Lecture Notes in Computer Science 4359

Commenced Publication in 1973
Founding and Former Series Editors:
Gerhard Goos, Juris Hartmanis, and Jan van Leeuwen

T0223295

Frank Geraets Leo Kroon
Anita Schoebel Dorothea Wagner
Christos Zaroliagis (Eds.)

Algorithmic Methods
for Railway Optimization

International Dagstuhl Workshop
Railway Optimization 2004
Dagstuhl Castle, Germany, June 20-25, 2004
Bergen, Norway, September 16-17, 2004
Revised Selected Papers

 Springer

Volume Editors

Frank Geraets
Deutsche Bahn AG, Konzernstrategie und Verkehrsmarkt (GSE)
10785 Berlin, Germany
E-mail: frank.geraets@bahn.de

Leo Kroon
Erasmus University Rotterdam, Rotterdam School of Management
NS Reizigers, Utrecht, The Netherlands
E-mail: L.Kroon@rsm.nl

Anita Schoebel
University of Göttingen, Institute for Numerical and Applied Mathematics, Germany
E-mail: schoebel@math.uni-goettingen.de

Dorothea Wagner
University of Karlsruhe, Department of Computer Science
P.O. Box 6980, 76128 Karlsruhe, Germany
E-mail: wagner@ira.uka.de

Christos Zaroliagis
University of Patras, Dept of Computer Eng. & Informatics
26500 Patras, Greece
E-mail: zaro@ceid.upatras.gr

Library of Congress Control Number: Applied for

CR Subject Classification (1998): F.2, E.1, G.2, I.2.8, I.3.5, G.1

LNCS Sublibrary: SL 1 – Theoretical Computer Science and General Issues

ISSN	0302-9743
ISBN-10	3-540-74245-X Springer Berlin Heidelberg New York
ISBN-13	978-3-540-74245-6 Springer Berlin Heidelberg New York

Springer is a part of Springer Science+Business Media

springer.com

© Springer-Verlag Berlin Heidelberg 2007
Printed in Germany

Typesetting: Camera-ready by author, data conversion by Scientific Publishing Services, Chennai, India
Printed on acid-free paper   SPIN: 12108947   06/3180   5 4 3 2 1 0

Preface

Algorithmic methods have reached a state of maturity as a consequence of decades of research. Despite its success, the current state of algorithmic research still faces severe difficulties, or cannot cope at all, with highly complex and data intensive real-world applications in large-scale networks. A prominent example is given by *railway networks*, which are used to model the complex systems of railway transportation. The complexity and size of optimization problems arising in railway transportation still pose challenges for algorithmic research.

This volume deals with problems arising in *railway optimization*, i.e., with planning and scheduling problems over several time horizons. Different challenging problems from the railway world are discussed from the point of view of computer science, algorithms, operations research, and discrete mathematics.

The first part of the volume consists of state-of-the-art papers that were selected after an open call that followed a Dagstuhl Seminar on *Algorithmic Methods for Railway Optimization* in June 2004. We had 12 submissions that underwent the standard peer review process, out of which 8 were selected for publication in this volume.

The second part consists of the nine accepted papers in the 4th Workshop on Algorithmic Methods and Models for Optimization of Railways (ATMOS 2004) that took place in Bergen, Norway, September 2004. The series of ATMOS workshops constitute a forum to present and discuss models, algorithms, and results related to railway optimization problems. ATMOS addresses researchers and practitioners working in computer science, discrete optimization, algorithms, or operations research. The ATMOS contributions come from all these fields and reflect the interdisciplinary character of railway optimization.

Within both parts we ordered the papers according to the hierarchical planning process in railway companies, which is roughly divided into the following four groups:

Network and Line Planning. This concerns the construction of the physical network (tracks and stations), the planning of the train lines along with their frequencies, and the design of the tariff system with the prices for the passengers.

Timetabling and Timetable Information. This concerns the construction of timetables, including scheduling and re-scheduling aspects, as well as timetable information systems.

Rolling Stock and Crew Scheduling. This concerns the scheduling of rolling stock and personnel to trains in order to carry out the train schedules in the timetable.

Real-Time Operations. This concerns the online reaction in the case of unexpected disruptions or delays. The delay management problem and its online version is a central issue, along with the improvement of dispatching systems.

Part I starts with three papers concerning the second group (Timetabling and Timetable Information). More specifically:

- In *"The Modeling Power of the PESP: Railway Timetables and Beyond,"* Christian Liebchen and Rolf H. Möhring show how various phases of the railway planning process can be integrated with periodic timetabling in order to achieve an additional optimization potential.
- In *"Cyclic Railway Timetabling: A Stochastic Optimization Approach,"* Leo Kroon, Rommert Dekker, and Michiel Vromans present a stochastic optimization model that can be used to find an optimal allocation of the running time supplements of a number of trains on a common infrastructure in order to minimize the average delays of these trains.
- In *"Timetable Information: Models and Algorithms,"* Matthias Müller-Hannemann, Frank Schulz, Dorothea Wagner, and Christos Zaroliagis give an overview of models and efficient algorithms for optimally solving timetable information problems under single or multiple criteria.

The next three papers concern the third group (Rolling Stock and Crew Scheduling). More specifically:

- In *"Estimates on Rolling Stock and Crew in DSB S-tog Based on Timetables,"* Michael Folkmann, Julie Jespersen, and Morten N. Nielsen describe two models for estimating the requested amount of rolling stock and crew required for operating a given timetable.
- In *"A Capacity Test for Shunting Movements,"* John van den Broek and Leo Kroon describe a capacity test for checking at an early stage of the planning process, whether the capacity of the infrastructure between the platform tracks and the shunting areas is sufficient for facilitating all the shunting movements that have to be planned in between the already timetabled train movements.
- In *"Railway Crew Pairing Optimization,"* Lennart Bengtsson, Rastislav Galia, Tomas Gustafsson, Curt Hjorring, and Niklas Kohl investigate the crew pairing problem that arises at major railways. Even though it is similar to the well-studied airline crew pairing problem, the size and complexity of the railway operation necessitate tailored optimization techniques.

The final two papers concern the fourth group (Real-Time Operations). More specifically:

- In *"Integer Programming Approaches for Solving the Delay Management Problem,"* Anita Schöbel presents path-based and activity-based integer programming models for the delay management problem and shows the equivalence of these formulations, and how to solve them in special cases.
- In *"Decision Support Tools for Customer-Oriented Dispatching,"* Claus Biederbick and Leena Suhl present decision support tools to be used by dispatchers in order to achieve customer orientation.

Part II starts with two papers in the first group (Network and Line Planning). More specifically:

- In *"An Integrated Methodology for the Rapid Transit Network Design,"* Gilbert Laporte, Ángel Marín, Juan A. Mesa, and Francisco A. Ortego present a new model for designing a line concept which is able to compete with private transportation systems.
- In *"A Simulation Approach for Fare Integration,"* Domenico Gattuso and Giuseppe Musolino deal with the problem of tariff planning and show their results within a case study in the region of Calabria.

The next four papers concern the second group (Timetabling and Timetable Information). More specifically:

- In *"Intelligent Train Scheduling on a High-Loaded Railway Network,"* Antonio Lova, Pilar Tormos, Federico Barber, Laura Ingolotti, Miguel A. Salido, and Montserrat Abril present an interactive application to assist planners when adding new trains on a complex railway network.
- In *"Platform Assignment,"* Sabine Cornelsen and Gabriele Di Stefano suggest an approach which is able to assign platforms to trains in such a way that a conflict-free realization of the timetable is possible, assuming a fixed timetable.
- In *"Finding All Attractive Train Connections by Multi-Criteria Pareto Search,"* Matthias Müller-Hannemann and Mathias Schnee discuss algorithms for timetable information systems that are able to find reasonable routes under various objectives.
- In *"The Railway Traveling Salesman Problem,"* Georgia Hadjicharalambous, Petrica Pop, Evangelia Pyrga, George Tsaggouris, and Christos Zaroliagis present a direct integer programming approach, based on timetable information, in order to find a tour through a railway system minimizing the overall time of the journey.

The next paper concerns the third group (Rolling Stock and Crew Scheduling). More specifically:

- In *"Rotation Planning of Locomotive and Carriage Groups with Shared Capacities,"* Taïeb Mellouli and Leena Suhl present a multi-layer multi-commodity network flow model which is able to handle various complex restrictions with respect to the planning of the circulation of locomotive and carriage groups.

The final two papers concern the fourth group (Real-Time Operations). More specifically:

- In *"An Estimate of the Punctuality Benefits of Automatic Operational Scheduling,"* Rien Gouweloos and Maarten Bartholomeus investigate the stability of railway systems and show the effects that suboptimal re-sequencing decisions have on the punctuality of the system.

– In *"Online Delay Management on a Single Train Line,"* Michael Gatto, Riko Jacob, Leon Peeters, and Peter Widmayer present a first theoretical investigation of online delay management problems and show that the special case of online delay management on a line is an extension of the ski-rental problem.

We would like to thank all those who submitted papers for consideration, as well as the referees for their invaluable contribution. We are grateful to Robert Görke for handling all technical issues in the preparation of this volume, especially for converting files from different sources into the current form. Finally, we acknowledge the support of the Human Potential Programme of EC under contract no. HPRN-CT-1999-00104 (project AMORE), and of the Future and Emerging Technologies Unit of EC (IST priority – 6th FP) under contract no. FP6-021235-2 (project ARRIVAL).

May 2007

Frank Geraets
Leo Kroon
Anita Schöbel
Dorothea Wagner
Christos Zaroliagis

Organization

Program Committee ATMOS 2004

Camil Demetrescu	University of Rome "La Sapienza," Italy
Oli B.G. Madsen	Techn. Univ. of Denmark, Lyngby, Denmark
Gabriele Di Stefano	University of L'Aquila, Italy
Anita Schbel (Co-chair)	Georg August Univ. of Göttingen, Germany
Leena Suhl	University of Paderborn, Germany
Frank Geraets (Co-chair)	Deutsche Bahn, Berlin, Germany
Gerhard J. Woeginger	TU Eindhoven, The Netherlands

List of Contributors

Montserrat Abril
DSIC
Universidad Politecnica de Valencia
Spain
mabril@dsic.upv.es

Federico Barber
DSIC
Universidad Politecnica de Valencia
Spain
fbarber@dsic.upv.es

Maarten Bartholomeus
HollandRailconsult
Postbus2855
3500GW, Utrecht
The Netherlands
mgpbartholomeus@hr.nl

Lennart Bengtsson
Jeppesen AB
Odinsgatan 9
Göteborg, Sweden
lennart.bengtsson@jeppesen.com

Claus Biederbick
University of Paderborn
Decision Support&OR Lab and
International Graduate School
for Dynamic Intelligent Systems
Warburger Str. 100
33098 Paderborn, Germany
biederbick@dsor.de

John van den Broek
Dept. of Mathematics and Computer
Science
Eindhoven University of Technology
NS Reizigers, Utrecht
The Netherlands
j.j.j.v.d.broek@tue.nl

Sabine Cornelsen
Universität Konstanz
Fachbereich Informatik &
Informationswissenschaft
Germany
cornelse@inf.uni-konstanz.de

Rommert Dekker
Rotterdam School of Economics
Erasmus University Rotterdam
3000 DR, Rotterdam
The Netherlands
R.Dekker@few.eur.nl

Michael Folkmann
Danish State Railways (DSB)
S-tog a/s, Production Planning
Kalvebod Brygge 32
1560 Copenhagen V, Denmark
mfolkmann@s-tog.dsb.dk

Rastislav Galia
Jeppesen AB
Odinsgatan 9
Göteborg, Sweden
rastislav.galia@jeppesen.com

Michael Gatto
Institute of Theoretical Computer
Science
ETH Zürich, Switzerland
gattom@inf.ethz.ch

Domenico Gattuso
Mediterranea University of Reggio
Calabria
Department of Computer Science,
Mathematics, Electronics and
Transportation
Feo di Vito, 89100 Reggio Calabria
Italy
domenico.gattuso@unirc.it

Rien Gouweloos
AtosConsulting
Papendorpseweg93
3528BJ, Utrecht
The Netherlands
rien.gouweloos@atosorigin.com

Tomas Gustafsson
Jeppesen AB
Odinsgatan 9
Göteborg, Sweden
tomas.gustafsson@jeppesen.com

Georgia Hadjicharalambous
Computer Technology Institute
P.O. Box 1122
26110 Patras, Greece
hadjicha@ceid.upatras.gr

Curt Hjorring
Jeppesen AB
Odinsgatan 9
Göteborg, Sweden
curt.hjorring@jeppesen.com

Laura Ingolotti
DSIC
Universidad Politecnica de Valencia
Spain
lingolotti@dsic.upv.es

Riko Jacob
Institute of Theoretical Computer
Science
ETH Zürich, Switzerland
rjacob@inf.ethz.ch

Julie Jespersen
Danish State Railways (DSB)
S-tog a/s, Production Planning
Kalvebod Brygge 32
1560 Copenhagen V, Denmark
jjespersen@s-tog.dsb.dk

Niklas Kohl
DSB Planning
Sølvgade 40
1349 Copenhagen K, Denmark
niko@dsb.dk

Leo G. Kroon
NS Reizigers
Department of Logistics
3500 HA, Utrecht
The Netherlands
and
Rotterdam School of Management
Erasmus University Rotterdam
3000 DR, Rotterdam
The Netherlands
L.Kroon@rsm.nl

Gilbert Laporte
Canada Research Chair
in Distribution Management
HEC Montréal
Canada
gilbert@crt.umontreal.ca

Christian Liebchen
TU Berlin
Institut für Mathematik,
Straße des 17. Juni 136
10623 Berlin, Germany
liebchen@math.tu-berlin.de

Antonio Lova
DEIOAC
Universidad Politecnica de Valencia
Spain
allova@eio.upv.es

Ángel Marín
Departamento de Matemática
Aplicada y Estadística
Universidad Politécnica de Madrid
Spain
amarin@dmae.upm.es

Taïeb Mellouli
Department of Management
Information Systems and
Operations Research
Martin-Luther-Universität
Halle-Wittenberg
Universitätsring 3
06108 Halle (Saale), Germany
mellouli@wiwi.uni-halle.de

Juan A. Mesa
Departamento de Matemática
Aplicada II
Universidad de Sevilla
Spain
jmesa@us.es

Rolf H. Möhring
TU Berlin
Institut für Mathematik,
Straße des 17. Juni 136
10623 Berlin, Germany
moehring@math.tu-berlin.de

Matthias Müller–Hannemann
Darmstadt University of Technology
Department of Computer Science
Hochschulstraße 10
64289 Darmstadt, Germany
muellerh@algo.
informatik.tu-darmstadt.de

Giuseppe Musolino
Mediterranea University of Reggio
Calabria
Department of Computer Science,
Mathematics, Electronics and
Transportation
Feo di Vito, 89100 Reggio Calabria
Italy
giuseppe.musolino@unirc.it

Morten N. Nielsen
Danish State Railways (DSB)
S-tog a/s, Production Planning
Kalvebod Brygge 32
1560 Copenhagen V, Denmark
monnielsen@s-tog.dsb.dk

Francisco A. Ortega
Departamento de Matemática
Aplicada I
Universidad de Sevilla
Spain
riejos@us.es

Leon Peeters
Institute of Theoretical Computer
Science
ETH Zürich, Switzerland
leon.peeters@inf.ethz.ch

Petrica Pop
Computer Technology Institute
P.O. Box 1122
26110 Patras, Greece
ppop@ceid.upatras.gr

Evangelia Pyrga
Department of Computer Engineering
and Informatics
University of Patras
26500 Patras, Greece
and
Computer Technology Institute
P.O. Box 1122
26110 Patras, Greece
pirga@ceid.upatras.gr

Miguel A. Salido
DCCIA
Universidad de Alicante
Spain
masalido@dccia.ua.es

Mathias Schnee
Darmstadt University of Technology
Department of Computer Science
Hochschulstraße 10
64289 Darmstadt, Germany
schnee@algo.
informatik.tu-darmstadt.de

Anita Schöbel
Institute for Numerical
and Applied Mathematics
Georg-August University
Göttingen, Germany
schoebel@math.uni-goettingen.de

Frank Schulz
Universität Karlsruhe
Department of Computer Science
P.O. Box 6980
76128 Karlsruhe, Germany
fschulz@ira.uka.de

Gabriele Di Stefano
Università dell'Aquila
Dipartimento di Ingegneria Elettrica
Italy
gabriele@ing.univaq.it

Leena Suhl
University of Paderborn
Decision Support&OR Lab and
International Graduate School
for Dynamic Intelligent Systems
Warburger Str. 100
33098 Paderborn, Germany
suhl@dsor.de

Pilar Tormos
DEIOAC
Universidad Politecnica de Valencia
Spain
ptormos@eio.upv.es

George Tsaggouris
Computer Technology Institute
P.O. Box 1122
26110 Patras, Greece
and
Department of Computer Engineering
and Informatics
University of Patras
26500 Patras, Greece
tsaggour@ceid.upatras.gr

Michiel J.C.M. Vromans
ProRail, Network Planning
3500 GA, Utrecht
The Netherlands
Michiel.Vromans@prorail.nl

Dorothea Wagner
Universität Karlsruhe
Department of Computer Science
P.O. Box 6980
76128 Karlsruhe, Germany
wagner@ira.uka.de

Peter Widmayer
Institute of Theoretical Computer
Science
ETH Zürich, Switzerland
widmayer@inf.ethz.ch

Christos Zaroliagis
Computer Technology Institute
P.O. Box 1122
26110 Patras, Greece
and
Department of Computer
Engineering and Informatics
University of Patras
26500 Patras, Greece
zaro@ceid.upatras.gr

Table of Contents

Part I

State of the Art

The Modeling Power of the Periodic Event Scheduling Problem: Railway Timetables — and Beyond*

Christian Liebchen and Rolf H. Möhring

TU Berlin, Institut für Mathematik, Straße des 17. Juni 136, D-10623 Berlin
{liebchen,moehring}@math.tu-berlin.de

Abstract. In the planning process of railway companies, we propose to integrate important decisions of network planning, line planning, and vehicle scheduling into the task of periodic timetabling. From such an integration, we expect to achieve an additional potential for optimization.

Models for periodic timetabling are commonly based on the Periodic Event Scheduling Problem (PESP). We show that, for our purpose of this integration, the PESP has to be extended by only two features, namely a linear objective function and a symmetry requirement. These extensions of the PESP do not really impose new types of constraints. Indeed, practitioners have already required them even when only planning timetables autonomously without interaction with other planning steps. Even more important, we only suggest extensions that can be formulated by mixed integer linear programs.

Moreover, in a selfcontained presentation we summarize the traditional PESP modeling capabilities for railway timetabling. For the first time, also special practical requirements are considered that we proove not being expressible in terms of the PESP.

1 Introduction

Traditionally, the planning process of railway companies is subdivided into several tasks. From the strategic level down to the operational level, the most prominent subtasks are network planning, line planning, timetable generation, vehicle scheduling, crew scheduling, and crew rostering, see Figure 1.

For a detailed description of these planning steps, as well as for an overview of solution approaches, we refer to Bussieck, Winter, and Zimmermann [4]. Notice that network planning and line planning are of course part of the strategic planning process of public transportation companies. In contrast, vehicle scheduling and crew scheduling are of operational nature. In between, timetabling forms the linkage between service and operation. An important reason for the division into at least five subtasks is the high complexity of the overall planning process ([4], [7]).

---

* Supported by the DFG Research Center "Mathematics for key technologies" in Berlin.

F. Geraets et al. (Eds.): Railway Optimization 2004, LNCS 4359, pp. 3–40, 2007.

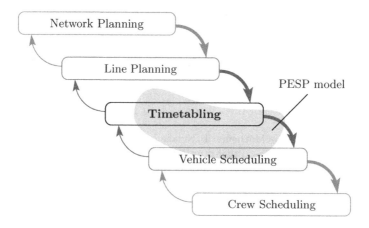

Fig. 1. Planning phases covered by the PESP beforehand

During the last years, a trend towards the integration of several planning steps has emerged. For example, vehicle and crew scheduling were successfully combined by Borndörfer, Löbel, and Weider [3] and by Haase, Desaulniers, and Desrosiers [8]. Similarly, a combination of line planning and network planning is the objective of Borndörfer, Grötschel, and Pfetsch [2]. Periodic timetabling has also served as a starting point for such attempts. Kolonko and Engelhardt-Funke [6] consider investments into infrastructure by using multi-criteria optimization. Nachtigall [20] computes timetables that require only few rolling stock for a specific vehicle schedule. Lindner [16] integrates the choice of rolling stock types in a non-linear model. Liebchen and Peeters [14] provide a linear model that serves as a good approximation for minimizing rolling stock while optimizing periodic timetables.

In this paper, we demonstrate how periodic timetable construction can be combined with other planning steps. Further, we incorporate other practical conditions on timetables such as timetable symmetry, line planning, and even infrastructure decisions. We show that this can in fact be achieved with only slight variations of the commonly used model for periodic timetable construction, the PESP model introduced by Serafini and Ukovich [30] in 1989. The variations keep much of the properties of the PESP model and are again mixed integer programs over a feasibility domain with essentially the same structure as the original PESP. In particular, all of the valid inequalities for the PESP stay valid, and some of the new formulations even speed up the solution time with standard MIP solvers. But there have also been proposed other solution techniques for PESP instances: constraint programming ([29]) and genetic algorithms ([21]). Hence, in this paper we will restrain ourselves to the pure modeling capabilities of the general PESP model—with only two small exceptions. But these exceptions have already been asked explicitly by practitioners for their own sake.

In the discussion of these modeling features, we will also lay out large parts of the map of the borderline between what still fits into the traditional PESP model, and what requires new features, and at which cost. To this end, we also review the traditional PESP modeling issues, thus altogether providing a selfcontained presentation of the PESP modeling capabilities and its extensions to symmetry, line planning, and network planning. Any of our suggestions for integrating these features can be formulated as a MIP, in particular not involving any quadratic terms.

The paper is organized as follows. Section 2 introduces the PESP. It presents its main formulations as a graph theoretic potential problem and as a mixed integer program, and reports on its complexity and a useful characterization of periodic timetables.

Section 3 discusses requirements for cyclic timetables that can be met by the PESP. These include simple requirements such as collision-free traffic on single tracks and headway between successive trains, but also more sophisticated ones such as bundling of lines, train coupling and sharing, fixed events in connection with hierarchical planning, and also disjunctive constraints and soft constraints.

Section 4 is devoted to timetable requirements that are beyond the scope of the traditional PESP, such as balanced reduction of service and symmetry of timetables. We show that the PESP or its MIP model only needs to be extended slightly in order to accommodate symmetry requirements.

Finally, in Section 5, we consider the integration of aspects of other planning steps into periodic timetable construction, in particular vehicle scheduling (minimization of rolling stock), line planning (simultaneous construction of line plan and timetable), and network planning (making infrastructure decisions). This integration makes essential use of the flexibility of the PESP, in particular disjunctive constraints, uses symmetry and, as a new technique, integrates aspects of graph techniques into the PESP in order to handle line planning.

All model features are illustrated by examples from our practical experience with timetable construction at Deutsche Bahn AG, S-Bahn Berlin GmbH, and BVG (Berlin Underground).

2   The Periodic Event Scheduling Problem (PESP)

In 1989, Serafini and Ukovich [30] introduced the Periodic Event Scheduling Problem (PESP), by which periodic timetabling instances may be formulated in a very compact way. Since then, this model has been widely used ([29,18,24,16,26]). In the Periodic Event Scheduling Problem (PESP), we are given a period time T and a set V of events, where an event models either the arrival or the departure of a directed traffic line at a certain station. Furthermore, we are given a set of constraints A. Every constraint $a = (i,j)$ relates a pair of events i, j by a lower bound ℓ_a and an upper bound u_a.

A solution of a PESP instance is a node assignment $\pi : V \mapsto [0, T)$ that satisfies

$$(\pi_j - \pi_i - \ell_a) \bmod T \leq u_a - \ell_a, \ \forall a = (i,j) \in A, \tag{1}$$

or $\pi_j - \pi_i \in [\ell_a, u_a]_T$ for short. We call a feasible node potential π a feasible *timetable*. Notice that we can scale an instance such that $0 \leq \ell_a < T$, and for the span $d_a := u_a - \ell_a$ of a *feasible interval* $[\ell_a, u_a]_T$ we may assume w.l.o.g. $d_a < T$.

Furthermore, for every fixed event i_0, every fixed point of time $t_0 \in [0, T)$, and every feasible timetable π there exists an equivalent timetable π' with $\pi'_{i_0} = t_0$. This is achieved by performing the simple shift $\pi'_i := (\pi_i - (\pi_{i_0} - t_0)) \bmod T$. Let us denote by $D = (V, A, \ell, u)$ the *constraint graph* modeling a PESP instance.

There are several practical aspects of periodic timetabling which profit from the presence of a linear objective function of the form

$$\sum_{a=(i,j) \in A} w_a \cdot (\pi_j - \pi_i - \ell_a) \bmod T,$$

with weights w_a. In our opinion, the most striking one is the integration of central aspects of vehicle scheduling, cf. section 5.1.

Another perspective on periodic scheduling can be obtained by considering tensions instead of potentials. In a straightforward way, define for a given node potential π its *tension*

$$\hat{x}_a := \pi_j - \pi_i, \forall a = (i, j) \in A.$$

We call a set of edges $C \subseteq A$ an *oriented cycle* if re-orienting a subset of its edges yields a directed circuit. The *incidence vector* γ_C of an oriented cycle C is a vector in $\{-1, 0, 1\}^A$, where the entry minus one indicates a backward arc of the oriented cycle. The cycle space $\mathcal{C}$ of a directed graph D is defined as

$$\mathcal{C} := \text{span}\{\gamma_C \mid C \text{ oriented cycle in } D\}.$$

Recall that a vector $\hat{x}$ is a tension (or potential difference), if and only if for some cycle basis B of $\mathcal{C}$, and each of its oriented cycles $C \in B$ with incidence vectors γ_C it holds that $\gamma_C \hat{x} = 0$ (e.g. [1]). This yields the following MIP formulation

$$
\left.
\begin{array}{ll}
\min c^t(\hat{x} + pT) & \min c^t x \\
\text{s.t. } \Gamma \hat{x} = 0 & \text{s.t. } \Gamma(x - pT) = 0 \\
\ell \le \hat{x} + pT \le u \quad \text{or} & \ell \le x \le u \\
p \in \mathbb{Z}^A, & p \in \mathbb{Z}^A,
\end{array}
\right]
\tag{2}
$$

where $\Gamma \in \{-1, 0, 1\}^{(|A| - |V| + 1) \times |A|}$ denotes the cycle-arc incidence matrix (*cycle matrix*) of some cycle basis of the directed graph D. The x variables are in fact a *periodic tension*, which we formally define for a given node potential π to be

$$x_{ij} := (\pi_j - \pi_i - \ell_{ij}) \bmod T + \ell_{ij}.$$

Sometimes, it is useful to define *slack variables* $\tilde{x}_a := x_a - \ell_a$.

Recall that cycle matrices are totally unimodular ([28]). This is the main observation to prove the following lemma.

Lemma 1 ([23]). *Let $\mathcal{I}$ denote an instance of PESP with integral vectors ℓ and u and an integer period time T. If $\mathcal{I}$ admits some feasible timetable $\pi \in [0, T)^V$, then it also admits an* integral *feasible timetable $\pi' \in \{0, \ldots, T - 1\}^V$.*

Already Serafini and Ukovich made the following simple but useful observation.

Lemma 2 (Serafini and Ukovich [30]). *If we relax the requirement* $\pi \in [0,T)^V$ *to* $\pi \in \mathbb{Q}^V$, *then for every spanning tree* H *and every feasible timetable* π *there exists an equivalent feasible timetable* π' *which induces* $p_a = 0$ *for* $a \in H$.

Notice that we may interpret the remaining non-zero integer variables as the representants of the elements of a (strictly) fundamental cycle basis. A generalization to integral cycle bases yields many variants of problem formulation 2, some of which are easier to solve for MIP solvers ([12]).

Periodic tensions can be characterized similarly to classic aperiodic tensions.

Lemma 3 (Cycle Periodicity Property). *A vector* $x \in \mathbb{Q}^A$ *is a periodic tension, if and only if for every cycle* C *with incidence vector* $\gamma_C \in \{-1,0,1\}^A$, *there exists some* $z_C \in \mathbb{Z}$, *such that*

$$\gamma_C x = z_C T. \tag{3}$$

The PESP is $\mathcal{NP}$-complete, since it generalizes Vertex Coloring ([23]). To see this, orient the edges of a Coloring instance arbitrarily and assign feasible periodic intervals $[1, T-1]_T$ to each of them. Solution methods for the PESP include Constraint Programming ([29]), Genetic Algorithms ([21]), and of course integer programming techniques. For a computational study in that these substantially different approaches are compared to each other, we refer to [15]. For the MIP approach, a very important ingredient is

Theorem 1 (Odijk [24]). *An integer vector* p *allows a feasible solution for the MIP* (2), *if and only if for every oriented cycle* C *of the constraint graph, the following* cycle inequalities *hold*

$$\underline{p}_C := \left\lceil \frac{1}{T}\left(\sum_{a \in C^+} \ell_a - \sum_{a \in C^-} u_a \right) \right\rceil \leq \sum_{a \in C^+} p_a - \sum_{a \in C^-} p_a \leq \left\lfloor \frac{1}{T}\left(\sum_{a \in C^+} u_a - \sum_{a \in C^-} \ell_a \right) \right\rfloor =: \overline{p}_C, \tag{4}$$

where C^+ *and* C^- *denote the forward and the backward arcs of the cycle* C.

We close this section by listing other totally different practical applications which can be modeled via the PESP ([30]). The most prominent ones are the scheduling of systems of traffic lights, and periodic job shop scheduling.

3   Timetabling Requirements Covered by the PESP

This section gives a broad overview of the timetable modeling capabilities of the PESP. Contrary to the following sections, practical requirements to be modeled are limited to those arising in periodic timetabling. Nevertheless, there are many facts we have to discuss in order to give a self-contained overview.

However, let us start by naming two facts which are definitely beyond the scope of the PESP: routing of trains through stations or even alternative tracks, and routing of the passenger flow. Hence, throughout this paper we assume fixed routes for both trains and passengers. A short motivation for these assumptions will be given at the beginning of Section 4.

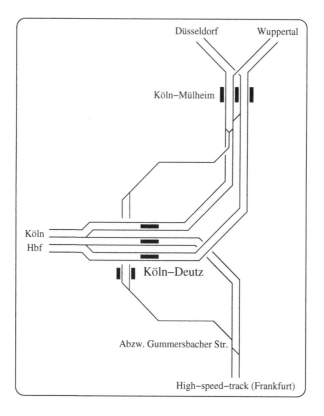

Fig. 2. Track map of Köln-Deutz (Cologne) — based on [11]

For the vast majority of practical requirements to be modeled, we provide examples which are close to practice. However, in particular time and track information might not always reflect practice exactly. Depending on the fact to be modeled, we provide a track map, a line plan, a visualization[1] of the timetable of a given track by means of a time-space diagram, and last but not least the resulting PESP subgraph. For readers not familiar with the first three types of charts, we refer to any textbook on railway engineering.

Most of our real-world examples are taken from the surroundings of the station Köln-Deutz (Cologne), which is part of the German ICE/IC-network. Figure 2 displays the general track map of Köln-Deutz. Unless stated otherwise, we assume a period time of $T = 60$ minutes.

3.1 Elementary Requirements

Both, for sake of completeness and in order to introduce the notation used in the following figures, we start by modeling the three most elementary actions within public transportation networks: trips, stops, and changeovers.

---

[1] In German: "Bildfahrplan."

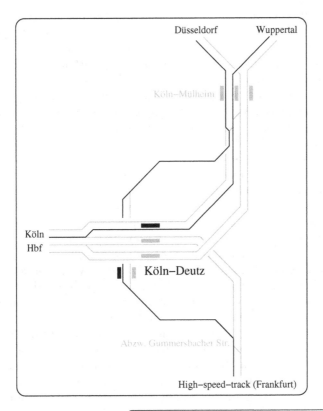

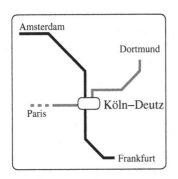

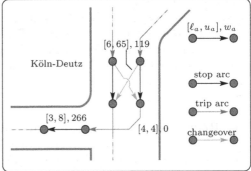

Fig. 3. Modeling elementary requirements: (a) two disjoint routes of lines serving Köln-Deutz; (b) the corresponding line plan; (c) PESP constraints modeling running activities, stopping activities, and changeover activities

In Figure 3 (a), we highlight the tracks used by two lines which cross at Köln-Deutz. The lines themselves are given in Figure 3 (b). Finally, we provide the constraint graph which models running, stopping, and changeover activities of these lines at Köln-Deutz in Figure 3 (c) as PESP constraints. For instance, the trip arc with the constraint $[4,4]_{60}$ ensures a trip time of precisely four minutes from Köln-Deutz to Köln Hbf. Within Köln Hbf, the minimum stopping time is set to three minutes such that passengers can board and alight the train. Finally, the increase of travel time for passengers that stay within the train is bounded by additional five minutes, providing an upper bound of $3 + 5 = 8$.

Notice that we ensure changeover quality by linearly penalizing changeover times which exceed a certain minimal changeover time required for changing platforms. In our example, a minimal changeover time of six minutes is assumed when connecting from Dortmund to Frankfurt. Using this approach, changeover arcs typically have a wide span.

An alternative way of modeling changeovers is to require some important ones not to exceed a maximal amount of effective waiting time. Then, we end up with rather small spans for changeover arcs. Schrijver and Steenbeek [29] follow this approach, which seems to be very suitable for constraint programming solvers.

Stopping arcs typically have very small span. In rather unimportant stations, in general it is a good choice to fix the span to zero, in particular if there is neither a junction of tracks, nor a single track, nor any changeovers.

Just as trip arcs, stopping arcs with span zero constitute redundancies which can be eliminated very efficiently in a preprocessing step. For example, one can contract any *fixed arc*, i.e. having zero span, together with its target node. Doing so, the arcs which were incident with the contracted target node only have to be redirected to the source node of the contracted arc, after having shifted their feasible intervals appropriately. Moreover, an arc being (anti-) parallel to another one can eliminated, if its feasible interval is a superset of the other arc. In addition to nodes with degree at most two, Lindner [16] gives further situations in which the graph can be simplified.

If there are several lines using the same track into the same direction, sometimes a balanced service might be required. For n lines, this can easily be achieved by introducing arcs with feasible interval $[\frac{T}{n}, T - \frac{T}{n}]_T$ between any unordered pair of events that represent the departure at the first station of the common track. Certainly, strict balancedness may be relaxed by increasing the feasible interval.

Safety Requirements. If, in contrast to the previous discussion, there is no need for a balanced service, then at least a minimal headway h between any two of them has to be ensured. In the easiest case, the lines are operated with the same type of trains, and their running time is fixed. Then, we can sufficiently separate any two lines by introducing constraints similar to the above ones, having feasible interval $[h, T - h]_T$. These can be inserted either at the beginning or at the end of their common track. The more sophisticated constellation of trains involving different speeds will be discussed in Section 3.2.

But two trains may also use the same track in opposite directions. This is mainly the case for single tracks, see Figure 4 (a). Obviously, a train may not enter the single track until the train of the opposite direction has left it. In Figure 4 (b), we give a timetable visualization that is extremely useful in particular for single tracks. We assume a fixed local signaling, and the grey boxes visualize

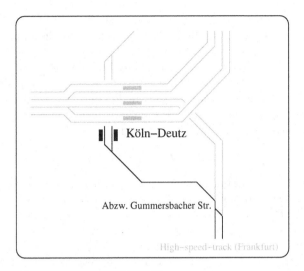

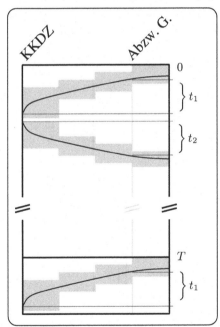

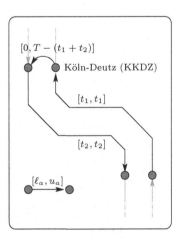

Fig. 4. Modeling single tracks: (a) a single track south of Köln-Deutz; (b) visualization of a feasible timetable for that single track; (c) PESP constraints ensuring safety distance for single track

the time a train blocks a certain part of the track. Surprisingly, there is only one single constraint needed to prevent two trains of opposite directions from colliding within the single track, as can be seen in Figure 4 (c). To that end, consider the western entry point to the single track. A train may only enter the single track after a train of the opposite direction has left ($\ell_a = 0$). But it also must have left the single track before the next train of the opposite direction may enter the single track ($u_a = T - (t_1 + t_2)$).

Note that so far we did not care about any buffer times and blocking times when setting the feasible interval to $[0, T - (t_1 + t_2)]_T$. Assuming a *minimal crossing time* b at both endpoints of the single track, i.e. the time that has to pass from a train leaving the single track until a train in opposite direction may enter, we obtain the following feasible interval

$$[b, T - (t_1 + t_2 + b)]_T.$$

Again, if there are several lines that have to be scheduled on a single track, one constraint for every unordered pair of opposite directions is needed.

Some authors ([9]) consider situations at crossings, where trains are shortly using the track of the opposite direction (cf. Figure 5), as another modeling feature. But this is just a special case of single tracks, if the network is modeled at an appropriate granularity. Abzw. Gummersbacher Straße has to be split into a northern station and a southern station which are linked by an eastern and a western track, where the western track can be traversed in both directions.

Fig. 5. Crossing of track of the opposite direction south of Köln-Deutz

3.2 More Sophisticated Requirements

Whereas the practical requirements discussed in the previous section might arise in almost every railway network, the following aspects are of a more specialized nature.

Fixed Events. When planning a timetable hierarchically, e.g. from international trains down to local trains, one has to consider the fixed settings of previous hierarchies without replanning their times. Hence, the capability to fix an event to a certain point of time is another important modeling feature.

Fortunately, due to the periodic nature of the PESP, we may shift every feasible timetable such that a fixed event i_0 is fixed to a desired point in time $t_0 \in [0, T)$, i.e. $\pi_{i_0} = t_0$, and the objective value remains unchanged. By defining one of the events to be fixed as a kind of "anchor" event, we can easily relate the other events i_j to be fixed to certain points of time t_j by introducing arcs $a_j = (i_0, i_j)$ with $\ell_{a_j} = u_{a_j} = t_j - t_0$.

Bundling of Lines. Hierarchical planning gives rise to a further challenging aspect of timetabling. Notice that if a track is used by trains of different speeds, the capacity of that track significantly depends on the ordering of the trains. The first two parts of Figure 6 visualize this effect. In the first scenario, slow and fast trains alternate, which implies that only two hourly lines of each of the two train types can be scheduled. However, if lines are bundled with respect to their speeds, three lines of the same two types of trains can be scheduled without having to invest into infrastructure, cf. Figure 6 (b).

On the one hand, when only planning the high-speed lines in the first step of a hierarchical approach, it may happen that decisions on a higher level result in infeasibility on a lower level. On the other hand, hierarchical decomposition might have been chosen because an overall planning was considered to be too complex.

In order to keep the advantage of decomposition but limit the risk of infeasibility on lower levels, we propose to only bundle the lines of the current level of hierarchy. Figure 6 (c) gives the complete set of lines which should be operated on the track in question. In Figure 6 (d), we provide the PESP graph for the ICE/IC network. To bundle the three active lines, we introduce an artificial event and require each of the departure events to be sufficiently close to that artificial event. Hereby, the departure events will be close to each other as well.

In particular, we must not choose one of the existing events as "anchor", because this would predict the corresponding line to be the head of the sequence of bundled lines. This must definitively be avoided, because — contrary to assumptions made by Krista [9] — the ordering of lines is indeed a major result of timetabling. Finally, based on profound estimates on passengers' behaviour the management has to decide whether it is more important to operate as many trains as possible—and hereby bundle the trains of the same type—or whether a balanced service within the different types of trains should be preferred.

Train Coupling/Train Sharing. During the last decade, in railway passenger traffic a trend emerged towards train units which can easily be coupled and shared. Doing so, more direct connections can be offered without increasing the capacity of some bottleneck tracks.

In Figure 7 (a), we display a line which is operated by two coupled train units between Berlin and Hamm. They split in Hamm to serve the two major routes of the Ruhr area, hereby offering direct connections from Berlin to the most important cities of that region. Still, this line occupies for example the high-speed track between Berlin and Hannover only once per hour.

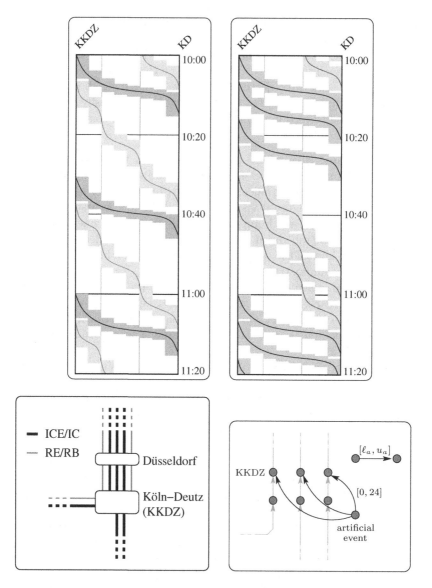

Fig. 6. Bundling of lines: (a) poor capacity if slow and fast trains are alternating; (b) capacity increase by bundling trains of the same type; (c) complete line plan for all the types of lines; (d) PESP constraints ensuring enough capacity for RE/RB lines already when planning only ICE/IC lines within the first step of a hierarchical planning

In Figure 7 (b), we provide PESP constraints which ensure the time for splitting the two train units in Hamm to be at least five minutes. Furthermore, for the two departing trains, a safety distance of four minutes is guaranteed.

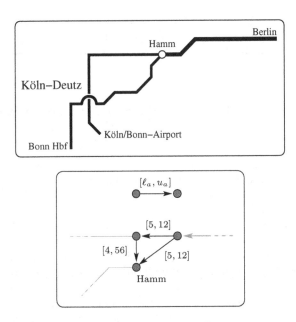

Fig. 7. Modeling train sharing: (a) line plan for the line Berlin-Hamm-{Bonn Hbf | Köln/Bonn-Airport}; (b) PESP constraints ensuring safety distance and time to split train units, but not specifying the ordering of departures

Notice that we do not need to specify which train should leave Hamm first. This decision will be made implicitly, and in an optimized way, by the PESP solver.

Variable Trip Times. As long as trip times are fixed, a usual safety constraint prevents two identical trains from overtaking each other. With h being the minimal headway for the track, we put an arc with feasible interval $[h, T - h]_T$ between the two events of entering the common track. If the line at the tail of the constraints is by f time units faster than the line at the tail of the constraints, overtaking can be prevented by modifying the constraint to $[h + f, T - h]_T$. This can be understood easily by having again a look at the corresponding situation in Figure 6 (a).

But this is no longer guaranteed if the model includes variable trip times. Even ensuring the minimal headway at the end of the track, too, does no longer prevent overtaking (even of trains of the same type) if the span in the trip times is at least twice the safety distance h, i.e. $u_a - \ell_a \geq 2h$. Schrijver and Steenbeek [29], Lindner [16], and Kroon and Peeters [10] tackle this phenomenon by adding extra constraints on the integer variables of the MIP formulations. Hereby, they leave the PESP model. In addition, Kroon and Peeters [10] provide some sufficient conditions on trip times, safety distance, and on the degree of flexibility of the trip times that prevent trains from overtaking.

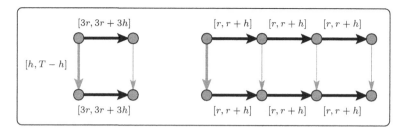

Fig. 8. Overtaking and variable trip times: (a) standard granularity does not prevent overtaking; (b) finer granularity prevents overtaking

In order to stay within the PESP model, we propose to subdivide[2] an initial trip arc into new smaller ones such that $u_a - \ell_a < 2h$ for every new trip arc. For an example, we refer to Figure 8, where **bold** arcs represent arcs of the spanning tree for which we set $p_a = 0$, cf. Lemma 2, and $3r$ is the minimum running time for the track.

Although this might seem to expand the model, the approach behaves rather well. More precisely, in every feasible timetable, the integer variables which we have to introduce for our additional arcs are in fact fixed to zero. This can simply be seen by applying the cycle inequalities (4) to any of the three squares in Figure 8 (b),

$$\underline{p} = \left\lceil \frac{1}{T}(r + h - (T - h) - (r + h)) \right\rceil = \left\lceil \frac{h - T}{T} \right\rceil = 0,$$

$$\overline{p} = \left\lfloor \frac{1}{T}((r + h) + (T - h) - h - r) \right\rfloor = \left\lfloor \frac{T - h}{T} \right\rfloor = 0.$$

Notice that the corresponding bounds for the initial formulation are only -1 and 1. But this is very natural, because there are three different types of timetables possible, of which we have to cut off two. The value one, for instance, models the fact that the second (lower) train is overtaking the first (upper) train.

Although we showed that the inconveniences caused by flexible running times can be overcome, we will assume fixed running times throughout the remainder of this paper.

3.3   General Modeling Capabilities

There are also important non-timetabling features which can be modeled by the PESP in a very elegant way. The types of such constraints are disjunctive constraints and soft constraints. Although they were originally introduced for their own sake, they turn out to be very useful for even more specialized requirements, which practitioners required to be modeled.

---

[2] This approach has also been discussed with Peeters [25,26] several years ago.

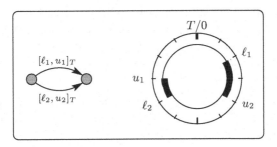

Fig. 9. Disjunctive constraints

Disjunctive Constraints. The feasible region of MIPs are commonly given as the intersection of finitely many half-spaces, plus some integrality conditions. If disjunctive constraints have to be modeled, usually artificial integer variables are introduced. However, the PESP offers a much more elegant way.

When introducing the PESP, Serafini and Ukovich [30] already made the important observation that the intersection of two PESP constraints is not always again a single PESP constraint. Rather, the feasible interval for a tension variable can become the *union* of two PESP constraints, e.g.

$$\pi_j - \pi_i \in [\ell_1, u_1]_T \cap [\ell_2, u_2]_T \Leftrightarrow \pi_j - \pi_i \in [\ell_1, u_2]_T \cup [\ell_2, u_1]_T.$$

We illustrate their observation in Figure 9. Nachtigall [20] observed that any union of k PESP constraints can be formulated as the intersection of at most k PESP constraints.

As an immediate practical application of disjunctive constraints, we consider optional operational stops. Long single tracks with no stop may cause the timetable of a line to be fixed within only small tolerances. In such a situation, Deutsche Bahn AG considers the option of letting the ICE/IC trains of one direction stop somewhere, although there is no ICE/IC station. In the current timetable, this takes places on the line between Stuttgart and Zurich, at Epfendorf.

If we want periodic timetable optimization to be competitive, we should enable the PESP to introduce an additional stop as well. We do so by introducing a pair of disjunctive constraints. The first constraint is a usual stop arc a_1. We set the lower bound ℓ_{a_1} to zero, which models the option of not introducing an additional stop. The upper bound u_{a_1} is set to the sum of the minimal increase b of travel time occurring from braking and accelerating, plus the maximal amount of stopping time s at the station. For the effected increase $\tilde{x}_a$ of travel time, this translates to

$$\tilde{x}_a \in \{0\}_T \cup [b, b + s]_T,$$

which is a disjunctive constraint. Notice that additional waiting time should be penalized in this situation similarly to an extension of a regular service stop. Moreover, if there are other lines operating on the same track, we have to take precautions that were discussed in the paragraph on variable trip times. However, optional operational stops make most sense within long single tracks. But there, in many cases there are not several lines using that large bottleneck.

Obviously, the introduction of an additional stop can also be due to the construction of a new station. Since such decisions are a part of network planning, we postpone this discussion until Section 5.3.

Soft Constraints. Nachtigall [19] investigated the combination of two antiparallel arcs $a_1 = (i, j)$ and $a_2 = (j, i)$. If they have an identical coefficient in the objective function and if neither of them can become infeasible for any vector π, or x resp., then they model a *soft constraint*.

Classically, if a certain tension value x_a does not satisfy a given PESP constraint $[\ell_a, u_a]_T$, one would declare the complete timetable as infeasible. But sometimes, it can be an alternative only to produce a significant penalty in the objective function, if a constraint is not satisfied.

To that end, we relax the upper bound of the original constraint to $\ell + T - 1$—we may assume the instance being scaled such that the precondition of Lemma 1 is satisfied. Further, we introduce a new antiparallel arc with feasible interval according to Figure 10. Then, these two constraints yield a piecewise constant

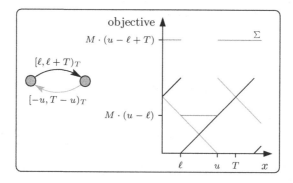

Fig. 10. Soft constraints

behavior of the objective function, which serves as an indicator for the violation of the original constraint, but without guaranteeing feasibility. For an initial constraint $x_a \in [\ell_a, u_a]$ consider the corresponding pair of artificial constraints a_1 and a_2—each of these having having cost coefficient M. They contribute to the objective function

$$M \cdot (x_{a_1} + x_{a_2}) = \begin{cases} M \cdot (u - \ell) & \text{if } x_{a_1} \in [\ell_a, u_a]_T, \text{ and} \\ M \cdot (u - \ell + T) & \text{otherwise,} \end{cases}$$

hereby indicating whether the original constraint a is satisfied for the tension vector x.

In our cooperation with Berlin Underground, we were asked to construct a timetable that, among the top 50 most important connections, maximizes the number of connections having a waiting time of at most five minutes. In fact, soft constraints are well-suited for letting MIP solvers produce a timetable being optimal subject to this kind of objective function.

4 Timetabling Requirements *Not* Covered by the PESP

Although the most important practical requirements for a periodic timetable can be modeled within the PESP, we are still aware of some special features for which the PESP fails. To the best of our knowledge this is the first time that practical requirements of timetabling are proven to be beyond the scope of the PESP.

First, one may think of situations in which it is not fixed which trains are operated on which track, for example within stations. Consider a station having two tracks in the same direction and three lines serving that direction. Then, we cannot decide a priori which pair of lines shall be within the station at the same time, hence omitting the sequencing constraint between these two lines. This observation is the motivation for the DONS system to be subdivided into CADANS, covering the timetabling step, and STATIONS, covering the routing aspect ([31]).

Notice that this requirement even affects the strategies for parking trains at terminus stations. Consider the following example, which has been inspired by the situation at Warschauer Straße of the Berlin fast train network, although there are further alternatives within that station. Within 20 minutes, two lines end at that station, both sharing the same track for arrival and departure, cf. Figure 11. For instance, arcs a_1 and a_2 measure the time that the two trains stay within this terminus station. We assume a turnover to take at least four

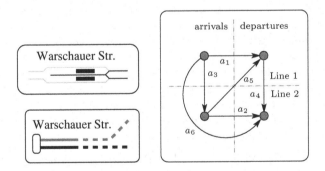

Fig. 11. Routing within terminus stations

minutes at the platform, or at least eight minutes when visiting the parking facility. Further, any arriving and departing trains block the platform for 59 seconds before and after their arrival and departure, respectively. To ensure that all passengers alighted before driving to the parking facility, the platform position is blocked for one more minute.

Proposition 1. *For every set of PESP constraints either timetables which are operable are classified as infeasible, or timetables which are not operable are classified as feasible.*

Proof. We start by analyzing the two major strategies individually: both lines turn at the platform, or line 2 turns in the parking facility, w.l.o.g. Table 1 provides tight lower and upper bounds for the six arcs in Figure 11 (c) with respect to these two scenarios. More precisely, with a strategy specified, we have that for every arc $a = (i, j)$ and every value $t_a \in [u_a, \ell_a]$ there exists an operable

Table 1. Tight interval bounds for different turning strategies at Warschauer Straße

Arc	Both at platform		Line 2 to parking		not specified	
	ℓ_a	u_a	ℓ_a	u_a	ℓ_a	u_a
a_1	4	12	4	13	4	13
a_2	4	12	8	18	4	18
a_3	6	14	6	17	6	17
a_4	6	14	2	14	2	14
a_5	10	18	7	18	7	18
a_6	10	18	6	18	6	18

timetable π such that

$$(\pi_j - \pi_i - \ell_a) \bmod T = t_a - \ell_a.$$

Further, by simple case inspection one can verify that every operable timetable which implements that specific strategy respects each of the given bounds. Hence, in order to provide general PESP constraints which characterize the operable timetables without having specified any parking strategy a priori, the feasible intervals must include the feasible intervals of both scenarios.

However, there exists a vector π which respects the six PESP constraints thus obtained (see the last two columns of Table 1), but which does *not* encode an operable timetable, because the two trains would be at the platform at the same time: line 1 arrives at minute 00 and departs only at minute 13, although line 2 already arrives at minute 06 and departs at minute 16. But for each of the $\binom{4}{2}$ potential differences between these four events there also exist operable timetables that attain the very same tension value.   □

Hence we cannot establish a set of PESP constraints that precisely identifies practically operable timetables as feasible solutions.

Apart from the rather important routing requirement, which unfortunately is simply out of scope for the PESP, we will analyze a very special situation in more detail, namely the balanced reduction of service. Finally, we will introduce the important notion of *symmetry*. On the one hand, symmetry slightly exceeds the original PESP, but on the other hand, when added explicitly, gives rise to a mechanism to include important aspects of line planning into the very same planning step as periodic timetabling and vehicle scheduling.

4.1 Balanced Reduction of Service

The Berlin fast train company (S-Bahn Berlin GmbH) aims at operating only one timetable for one whole day. The late evening service differs from the rush hour only in that some trains are omitted. Hence, the timetable must respect the available capacity during the rush hour, and it has to offer a balanced service in the late evening as well.

From a pure operations point of view, it could seem strange to sidestep an intraday change of the timetable structure. It is for sure that the information technology available in the 21st century could cope with this. But it is still the policy of the company. It is given as a motivation that customers really expect to have only one single timetable to be kept in mind for their station.

Consider the approximately 10 km long track from Zoo station to Berlin East station. On it, a minimal headway of 2.5 minutes has to be respected. The period time is 20 minutes and eight[3] lines (having identical train types) per period and direction have to be scheduled. In the late evening service, there are four trains every 20 minutes, two of them being fixed to a 10 minutes time lag. We call these two lines *core-lines*.

Of course it would be ideal to have a five minutes time lag between two consecutive trains in the evening. But this is impossible because one of the evening trains is required to serve Potsdam every 10 minutes together with a rush hour train. Hence, one should ensure that the maximal time lag between two consecutive trains does not exceed 7.5 minutes.

But this simple requirement cannot be covered by the PESP. Consider the two types of timetables given in Table 2. Timetables of type 1 satisfy our requirement

Table 2. Possible timetables for the late evening service from Zoo station to Berlin East station. This table only shows the core-lines that are actually running in the evenings. Each of the – entries is a joker for a rush-hour train.

Timetable	Departure times (T = 20 minutes)								
Type 1	0.0	–	–	7.5	10.0	12.5	–	–	(20.0)
Type 2	0.0	2.5	–	7.5	10.0	–	–	–	(20.0)

by bounding the maximum distance between two consecutive trains to 7.5 minutes, but type 2 does not because there we have a gap of 10 minutes.

Proposition 2. *For every set of PESP constraints either timetables of both types are feasible, or timetables of both types are infeasible.*

Proof. There are two types of constraints to be analyzed:

  i. one constraint between the two non-core lines,
 ii. four constraints between one of the two core lines and one of the two non-core lines.

---

[3] One of them only serves as a free slot for occasional non-passenger trips.

Since we must not specify the sequence of the lines in advance, only symmetric constraints $[\ell, T - \ell]_T$ make sense. Moreover, all constraints of type (ii) have to be identical for the same reason.

To guarantee feasibility of type 1 timetables, we deduce $\ell \leq 5$ for the constraint of type (i) and $\ell \leq 2.5$ for the constraints of type (ii). But then, timetables of type 2 stay feasible as well. Hence, in order to cut off timetables of type 2, we have to increment one of the given bounds. But since they are tight, this would immediately cut off timetables of type 1 as well.   □

Notice, however, that other railway companies implement other strategies to attain a balanced reduction of service. We will present an approach which turns out to be easier for timetabling, but slightly more complex for operation and customers.

Consider the track Niederhöchststadt-Langen (Hessen) via Frankfurt Hbf of S-Bahn Frankfurt. Compare the regular service hourly pattern to the weak-traffic service hourly pattern, which are given in Table 3. For the weak-traffic service,

Table 3. Timetables for regular service and weak-traffic service between Niederhöchststadt and Langen (Hessen)[17]

	regular service				weak traffic	
Line	S4	S3	S4	S3	S4	S3
Bad Soden	–	20	–	50	–	50
Kronberg	09	–	39	–	24	–
Niederhöchststadt	14	29	44	59	29	59
Langen (Hessen)	56	11	26	41	11	41
Darmstadt Hbf	–	25	–	55	–	55

every second train is omitted. To prevent a 45 minutes gap every hour, one of the two lines is shifted by 15 minutes and uses the slot of the train of the other line, which has just been skipped.

If we assume none of the lines to share a track with other lines outside their common part, then we can easily deduce a feasible timetable for the weak-traffic service from a periodic timetable, which is feasible for the regular service. In case of single tracks along the peripherical segments, the only thing to be ensured is that the shift of 15 minutes appears simultaneously for the two directions. Hereby, every meeting point for the weak-traffic service is already a meeting point for the regular service — hence, single tracks stay respected. Trivially, along the common track no conflicts will appear either.

4.2   Symmetry of a Periodic Timetable

Throughout our discussion of symmetry, we assume that for every directed line there exists another directed line serving the same stations just in opposite order. Moreover, the concept of symmetry makes only sense, if for every traffic line, the running and stopping times of its two opposite directions are the same. Also for the minimum headways and other operational constraints we require them to be identical in both directions. Furthermore, the passenger flow is assumed to be symmetric.

First, observe that in every periodic timetable with period time T, every train meets some train of the opposite direction of its line twice within the period time. In general, every line can have different times for these meetings.

A periodic railway timetable is called *symmetric with* (global) *axis s*, if at time s every train in the network meets a train of the opposite direction of its line. From the above considerations we deduce that we may assume w.l.o.g $s \in [0, \frac{T}{2})$.

For the arrival or departure event of a directed line at a certain station, we denote by its *complementary event* the departure or arrival, resp., of the opposite line at the same station. In the sequel, we provide two characterizations of symmetric timetables.

Lemma 4. *A timetable is symmetric with axis s, if and only if for every pair i and $\bar{i}$ of complementary events there holds*

$$\frac{(\pi_i + \pi_{\bar{i}}) \bmod T}{2} = s. \tag{5}$$

Proof. Let i and $\bar{i}$ be any two complementary events. By definition, they are part of the two opposite directions of the same line. Moreover, they are located in the same station S.

In a symmetric timetable, the trains of the two opposite directions meet at times s and $s + \frac{T}{2}$. Consider two virtual events j and $\bar{j}$ of passing the meeting point M. As the trains meet there, we have $\pi_j = \pi_{\bar{j}} \in \{s, s + \frac{T}{2}\}$.

We assumed the travel times of two opposite trains to be identical and denote the travel time between S and M by t. Hence, w.l.o.g.

$$(\pi_i + \pi_{\bar{i}}) \bmod T = ((\pi_j + t) + (\pi_{\bar{j}} - t)) \bmod T = (2 \cdot \pi_j) \bmod T. \qquad \square$$

To define a counterpart of condition (5) for the tension formulations (2), we define two arcs $a = (i, j)$ and $\bar{a} = (\bar{j}, \bar{i})$ to be *complementary*, if $\{i, \bar{i}\}$ and $\{j, \bar{j}\}$ are complementary, and we have $\ell_a = \ell_{\bar{a}}$ and $u_a = u_{\bar{a}}$. With these definitions at hand, we are able to define a symmetric instance of PESP: A constraint graph is called *symmetric*, if every arc connects either two complementary events, or if for every arc $a \in A$ there exists some complementary arc $\bar{a} \in A \setminus \{a\}$.

Lemma 5. *Consider an instance of PESP that is modeled by a connected symmetric constraint graph. Let π be a feasible timetable with corresponding periodic tension x. There exists some $s \in [0, \frac{T}{2})$ such that Condition (5) holds for every pair of symmetric events, if and only if every pair of complementary arcs a and $\bar{a}$ fulfills*

$$\tilde{x}_a = \tilde{x}_{\bar{a}}. \tag{6}$$

Proof. "⇒": Let $a = (i,j)$ and $\bar{a} = (\bar{j}, \bar{i})$ denote two complementary arcs of the constraint graph. Then, we have

$$
\tilde{x}_a = x_a - \ell_a \overset{(2)}{=} (\pi_j - \pi_i - \ell_a) \bmod T
$$

$$
\overset{(5)}{=} (2s - \pi_{\bar{j}} - (2s - \pi_{\bar{i}}) - \ell_a) \bmod T
$$

$$
= (\pi_{\bar{i}} - \pi_{\bar{j}} - \ell_{\bar{a}}) \bmod T = x_{\bar{a}} - \ell_{\bar{a}} = \tilde{x}_{\bar{a}}.
$$

"⇐": Let x be the periodic tension of some feasible timetable π. We show that there exists one global symmetry axis s such that Condition (5) is satisfied for π.

We compute s from an arbitrary fixed event, say i,

$$
s := \frac{(\pi_i + \pi_{\bar{i}}) \bmod T}{2}.
$$

Now, we consider an arbitrary pair of complementary events j and $\bar{j}$. Since D is connected and symmetric, there exists a path P from i to j or $\bar{j}$ that only contains arcs a such that $\bar{a} \in A \setminus \{a\}$. We assume w.l.o.g. that P starts at i and ends at j. By setting

$$
x_P := \sum_{a \in P^+} x_a - \sum_{a \in P^-} x_a,
$$

we obtain $\pi_j = (\pi_i + x_P) \bmod T$. As for every $a \in P$ there exists its complementary arc $\bar{a} \in A \setminus \{a\}$, the complementary path $\overline{P}$ of P from $\bar{j}$ to $\bar{i}$ is well-defined. Equation (6) ensures $x_{\overline{P}} = x_P$.

In total, we obtain

$$
\frac{(\pi_j + \pi_{\bar{j}}) \bmod T}{2} = \frac{(\pi_i + x_P + \pi_{\bar{i}} - x_{\overline{P}}) \bmod T}{2} = \frac{(\pi_i + \pi_{\bar{i}}) \bmod T}{2} = s. \qquad \square
$$

Remark 1. If the line plan of a traffic network is connected and the constraint graph is symmetric, we are able to give an even more compact characterization of symmetry. Then, a feasible tension encodes a symmetric timetable, if and only if Condition (6) is satisfied for changeover arcs and stopping arcs. In fact, in the proof of Lemma 5 we can then find a path that only uses such arcs, plus trip arcs, which we assume to have zero span.

Surely, one can introduce a certain tolerance Δ on the symmetry requirement. But notice that in this case, condition (6) has to be blown up by a new integer variable.

Example 1 (Deutsche Bahn AG). Figure 12 shows two real-world timetable queries for opposite directions. These are representative for large parts of central European countries, such as Germany and Switzerland, which are operated with symmetry axis zero within only minor tolerances. Hence, if not stated otherwise we assume $s = 0$ throughout this paper for ease of notation.

We check the three characterizations of symmetry. Most striking, the changeover waiting time is almost the same in both directions, cf. Remark 1 and

Station/Stop	Date	Time	Platform	Products	Comments
Berlin Zoologischer Garten	05.06.03	dep 09:54	4	ICE 952	InterCityExpress
Wolfsburg		dep 10:54			BordRestaurant
Hannover Hbf		dep 11:31			
Bielefeld Hbf		dep 12:24			
Hamm(Westf)		dep 12:54			
Hagen Hbf		dep 13:25			
Wuppertal Hbf		dep 13:42			
Köln-Deutz		dep 14:11			
Köln Hbf	05.06.03	arr 14:14	6		
Köln Hbf	05.06.03	dep 15:13	8	ICE 14	InterCityExpress
Aachen Hbf		dep 15:52			Onboard meeting place
Aachen Süd(Gr)					
Liege-Guillemins					
Bruxelles-Midi	05.06.03	arr 17:46			

Duration: 7:52; runs daily

All information is issued without liability. Software/Data: HAFAS 5.00.DB.4.5 - 20.05.03 [5.00.DB.4.5/v4.05.p0.13_data:59e79704]

Station/Stop	Date	Time	Platform	Products	Comments
Bruxelles-Midi	05.06.03	dep 12:16		ICE 15	InterCityExpress
Liege-Guillemins		dep 13:28			Onboard meeting place
Aachen Süd(Gr)					
Aachen Hbf		dep 14:10			
Köln Hbf	05.06.03	arr 14:46	3		
Köln Hbf	05.06.03	dep 15:47	2	ICE 953	InterCityExpress
Köln-Deutz		dep 15:51			BordRestaurant
Wuppertal Hbf		dep 16:17			
Hagen Hbf		dep 16:35			
Hamm(Westf)		dep 17:10			
Bielefeld Hbf		dep 17:37			
Hannover Hbf		dep 18:31			
Wolfsburg		dep 19:05			
Berlin Zoologischer Garten	05.06.03	arr 20:02	1		

Duration: 7:46; runs Mo - Fr, not 29. May, 9. Jun, 21. Jul, 15. Aug, 11. Nov
Hint: Prolonged stop

All information is issued without liability. Software/Data: HAFAS 5.00.DB.4.5 - 20.05.03 [5.00.DB.4.5/v4.05.p0.13_data:59e79704]

Fig. 12. Symmetric timetables in practice

Equation (6). To check Condition (5), we consider the arrival of ICE 952 in Köln Hbf and the complementary departure of ICE 953. The two events sum up to $(14 + 47) \bmod 60 \approx 0$, and the same can be observed for the Brussels trains. Finally, notice that the Berlin line has one of its meeting points between Köln-Deutz and Wuppertal Hbf, at minute zero, of course. To that end, we have to know that the trains from Berlin arrive at Köln-Deutz at minute 09, which is two minutes before its departure at minute 11.

Some practitioners consider the changeover condition in Remark 1 to be an important advantage of symmetric timetables. Even though this might depend on personal preferences, we do *not* consider this really to be a striking argument for symmetry. Actually, there are examples which prove that symmetric timetables are only suboptimal, even if the input data is symmetric ([13]).

Apparently there are not yet many discussions of symmetric timetables available. But among further motivations for symmetry, as they can be found in [13], the most convincing one seems to be that symmetry halves the complexity of an instance. This can in particular be useful if there are complex interfaces to international trains or to regional traffic, and when planning is performed manually. However, this argument should become less important in the future, as we think that PESP solvers achieve some more progress in performance, and hence find their way into practice.

With the following theorem, we are able to prove a conjecture that has been stated in [13].

Theorem 2. *Symmetry of periodic timetables cannot be guaranteed by only using PESP constraints* (1).

Proof. Consider the PESP instance in Figure 13 (b). The PESP constraints that relate the two opposite directions of the line to be considered model the two single tracks of the track map that is shown in Figure 13 (a). The minimum

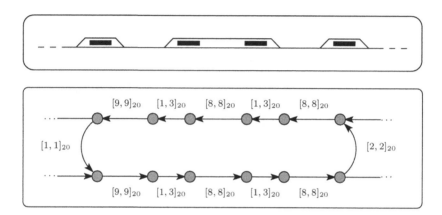

Fig. 13. A track map (a) on which an instance of PESP (b) does not admit any integral symmetric solution

crossing times (cf. Section 3.1) that apply to a certain single track depend on the infrastructure and the signaling system. For the western single track, we assume minimum crossing times of one time unit at both of its endpoints, for the eastern single track we assume two time units at both of its endpoints. Hence, given a period time of $T = 20$ time units and the indicated one-way running times of eight and nine time units, the single track constraints become tight.

Summing up the lower bounds of the constraints of the directed cycle yields 57, summing up the upper bounds provides 63. Hence, there exist feasible timetables. Moreover, due to the cycle periodicity property (Lemma 3), we know that in each of the feasible solutions the tension values sum up to 60. Hence, a slack of three time units has to be distributed on the four arcs with positive span.

In every symmetric feasible timetable, both of the directions obtain 1.5 time units of slack, hereby implying non-integral tension values. In contrast, by Lemma 1 every feasible system of PESP constraints (1) admits a feasible integral timetable.   □

Hence, we will have to add non-PESP constraints to the MIP formulations of a PESP instance in order to ensure symmetry. This is really required in practice, because in particular with national railway companies, we gained the experience that the symmetry requirement is really a knockout criterion.

To summarize, besides a linear objective function, symmetry is the second important requirement arising in the practice of periodic railway timetabling, by which the initial PESP model should be extended. Fortunately, in computations on real-world data sets it has been observed that MIP solvers may profit from the addition of symmetry constraints, in particular in formulation (6) ([13]). Such a generalized MIP model even inherits large parts of the structure of a pure PESP model. Most important, the cycle inequalities (4) remain valid.

5 Further Planning Steps Covered by the PESP

In the following, we will demonstrate that the modeling capabilities of the PESP are not limited only to periodic timetabling. Rather, central aspects of both preceeding and succeeding planning steps in the sense of Figure 1 can be integrated.

We start this discussion with the well-established technique of minimizing the number of vehicles required to operate a periodic timetable by penalizing waiting times of vehicles. Hereafter, we provide first ideas for the integration of important decisions of line planning. We close this section by proposing a way to model some specialized decisions arising in network planning.

5.1 Aspects of Vehicle Scheduling

Almost all companies in public transportation have in common that they want to minimize the amount of rolling stock required to serve their networks. Notice that the quality of the vehicle schedule for a fully periodic timetable, i.e. with no peak trips included, is largely determined by the timetable.

Consider for example the hourly line displayed in Figure 14 (a). Assume the minimal travel times between the two endpoints to be 235 minutes for each direction. Given strict minimal turnover times of 45 and 60 minutes, respectively, the minimal number of vehicles required to operate this line is precisely

$$N := \left\lceil \frac{1}{60}(235 + 235 + 45 + 60) \right\rceil = 10.$$

A timetable which lets the trains leave at the full hour from Frankfurt and Amsterdam can indeed be operated with only 10 trains, at least if the stopping times are extended only moderately. On the contrary, a timetable in which only the trains starting at Frankfurt depart at minute 00, but the trains from Amsterdam leave at minute 30 requires at least 11 vehicles. Hence, the amount of vehicles depends on the timetable.

We will analyze in which special cases pure PESP constraints are able to control the number of trains required. After that, we show that a linear objective function covers many more of the practical cases.

Proposition 3 (Nachtigall [20]). *Consider a fixed traffic line with period time T. If we assume trains always to serve only this line, and if we do not allow to insert additional stopping time, then there exist upper bounds u for the turnover activities, such that the only feasible timetables are those which can be operated with the minimal amount of trains.*

Proof. We present a proof of this simple fact, both in order to provide the notation used in the following paragraphs, and because it avoids modulo-notation.

Denote the endpoints of the line by A and B. Let ℓ_{AB} denote the minimal travel time from A to B, i.e. the sum of the minimal stopping and running times of the activities of this directed traffic line. Moreover, denote by ℓ_B the minimal amount of time a train has to stay in endpoint B between two consecutive trips.

The minimal number N of trains required to operate this line is precisely

$$N = \left\lceil \frac{\ell_{AB} + \ell_B + \ell_{BA} + \ell_A}{T} \right\rceil.$$

From the cycle periodicity property (3) we know that every feasible timetable x fulfills

$$x_{AB} + x_B + x_{BA} + x_A = zT, \tag{7}$$

for some $z \in \mathbb{Z}$. Hence, we must ensure $z = N$. To that end, consider the slack

$$\sigma := NT - (\ell_{AB} + \ell_B + \ell_{BA} + \ell_A) \tag{8}$$

of this traffic line, implying $(x_A - \ell_A) + (x_B - \ell_B) = \sigma$. But since $\sigma < T$, by setting

$$u_A := \ell_A + \sigma \tag{9}$$

we even ensure $x_{AB} + x_B + x_{BA} + x_A < (N+1)T$.   □

Let us now analyze the case in which additional stopping times may be inserted, i.e. $u_{AB} > \ell_{AB}$. We will show that together with the constraints (9), some timetables which require an additional train may become feasible.

On the one hand, consider a timetable for which we have $x \equiv \ell$ for all activities, except for the turnover time in one endpoint. This timetable can still be operated with the minimal number of trains, showing that decreasing the value (9) for u_A would cut off timetables we seek for.

On the other hand, assume $x_{AB} = u_{AB}$ and $x_{BA} = u_{BA}$. If

$$(u_{AB} - \ell_{AB}) + (u_{BA} - \ell_{BA}) + \sigma \geq T, \tag{10}$$

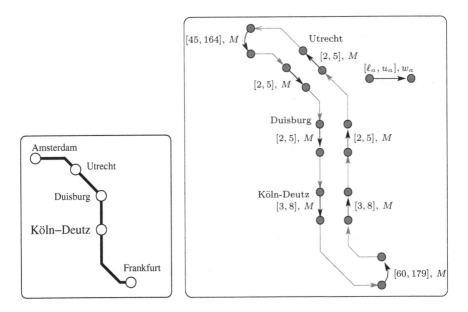

Fig. 14. Modeling aspects of vehicle scheduling: (a) line plan; (b) PESP constraints measuring the number of trains required to operate the line

then we can extend x to a timetable that still respects (9), but which requires at least one additional train. For instance, if inequality (10) is tight, then for $x \equiv u$ we have

$$
\begin{aligned}
x_{AB} + x_B + x_{BA} + x_A &= u_{AB} + u_B + u_{BA} + u_A \\
&\overset{(9)}{=} (u_{AB} - \ell_{AB}) + (\ell_B + \sigma) + (u_{BA} - \ell_{BA}) + \\
&\quad + (\ell_A + \sigma) + \ell_{AB} + \ell_{BA} \\
&\overset{(10)}{=} T + \sigma + \ell_{AB} + \ell_B + \ell_{BA} + \ell_A \\
&\overset{(8)}{=} (N + 1)T.
\end{aligned}
$$

The above dilemma is our main motivation for the need of a linear objective function. Such a function takes advantage of equation (7): By assigning a value M to the arcs modeling a traffic line, every additional train pays $M \cdot T$ to the objective function value. Of course, if suffices to consider arcs with positive span, cf. Figure 14 (b). If the value for M is chosen relatively large compared to the passenger weights, the objective function essentially models the piecewise constant behavior of the cost of the rolling stock for operating the railway network.

From a more local perspective, we just penalize idle time of trains. But this can even be done without knowing a priori the circulation plan of the trains. Although a straight-forward exact model involves a quadratic objective function, Liebchen and Peeters [14] report that a simple linear relaxation in terms of the PESP yields results of high quality.

Very recently, Nyhave, Hove, and Clausen [22] proposed an integer linear model to precisely count the number of trains required to operate a timetable, even if trains are allowed to switch lines in their endpoints. This approach does not depend on additional assumptions as synchronization constraints or pre-defined time-windows for turnaround times, as they were used by Peeters [26]. In the sequel, we translate their ideas into the PESP plus some additional variables and constraints.

Consider a station S that is a terminus for the two lines 1 and 2. Denote by a_i and d_i the arrival and departure events in station S of line i. We introduce the following arcs

$$a_{11} = (a_1, d_1) \text{ and } a_{22} = (a_2, d_2),$$
$$a_{12} = (a_1, d_2) \text{ and } a_{21} = (a_2, d_1).$$

The effective waiting times for the trains in S are $\tilde{x}_{11} + \tilde{x}_{22}$ if trains stay on their lines, or $\tilde{x}_{12} + \tilde{x}_{12}$ if trains switch lines. Notice that $(a_{11}, a_{21}, a_{22}, a_{12})$ is an oriented cycle. In particular, there exists some $z \in \mathbb{Z}$ such that

$$x_{11} + x_{22} = x_{12} + x_{21} + zT.$$

In most cases, we have $\ell_{a_{11}} = \ell_{a_{12}}$ and $\ell_{a_{21}} = \ell_{a_{22}}$. Then, we even know that there exists some $r \in [0, T)$ and $b_1, b_2 \in \{0, 1\}$ such that

$$r = \tilde{x}_{11} + \tilde{x}_{22} - b_1 \cdot T = \tilde{x}_{12} + \tilde{x}_{21} - b_2 \cdot T.$$

Hence, in an optimal vehicle schedule the total effective waiting time in station S amounts to $r + \min\{b_1, b_2\} \cdot T$. To obtain a MIP-formulation, we introduce a new (rational) variable w and require

$$w \geq b_1 + b_2 - 1 \text{ and } w \geq 0.$$

Finally, station S contributes

$$M \cdot (r + w \cdot T)$$

to the objective function, where M again denotes the cost factor for vehicle waiting time.

5.2   Aspects of Line Planning

Our main idea for letting PESP solvers even take decisions of line planning is to combine — or match — pre-defined line-segments. To that end, we will make intensive use of disjunctive constraints. Unfortunately, we will only be able to ensure symmetric line plans if we require symmetry also within the stations where lines are matched.

We are aware of only one other approach for integrating the planning phases of line planning, timetabling and vehicle scheduling ([32]). Whereas that approach is based on the assumption that the line plan contains no cycles, our ideas do not require any restrictive assumptions on the topology of the network. Rather, we are able to keep even very important technical restrictions such as single tracks.

Notice that bad decisions at the level of line planning may cause very bad results also for vehicle scheduling. Consider the four line segments displayed in Figure 15. We assume a period time of $T = 60$ minutes and a minimal turnover time of 30 minutes at each of the four terminus stations. The time for a one-way trip from the matching station to one of the endpoints is indicated at the corresponding edge.

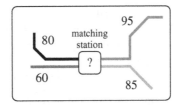

Fig. 15. Line segments where only one matching provides good vehicle schedules

In fact, the vehicle schedule is fixed due to the distinct endpoints. Combining the south-west segment with the north-east segment causes this line to require at least

$$\left\lceil \frac{1}{60}(60 + 95 + 30 + 95 + 60 + 30) \right\rceil = \left\lceil \frac{370}{60} \right\rceil = 7 \text{ trains.}$$

The other line of the same matching requires seven trains, too.

In contrast, the other matching implies seven trains only for the northern line consisting of the two top line segments. But the other line can be operated with only six trains. Hence, already the line plan has a major impact on the cost of operation. Claessens et al. [5] consider this phenomenon in their approach for constructing cost-optimal line plans.

However, they omit the important intermediate linking step of computing a timetable. Therefore, their approach must also consider possible constellations in which there is no feasible timetable using only six trains for the southern line. This would be the case, if there was a single track with travel time 25 minutes for every direction just at the end of the south-east segment. The same holds if it is required that the two lines together form an exact half-hourly service along the backbone of the network.

We consider a track that has to be served in the same direction by n directed lines which are operated by trains of identical type. We denote the matching station by S which resides between the two endpoints of the common track. We consider n line segments $L_1^a, \ldots, L_n^a$ which have station S as their common endpoint, and n line segments $L_1^d, \ldots, L_n^d$ having station S as their common starting point. Any (bipartite) perfect matching between the arriving and the departing line segments induces a line plan.

But from the perspective of timetabling, there are only n arrival events $a_1, \ldots, a_n$ as well as n departure events $d_1, \ldots, d_n$ visible. Hence, we must deduce only from their arrival times π_{a_i} and their departure times π_{d_j} which arriving line segment L_i^a should be matched with which departing line segment L_j^d.

This can be done in a canonical way, if we choose the matching station S such that it has only one track in the direction of the line segments we consider. If necessary, we add an artificial station in the middle of some track. Then, at most one train can be in S at the same time. Timetables respecting this constraint can be characterized very easily as follows.

Definition 1 (Alternating timetable). *For a fixed station S and a fixed direction, a periodic timetable π with n pairwisely different arrival times $0 \leq \pi_{a_1} < \cdots < \pi_{a_n} < T$ and n pairwisely different departure times $0 \leq \pi_{d_1} < \cdots < \pi_{d_n} < T$ is called* alternating *at S, if either $\pi_{a_i} \leq \pi_{d_i} < \pi_{a_{i+1}}$ for every $i = 1, \ldots, n$, or $\pi_{d_i} < \pi_{a_i} \leq \pi_{d_{i+1}}$ for every $i = 1, \ldots, n$, where we define $\pi_{\cdot_{n+1}} := \pi_{\cdot_1} + T$.*

Lemma 6. *A timetable π ensures that there is always at most one train at station S if and only if it is alternating at S.*

Hence, for an alternating periodic timetable, we combine the arriving line segment L_i^a with the departing line segment L_j^d, if and only if the latter marks the unique first possible departure. In the sequel, we will give PESP constraints ensuring every feasible timetable to be alternating at S. Thus, every feasible timetable will encode some unique matching and the associated line plan.

The first two sets of constraints ensure the minimal headway d in front of and behind the matching station S:

$$\forall i, j \in \{1, \ldots, n\} : \pi_{a_j} - \pi_{a_i} \in [d, T - d]_T, \tag{11}$$

$$\forall i, j \in \{1, \ldots, n\} : \pi_{d_j} - \pi_{d_i} \in [d, T - d]_T. \tag{12}$$

Notice that (11) and (12) can only be fulfilled if $0 \leq d \leq \frac{T}{n}$. Moreover, we relate arrival events to departure events by the following disjunctive constraints

$$\forall i, j \in \{1, \ldots, n\} : \pi_{d_j} - \pi_{a_i} \in [0, T - d + h]_T, \tag{13}$$

$$\forall i, j \in \{1, \ldots, n\} : \pi_{d_j} - \pi_{a_i} \in [d, T + h]_T, \tag{14}$$

where we denote by h the maximal stopping time for a train at station S. Together, these constraints (13) and (14) yield

$$(\pi_{d_j} - \pi_{a_i}) \bmod T \in [0, h] \,\dot\cup\, [d, T - d + h]. \tag{15}$$

Trivially, $0 \leq h < d$ is necessary for every feasible timetable π to be alternating at S.

Theorem 3. *Let π be a timetable respecting constraints (11) to (14). Then for every departure event d_j, there exists a* unique *arrival event a_i satisfying*

$$\pi_{d_j} - \pi_{a_i} \in [0, h]_T, \tag{16}$$

if and only if $h < (n + 1)d - T$.

Since $0 \leq h$, from $h < (n+1)d - T$ we conclude $\frac{T}{n+1} < d$.

Proof. "$\Rightarrow$": We assume $h \geq (n+1)d - T$. Since $d = \frac{T}{n}$ would imply $h \geq d$, we must only investigate the case that $d < \frac{T}{n}$. We will construct a timetable which respects the constraints (11) to (14), but which contradicts (16).

Define $\pi_{a_i} := (i-1)d$, for all $i = 1, \ldots, n$, and $\pi_{d_j} := j \cdot d$, for all $j = 1, \ldots, n$. By construction, all the constraints are satisfied. However, since $\pi_{a_n} + h < n \cdot d = \pi_{d_n}$, for departure π_{d_n} none of the arrival events fulfills (16), q.e.d.

"$\Leftarrow$": We assume there exists a timetable π having one departure event d_0 such that

$$\forall i = 1, \ldots, n : (\pi_{d_0} - \pi_{a_i}) \bmod T > h,$$

but which respects the constraints (11) to (14). We may assume w.l.o.g. that for the cyclic predecessor arrival a_1 of d_0 we have $\pi_{a_1} = 0$. As π is feasible, it satisfies (15). From our assumption, we conclude $d \leq \pi_{d_0}$ and $\pi_{d_0} + (d-h) \leq \pi_{a_2}$, and hence $\pi_{a_2} - \pi_{a_1} \geq 2d - h$. Event a_1 also takes place at time T. For notational convenience, we define $\pi_{a_{n+1}} := T$. With this notation, we have $\pi_{a_{i+1}} - \pi_{a_i} \geq d$, for all $i = 2, \ldots, n$. By the definition of $\pi_{a_{n+1}}$, we know that

$$\sum_{i=1}^{n} (\pi_{a_{i+1}} - \pi_{a_i}) = \pi_{a_{n+1}} - \pi_{a_1} = T.$$

Summing up the lower bounds yields $T \geq (n+1)d - h$, which contradicts the hypothesis of Theorem 3. $\qquad \square$

Corollary 1. *If $h < (n+1)d - T$, then every timetable which respects constraints* (11) *to* (14) *is an alternating timetable.*

In Figure 16, we provide an example for the easiest case, namely matching two lines. As usual, we assume the period time to be 60 minutes.

Remark 2. There are of course alternating periodic timetables in the case $d \leq \frac{T}{n+1}$. PESP solvers are able to detect even those, if we were able to pre-define sufficiently many empty slots. By an "empty slot" we understand an artificial line which we have to schedule in the same way as the original lines, hereby separating the lines before and after the empty slot.

In more detail, let us assume that $\frac{T}{n^*+1} < d \leq \frac{T}{n^*}$ for some $n^* > n$, and that h satisfies the assumptions of Theorem 3 for n^*. We then introduce $n^* - n$ artificial dummy arrival and departure events a_i and d_i, $i = n+1, \ldots, n^*$. To prevent the original line segments from being matched with an artificial event, we require $\pi_{d_i} - \pi_{a_i} \in [0, h]$ for all $i = n+1, \ldots, n^*$.

By construction, the only feasible timetables let the original arrivals and departures alternate. However, perfectly balanced timetables, i.e. $\pi_{a_i} := (i-1)\frac{T}{n}$, are infeasible under these settings if $n^* < 2n$, since they do not provide $n^* - n$ empty slots.

Recall that so far we have considered only one direction. Hence, there is no mechanism yet to bind the matching of one direction to that of the opposite

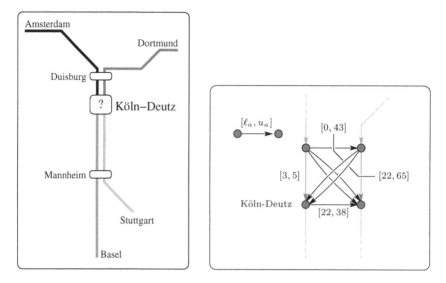

Fig. 16. Modeling aspects of line planning: (a) line segments; (b) PESP constraints ensuring the segments to be matched

direction. But the matchings of opposite directions must fulfill the symmetry assumption that we gave at the beginning of Section 4.2. Otherwise, the trains from direction A could pass the matching station S in order to continue towards B, but the trains from B pass S before continuing in direction C. Thus, it would not be possible to communicate the line plan in the way customers are used to, because it may no more be visualized by an undirected graph. However, limited asymmetries in operation are accepted in practice.

Example 2 (S-Bahn Berlin GmbH). We consider the line S2 serving the route Blankenfelde-Lichtenrade-Buch-Bernau. Between Lichtenrade and Buch, a ten minutes frequency must be offered, for the remaining parts 20 minutes suffice.

In the current timetable ([27]), this line is served in an asymmetric way. In order to cope with the single tracks (which are present at both endpoints) to limit the total amount of stopping time, and to ensure an efficient employment of the rolling stock, an asymmetric service is offered, and we present it in table 4.

In order to ensure symmetric line plans, we have to guarantee the following condition. If we combine the arrival event a_i with the departure event d_j in one direction, then in the opposite direction the complementary arrival event a'_j must be combined with the departure event d'_i. More precisely, when considering the corresponding tension variables $x_{a_i d_j}$ and $x_{a'_j d'_i}$, they must fulfill

$$x_{a_i d_j} \in [0, h] \Leftrightarrow x_{a'_j d'_i} \in [0, h]. \tag{17}$$

In fact, this condition is quite similar to the symmetry constraints (6). What makes things more complicated is the fact that we must not predict in advance

Table 4. Asymmetric service of line S2 (Berlin)

Blankenfelde	dep			10:09			arr o	11:14	
Lichtenrade	dep ↓	10:15 10:25		arr o	11:05 11:15
Buch	arr o	11:06 11:16		dep ↑	10:14 10:24
Bernau	arr o	11:21		dep			10:10

for which pairs (i, j) requirement (17) has to hold, and for which pairs it may be violated. Hence, we propose to guarantee property (17) for the matched pairs by imposing symmetry requirements on *every* pair of complementary junctions. But it is clear that this approach cuts off feasible timetables for symmetric line plans just because such timetables need not to be symmetric, see e.g. example 3.

Example 3 (S-Bahn Berlin GmbH). Consider the current timetable ([27]) of the ring subnetwork of S-Bahn Berlin GmbH, of which we provide an excerpt in table 5. Obviously, the line plan is symmetric. But the timetable is not symmetric.

Table 5. Symmetric line plan but asymmetric timetable

	Direction A					
Line	S45	S46	S8	S9	S47	S8
Origin	BFHS	BKW	BGA	BFHS	BSPF	BZN
Schöneweide dep ↓	xx:01	xx:06	xx:10	xx:13	xx:15	xx:18
Baumschulenweg arr o	xx:03	xx:09	xx:13	xx:16	xx:17	xx:21
Destination	BHMS	BGS	BPKR	BZOO	BWES	BPKR
	Direction B					
Line	S8	S46	S9	S47	S8	S45
Origin	BPKR	BGS	BZOO	BWES	BPKR	BHMS
Baumschulenweg dep ↓	xx:02	xx:06	xx:08	xx:13	xx:14	xx:19
Schöneweide arr o	xx:05	xx:08	xx:10	xx:15	xx:17	xx:21
Destination	BGA	BKW	BFHS	BSPF	BZN	BFHS

This can be seen by calculating the symmetry axes of lines S47 and S9 at station Schöneweide. Departure and arrival of line S47 sum up to 30, hence the trains of this line meet at times 5 and 15. For line S9 the sum yields 23, providing a symmetry axis of 1.5. An easier argument for asymmetry is that the sequence of the trains in direction B is not the inverse of the one in direction A.

There are two main objectives for the matching approach. First, we want to offer direct trips for as many passengers as possible. Second, the timetable should require only few trains for operation.

For the second criterion, in the case $h = 0$, no additional weight on arcs within the matching node is required in order to minimize the amount of rolling stock required to operate the timetable. In the case $h > 0$, one could put the vehicle weight on the arcs with feasible interval $[0, T - d + h]$. But this would no longer yield the desired exact piecewise-constant behavior of the objective, because some double counting can appear.

For maximizing the number of direct travelers, we consider the number of passengers w_{ij} starting their trip before the common track on a train covering line segment L_i^a, and finishing their trip after the common endpoint on a train covering line segment L_j^d. The value w_{ij} is added to the weight of the arc $a = (a_i, d_j)$ with $\ell_a = 0$ and $u_a = [0, T - d + h]$. The resulting cost coefficients in the objective function make even sense for pairs of line segments which are not matched, because long changeover times of many passengers are penalized.

Notice that the values w_{ij} are only well-defined if the two line segments do not serve a second matching station. This shows that the decisions to be taken within a matching station are of a rather local nature.

Summarizing, there are important scenarios in which the PESP can integrate relevant aspects of line planning into a model suited for timetabling and key issues of vehicle scheduling. This is in particular the case if symmetric timetables and balanced sequences along the common tracks, i.e. $d > \frac{T}{n+1}$, are requested for their own sake. Moreover, we observed that the larger the distance between two matching stations, the more reliable the passenger weight that we propose.

We think that fast train networks of European agglomerations, such as Frankfurt, Munich, or Paris (RER), are well-suited candidates for this approach. There, many passengers might have their origin or destination somewhere on the backbone route, and balanced sequences must be ensured due to the large number of lines per period.

5.3 Aspects of Network Planning

We propose to also model two questions which arise in network planning within the PESP: the extension of existing tracks, and thus lines, beyond their current endpoints, and the construction of faster tracks as substitution for existing ones. Taking into account that, in these questions, we have to select one option out of a small number of disjoint options, it is evident that we will make intensive use of disjunctive constraints, cf. Section 3.3. Recall that there, we already discussed the introduction of optinal additional stops. With appropriate weights that reflect amortisation—see below—these may also cover the construction of new stations along an existing track.

We only discuss the construction of faster tracks in detail. But the reader will have no difficulty to adapt our suggestions to the very similar task of the extension of tracks.

In Figure 17, we provide a constraint graph which offers the option of a new track between Aachen and Köln, being then part of the European high-speed line PBK (Paris-Brussels-Köln). We provide the status quo, with one intermediate stop, only for illustration purposes. In the future, we have the option to either use the current tracks, thus keeping a trip time of 38 minutes, or to establish the new high-speed track, hereby reducing the trip time down to 26 minutes.

To define appropriate weights for the arcs, we have to take into account three different types of objectives: The number of customers c who profit from a new

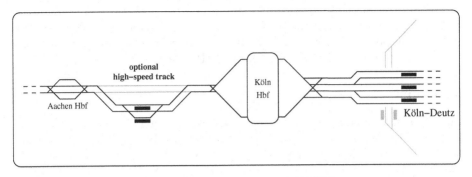

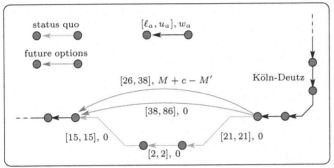

Fig. 17. Modeling aspects of network planning: (a) infrastructure including optional high-speed track; (b) PESP constraints taking into account the two infrastructural alternatives

track by shorter travel times, the trip times of the trains trains which may allow to reduce the number of trains requires (M, c.f. Section 5.1), and the cost M' of the investment. One can imagine that it is an absolutely non-trivial management decision to derive an hourly weight M' from the total cost of the investment.

Similarly to line planning, investments into infrastructure will only make sense if they are effected for both directions at the same time. Again, we ensure symmetric investments by requiring the timetable to be symmetric.

Let us now analyze the situation in which several lines have the option of using the same new, faster track. Of course, we want to ensure that infrastructure is only paid once in terms of the objective function. Hence, we have to partition the total cost onto all of the concerned lines. But what if in a solution of a PESP instance only one line is routed over the new track?

But a reasonable allocation of the total costs is only possible, if we know in advance how many lines will have to use the new track. Unfortunately, we are only able to ensure this with constraints of the types already introduced, if *all* the lines must use the same track. This would, e.g., be the case when analyzing two mutually exclusive variants of constructing a new track.

We can guarantee that all the lines use the same track simply by enforcing the same running time for each line. This is achieved by introducing constraints of type (6). But notice that in this case we cheat a bit, because those constraints no longer relate only pairs of complementary arcs to each other... Anyway, the MIP formulation of this even slightly more extended model incorporates many of the computational aspects of the pure PESP model.

6   Conclusion

Our discussion of the PESP model shows that it has a great modeling power and extendability. We have demonstrated that many non-standard requirements for periodic timetables and also important aspects of other – traditionally separate – planning phases can be integrated into the PESP. Figure 18 displays the gain by this modeling power over the traditional use of the PESP displayed in Figure 1.

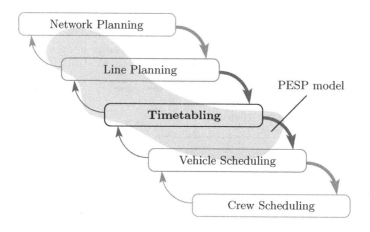

Fig. 18. Planning phases covered by the PESP with our contribution

Interestingly, this integration into the PESP has been possible without seemingly complicating it too much. In all cases, we obtained mixed integer programs that still have the characteristics of a PESP. Hence we believe that these extended models stay computationally tractable also for networks of relevant sizes. So far, our belief is confirmed by a confidential study for S-Bahn Berlin GmbH for two of its three major subnetworks.

We therefore hope that these models, through their integrative approach to vehicle scheduling, timetabling, line planning, and infrastructure planning, will eventually lead to better decision making in practice.

Acknowledgments

We want to thank the staff of Deutsche Bahn AG, S-Bahn Berlin GmbH, and Berliner Verkehrsbetriebe (BVG) for providing us with both real-world data and very detailed requirements of their specific periodic timetabling problems. Moreover, we thank the referees for their *very* detailed suggestions.

References

1. Bollobás, B.: Modern Graph Theory. Graduate Texts in Mathematics, vol. 184 (2nd printing) Springer, Heidelberg (2002)
2. Borndörfer, R., Grötschel, M., Pfetsch, M.E.: A path-based model for line planning in public transport. Technical Report 05-18, Zuse Institute, Berlin (2005)
3. Borndörfer, R., Löbel, A., Weider, S.: Integrierte Umlauf- und Dienstplanung im öffentlichen Nahverkehr. In: HEUREKA '02: Optimierung in Transport und Verkehr, Tagungsbericht, number 002/72. FGSV Verlag (in German) (2002)
4. Bussieck, M.R., Winter, T., Zimmermann, U.: Discrete optimization in public rail transport. Mathematical Programming B 79, 415–444 (1997)
5. Claessens, M.T., van Dijk, N.M., Zwaneveld, P.J.: Cost optimal allocation of rail passenger lines. European Journal of Operational Research 110(3), 474–489 (1998)
6. Engelhardt-Funke, O., Kolonko, M.: Analysing stability and investments in railway networks using advanced evolutionary algorithms. International Transactions in Operational Research 11, 381–394 (2004)
7. Grötschel, M., Löbel, A., Völker, M.: Optimierung des Fahrzeugumlaufs im öffentlichen Nahverkehr. In: Hoffmann, K.H., Jäger, W., Lohmann, T., Schunck, H. (eds.) Mathematik - Schlüsseltechnologie für die Zukunft, Springer, Heidelberg (in German) (1997)
8. Haase, K., Desaulniers, G., Desrosiers, J.: Simultaneous vehicle and crew scheduling in urban mass transit systems. Transportation Science 35(3), 286–303 (2001)
9. Krista, M.: Verfahren zur Fahrplanoptimierung am Beispiel der Synchronzeiten. Ph.D. thesis, Technische Universität Braunschweig (in German) (1997)
10. Kroon, L.G., Peeters, L.W.P.: A variable trip time model for cyclic railway timetabling. Transportation Science 37, 198–212 (2003)
11. Leuschel, I.: Der Fernverkehrsfahrplan 2003 der Deutschen Bahn AG. Eisenbahntechnische Rundschau (In German) 51(7–8), 452–464 (2002)
12. Liebchen, C.: Finding short integral cycle bases for cyclic timetabling. In: Di Battista, G., Zwick, U. (eds.) ESA 2003. LNCS, vol. 2832, pp. 715–726. Springer, Heidelberg (2003)
13. Liebchen, C.: Symmetry for periodic railway timetables. Electronic Notes in Theoretical Computer Science 92, 34–51 (2004)
14. Liebchen, C., Peeters, L.: Some practical aspects of periodic timetabling. In: Chamoni, P., Leisten, R., Martin, A., Minnemann, J., Stadtler, H. (eds.) Operations Research 2001, Springer, Heidelberg (2002)
15. Liebchen, C., Proksch, M., Wagner, F.H.: Performance of algorithms for periodic timetable optimization. In: Hickman, M. (ed.) Computer-Aided Transit Scheduling— Proceedings of the Ninth International Workshop on Computer-Aided Scheduling of Public Transport. Lecture Notes in Economics and Mathematical Systems, Springer, Heidelberg (to appear, 2005)
16. Lindner, T.: Train Schedule Optimization in Public Rail Transport. Ph.D. thesis, Technische Universität Braunschweig (2000)
17. Rhein-Main Verkehrsverbund: Fahrplanbuch Gesamtausgabe (gültig ab 14. Dezember 2003) (2003)
18. Nachtigall, K.: A branch and cut approach for periodic network programming. Hildesheimer Informatik-Berichte 29, Universität Hildesheim (1994)
19. Nachtigall, K.: Cutting planes for a polyhedron associated with a periodic network. Institutsbericht IB 112-96/17, Deutsche Forschungsanstalt für Luft- und Raumfahrt e.V (July 1996)

20. Nachtigall, K.: Periodic Network Optimization and Fixed Interval Timetables. Habilitation thesis, Universität Hildesheim (1998)
21. Nachtigall, K., Voget, S.: A genetic algorithm approach to periodic railway synchronization. Computers and Operations Research 23(5), 453–463 (1996)
22. Nielsen, M.N., Hove, B., Clausen, J.: Constructing periodic timetables using MIP— a case study from DSB S-train. International Journal of Operations Research, 1 (2005)
23. Odijk, M.A.: Construction of periodic timetables, Part 1: A cutting plane algorithm. Technical Report 94-61, TU Delft (1994)
24. Odijk, M.A.: A constraint generation algorithm for the construction of periodic railway timetables. Transportation Research B 30(6), 455–464 (1996)
25. Peeters, L.W.P.: Personal Communication (2000)
26. Peeters, L.W.P.: Cyclic Railway Timetable Optimization. Ph.D. thesis, Erasmus Universiteit Rotterdam (2003)
27. S-Bahn Berlin GmbH: S-Bahn-Fahrplan (gültig ab 16. Juni 2003) (2003)
28. Schrijver, A.: Theory of Linear and Integer Programming, 2nd edn. Wiley, Chichester (1998)
29. Schrijver, A., Steenbeek, A.G.: Dienstregelingontwikkeling voor Nederlandse Spoorwegen N.S. Rapport Fase 1, Centrum voor Wiskunde en Informatica (Oktober 1993)
30. Serafini, P., Ukovich, W.: A mathematical model for periodic scheduling problems. SIAM Journal on Discrete Mathematics 2(4), 550–581 (1989)
31. van den Berg, J.H.A., Odijk, M.A.: DONS: Computer aided design of regular service timetables. In: Murthy, T.K.S., Mellitt, B., Brebbia, C.A., Sciutto, G., Sone, S. (eds.) Computers in Railways IV (COMPRAIL)—vol. 2: Railway Operations. WIT Press (1994)
32. Völker, M.: Ein multikriterieller Algorithmus zur automatisierten Busliniennetzplanung. Lecture on the OR Workshop Optimierung im öffentlichen Nahverkehr (2003)

Cyclic Railway Timetabling:
A Stochastic Optimization Approach

Leo G. Kroon[1,2], Rommert Dekker[3], and Michiel J.C.M. Vromans[4]

[1] NS Reizigers, Department of Logistics,
3500 HA, Utrecht, The Netherlands
[2] Rotterdam School of Management
L.Kroon@rsm.nl
[3] Rotterdam School of Economics,
Erasmus University Rotterdam
3000 DR, Rotterdam, The Netherlands
R.Dekker@few.eur.nl
[4] ProRail, Network Planning,
3500 GA, Utrecht, The Netherlands
Michiel.Vromans@prorail.nl

Abstract. Real-time railway operations are subject to stochastic disturbances. However, a railway timetable is a deterministic plan. Thus a timetable should be designed in such a way that it can absorb the stochastic disturbances as well as possible. To that end, a timetable contains buffer times between trains and supplements in running times and dwell times. This paper first describes a stochastic optimization model that can be used to find an optimal allocation of the running time supplements of a single train on a number of consecutive trips along the same line. The aim of this model is to minimize the average delay of the train. The model is then extended such that it can be used to improve a given cyclic timetable for a number of trains on a common railway infrastructure. Computational results show that the average delay of the trains can be reduced substantially by applying relatively small modifications to the timetable. In particular, allocating the running time supplements in a different way than what is usual in practice can be useful.

1   Introduction

Punctuality of railway services is a highly important issue, since punctuality is considered as one of its main performance indicators. In the Netherlands, punctuality is defined as the percentage of trains that arrive with a delay of less than 3 minutes at one of the larger railway stations. In several other countries, a 5 minute margin is used, or only the delays at the end points of a line are taken into account. Delays of trains occur since real-time railway operations are subject to stochastic disturbances. However, the underlying railway timetable is a deterministic plan. Therefore, the stochastic disturbances in the operations should be taken into account in the design of a timetable as well as possible: the timetable

F. Geraets et al. (Eds.): Railway Optimization 2004, LNCS 4359, pp. 41–66, 2007.

should be *robust*. In order to cope with the disturbances in the real-time operations, a timetable contains buffer times between trains and supplements in the
running times and in the dwell times of the trains.

Many authors addressed the analysis and the improvement of the punctuality
of railway services: several relevant models have been developed to that end.
The main examples of these models are the following: (*i*) simulation models (see
Bergmark [1], König [10], Middelkoop and Bouwman [13], and Wahlborg [24]),
(*ii*) Max-Plus models (see Goverde [4], De Kort [11], and Soto Y Koelemeijer et
al. [21]), and (*iii*) analytical models (see Carey [3], Higgins and Kozan [7], and
Huisman and Boucherie [8]). Other relevant literature on stochastic methods for
the improvement of railway timetables is Hallowell and Harker [6], Schwanhäußer
[19], Mühlhans [14], and Petersen and Taylor [17]. However, a drawback of the
existing models is that they are mainly *evaluation* models and that, based on
these models, *optimization* of a timetable can only be achieved by trial-and-error:
the timetable is modified and then the evaluation model is used afterwards to
evaluate the effect of the modification. If necessary, these steps are repeated.

In contrast with the existing models, the current paper describes a *stochastic
optimization model* that can be used to modify a given cyclic timetable and,
at the same time, to evaluate the modified timetable by operating a number of
realizations of the trains in the timetable. We refer to Birge and Louveaux [2]
for more information on stochastic optimization. In our model, the trains are
operated as much as possible according to the modified timetable, but subject
to external stochastic disturbances. The main criterion that is used to modify
the timetable is minimization of the average delay of the trains. Note that other
criteria can be handled as well. The structure of the model is such that it is a
symbiosis of a *timetabling* model and a *simulation* model.

The first model in this paper generates a timetable for a single train that
is operated under stochastic external disturbances on a number of consecutive
trips along the same line. Here a trip is a movement of a train from one station
to the next. The model is used to allocate a fixed total amount of running time
supplement to the consecutive trips such that the average delay of the train
is minimal. The model is then extended to be applicable in a more complex
situation where several trains are operated according to a given cyclic timetable
and on a common railway infrastructure. These trains are also operated under
stochastic external disturbances. The extended model is used here to improve
the timetable with respect to the average delay of the trains by re-allocating
buffer times and time supplements. The application of the extended model to a
practical case shows that, within the model, the modification of a given timetable
may lead to a substantial reduction of the average delay of the trains.

This paper is structured as follows. Sect. 2 describes several aspects that are
relevant for the allocation of running time supplements. In Sect. 3 we describe
the above mentioned first stochastic optimization model. In Sect. 4, we prove
that, if the train runs over just *two* consecutive trips and if there is a finite
probability distribution of the disturbances, then the results of the stochastic
optimization model converge to the true optimum if the number of realizations

tends to infinity. In Sect. 5, we present the computational results related to the model of Sect. 3. Sect. 6 describes the above mentioned extended stochastic optimization model. Computational results that were obtained by applying this model to the railway corridor between Haarlem and Maastricht/Heerlen are described in Sect. 7. The paper is finished with conclusions in Sect. 8.

2 Running Time Supplements

2.1 A Trade-Off

To obtain a high punctuality of the railway services, it is desirable that trains are able to run faster than planned in order to make up for earlier delays. This means that the planned running times should be longer than the technically minimum running times. The difference between the planned running time and the technically minimum running time is the *running time supplement*.

Also other processes (e.g. halting at stations) may obtain time supplements in the planning. However, in this paper we mainly focus on the allocation of running time supplements. Note that these other process time supplements may be handled in the same way as the running time supplements.

In general, higher running time supplements lead to a better punctuality of the railway services. However, higher running time supplements also lead to higher *planned* running times. This means that the planned travel times of the passengers increase as well. Note that these planned travel times do not only depend on the total *amount* of running time supplements, but also on the *distribution* of the running time supplements among the trips in the timetable. Note further that running time supplements may even have a negative influence on the *realized* travel times. Indeed, each minute of running time supplement in the timetable brings the risk that it is not needed, since there are no disturbances. Furthermore, longer planned running times increase the block occupation times and therewith the track occupation rates. Additionally, longer planned running times require more personnel and rolling stock, hence they are negative for the efficiency of the railway system.

On the other hand, running time supplements add to the predictability of the realized travel times and to the reliability of the railway system as a whole. As a consequence, the total amount and the allocation of the running time supplements should be chosen by a trade-off between the above elements.

2.2 Application in Practice

In the Netherlands, running time supplements are currently (2007) approximately 5% of the technically minimum running times. This percentage is used nationwide for all types of passenger services. However, due to rounding -because of the integer character of the timetable- and local circumstances, the actual percentages may deviate from this percentage. Furthermore, cargo trains are usually planned 5 kilometer per hour below their maximum speed. Additionally, a running time supplement of 5% of the running times may be used for cargo trains.

On top of that, for a cargo train, the planned acceleration and deceleration times are based on a maximum total weight of the train. The difference between this maximum weight and the actual weight of the train acts as an extra running time supplement.

In Switzerland, running time supplements have several components (see Haldeman [5]). First, there is a proportional running time supplement, which equals 7% of the running time for passenger trains and 11% for cargo trains. Secondly, *special operational supplements* are added at highly utilized nodes. Additionally, one minute of supplement is added for each 30 minutes of running time. For trips with high average speeds, the supplements are even higher. In the United Kingdom running times are based on past performance on a railway section (see Rudolph [18]). Supplements are not explicitly defined here.

Leaflet 451-1 of the UIC (see UIC [22]) gives recommendations for running time supplements. It recommends a running time supplement to be the sum of a distance dependent supplement and a percentage of the technically minimum running time. The distance dependent supplement is 1.5 minute per 100 kilometer for locomotive-hauled passenger trains and 1 minute per 100 kilometer for multiple unit passenger trains. The running time dependent supplements may vary between 3% for relatively slow trains and 7% for faster trains. For locomotive-hauled trains, this percentage also depends on the weight of the train. For cargo trains, supplements are generally higher. The running time dependent supplement can be replaced by a second distance dependent supplement.

It can be concluded that in practice it is common to allocate the running time supplement on a certain trip to a large extent *in proportion* to the running time on that trip. In this paper, such an allocation is called a *proportional* allocation. However, this paper demonstrates that, from a punctuality point of view, it is better to allocate a somewhat larger part of the total running time supplement to the first trips of the complete route of a train. Indeed, a delay reduction on a certain trip does not only reduce the delay on the respective trip, but also on all subsequent trips. This means that the delay reduction is measured at all subsequent measuring points. Consequently, an early running time supplement is more effective than a late running time supplement. Therefore, one would expect to have a relatively large part of the running time supplements early on. But there is also a downside: if there are no early disturbances, then early supplements are lost. Hence they are useless in that case.

The stochastic optimization model described in this paper can be used to analyze this stochastic trade-off for a cyclic timetable. Also the choice between running time supplements and dwell time supplements can be supported by this model. However, for ease of presentation, we first focus in Sect. 3 on the allocation of a fixed amount of running time supplement to the consecutive trips of a single train. This allocation is done in such a way that the average arrival delay of the train is minimal. Thereafter, in Sect. 6, we describe a more complicated situation where several trains are operated on a common railway infrastructure.

3   A Single Train on a Single Line

In this section, we present a stochastic optimization model for allocating a fixed amount of running time supplement S to N consecutive trips of a single train on a single line. On each of the trips, the train is subject to external disturbances, possibly leading to a delay of the train. This delay is measured at the end of each trip. The objective is to minimize the average delay of the train.

This situation is illustrated in Fig. 1. Here the running time supplements on the trips t are denoted by the variables s_t. The external disturbances are denoted by the parameters δ_t, and the resulting delay is represented by the variables D_t. The figure represents the fact that the disturbances δ_t are partially compensated by the running time supplements s_t, so that they can "leave the train" again. The disturbances that cannot be compensated, since the running time supplements have been used completely, accumulate in the delays D_t.

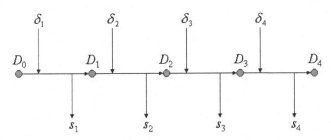

Fig. 1. Relation between disturbances, running time supplements, and delay

In this paper we assume that all running time supplement allocated to a trip can be used for recovering from a disturbance at the same trip. In other words, the disturbances are assumed to take place at the start of a trip (or at the preceding station), as is shown in Fig. 1. This assumption may be relaxed by splitting the trips into a number of smaller trips.

The stochastic optimization model contains a *planning* part for determining the running time supplements and an *evaluation* part for determining the resulting average delay of the train. To that end, at the same time as the running time supplements s_t are determined, R realizations of the train are operated along the N trips subject to externally generated disturbances. Let $\delta_{t,r}$ denote the disturbance incurred on trip t by realization r of the train. Furthermore, the resulting delay of realization r of the train by the end of trip t is denoted by $D_{t,r}$. Then the following relation describes the balance of the external disturbances $\delta_{t,r}$, the running time supplements s_t, and the delays $D_{t,r}$:

$$D_{t,r} = \max\{\, 0,\ D_{t-1,r} + \delta_{t,r} - s_t \,\} \text{ for } t = 1, \ldots, N;\ r = 1, \ldots, R. \quad (1)$$

Note that equation (1) is a mathematical representation of Fig. 1. It follows that, if $\delta_{t,r} > s_t$, then the delay of realization r of the train increases over trip t. If $\delta_{t,r} \leq s_t$, then the delay of realization r of the train may decrease over trip t. Note that equation (1) assumes that the train is not influenced by other trains: disturbances are assumed to be autonomous and external. A further consequence of equation (1) is that the amount of running time supplement $U_{t,r}$ that is actually used by realization r on trip t equals

$$U_{t,r} = \min\{\ D_{t-1,r} + \delta_{t,r},\ s_t\ \} \text{ for } t = 1, \ldots, N;\ r = 1, \ldots, R. \tag{2}$$

Obviously, equation (1) is not a linear equation. However, it can be linearized easily. Now the complete model can be described as follows:

$$\min D = \sum_{t=0}^{N} \sum_{r=1}^{R} w_t D_{t,r}/R \tag{3}$$

subject to

$$D_{t-1,r} + \delta_{t,r} - s_t \leq D_{t,r} \quad \text{for } t = 1, \ldots, N;\ r = 1, \ldots, R \tag{4}$$

$$\sum_{t=1}^{N} s_t \leq S \tag{5}$$

$$D_{t,r} \geq 0 \quad \text{for } t = 0, \ldots, N;\ r = 1, \ldots, R \tag{6}$$

$$s_t \geq 0 \quad \text{for } t = 1, \ldots, N \tag{7}$$

The objective function (3) indicates that the objective is to minimize the average weighted delay D. For $t = 1, \ldots, N$, the weight w_t indicates the weight of the delay at the end of trip t, depending on the number of involved passengers or on the status of the station at the end of trip t (e.g. an ordinary station or a transfer station). Constraints (4) and (6) together give the linearized version of equation (1) relating the delay at the end of trip t to the delay at the end of trip $t-1$. Next, constraint (5) expresses the fact that only a fixed amount of running time supplement is to be distributed among the trips. Finally, constraints (6) and (7) indicate that the variables are to be non-negative.

4   Convergence

In this section we consider the same model as in the previous section, but for the case of a single train that is operated over *two* consecutive trips. We assume that the probability distribution of the disturbances (δ_1, δ_2) has a finite set I of possible values. Each of these values (δ_1^i, δ_2^i) has a probability of occurrence p_i.

For this case we prove that the results of the stochastic programming model converge to the optimal allocation of the running time supplement if the number of realizations tends to infinity. Here we assume that the optimal allocation of the running time supplement is unique. The latter is not essential, but it simplifies the proof somewhat. The assumption holds e.g. if $|I|$ is odd and all disturbances (δ_1^i, δ_2^i) satisfy $\delta_1^i + \delta_2^i > S$, but also in many other situations. Further results on convergence in stochastic optimization can be found in Linderoth et al. [12].

4.1   Optimal Running Time Supplement

The running time supplement allocated to trip 1 is denoted by s. Then the running time supplement allocated to trip 2 equals $S - s$. Fig. 2 shows the partitioning of the positive (δ_1, δ_2) quadrant for a given value of s into the areas $A_1(s)$, $A_2(s)$, $A_3(s)$, and $A_4(s)$. For example, $A_1(s)$ is the area with relatively small disturbances on both trips. As a consequence, on both trips the delays can be compensated by the running time supplements. Similarly, $A_2(s)$ is the area with relatively small disturbances on the first trip and relatively large disturbances on the second trip. This results in delays on the second trip only.

The delay of the train by the end of trip t $(t = 1, 2)$ if the disturbances equal (δ_1^i, δ_2^i) is denoted by D_t^i. In that case, the total weighted delay of the train over the two trips is denoted by D^i. As a consequence, for a given value s of the running time supplement on the first trip, the following weighted delays are caused by the disturbances (δ_1^i, δ_2^i):

- If (δ_1^i, δ_2^i) in $A_1(s)$, then $D_1^i = 0$ and $D_2^i = 0$. Hence $D^i = 0$.
- If (δ_1^i, δ_2^i) in $A_2(s)$, then $D_1^i = 0$ and $D_2^i = \delta_2^i - (S - s)$. Hence $D^i = w_2(\delta_2^i - S + s)$.
- If (δ_1^i, δ_2^i) in $A_3(s)$, then $D_1^i = \delta_1^i - s$ and $D_2^i = 0$. Hence $D^i = w_1(\delta_1^i - s)$.
- If (δ_1^i, δ_2^i) in $A_4(s)$, then $D_1^i = \delta_1^i - s$ and $D_2^i = \delta_1^i + \delta_2^i - S$. Hence $D^i = w_1(\delta_1^i - s) + w_2(\delta_1^i + \delta_2^i - S) = (w_1 + w_2)\delta_1^i + w_2\delta_2^i - w_1 s - w_2 S$.

From the foregoing it follows that, for a given value s of the running time supplement on the first trip, the average weighted delay $D(s)$ of the train can be expressed as follows:

$$D(s) = \sum_{i \in A_2(s)} p_i w_2(\delta_2^i - S + s) + \sum_{i \in A_3(s)} p_i w_1(\delta_1^i - s) +$$
$$\sum_{i \in A_4(s)} p_i((w_1 + w_2)\delta_1^i + w_2\delta_2^i - w_1 s - w_2 S). \tag{8}$$

The minimization problem to be solved is to find a value s^* for the running time supplement on the first trip such that the average delay $D(s^*)$ is minimal.

It is not difficult to see that the average delay $D(s)$ is a continuous and convex piecewise linear function in s. Furthermore, (8) implies that, if s is not equal to one of the values δ_1^i and $S - s$ is not equal to one of the values δ_2^i, then a slight modification Δs of the running time supplement on the first trip gives the following modification $\Delta D(s)$ of the average delay on the two trips:

$$\Delta D(s) = \sum_{i \in A_2(s)} p_i w_2 \Delta s - \sum_{i \in A_3(s)} p_i w_1 \Delta s - \sum_{i \in A_4(s)} p_i w_1 \Delta s$$
$$= \Delta s \left(\sum_{i \in A_2(s)} p_i w_2 - \sum_{i \in A_3(s) \cup A_4(s)} p_i w_1 \right).$$

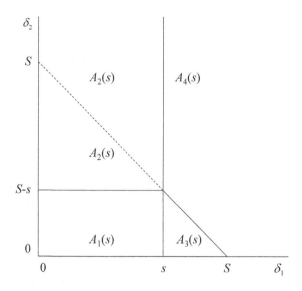

Fig. 2. Partitioning into the areas $A_1(s)$, $A_2(s)$, $A_3(s)$, and $A_4(s)$

It follows that the average delay is minimal if the running time supplement s on the first trip is such that, around s, the above expression changes from a negative value (decreasing average delay $D(s)$) to a positive value (increasing average delay $D(s)$). Hence, the optimal running time supplement s^* on the first trip is such that the expression

$$\sum_{i \in A_2(s)} p_i w_2 - \sum_{i \in A_3(s) \cup A_4(s)} p_i w_1$$

is negative for $s = s^* - \Delta s$ and is positive for $s = s^* + \Delta s$ for a sufficiently small value of Δs. It follows that s^* coincides with one of the values $\{ \delta_1^i, \ S - \delta_2^i \mid i \in I \}$. Note that here the assumption is used that there is a unique optimal allocation of the running time supplement.

4.2   Stochastic Optimization Model

Next, suppose that we have a random sample of R realizations of pairs of disturbances. Let R_i denote the number of occurrences of the pair (δ_1^i, δ_2^i) in this sample. Furthermore, let s denote the proposed value for the running time supplement on the first trip. Then, in the same way as in the previous section, it follows that the average delay $D_R(s)$ can be expressed as follows:

$$D_R(s) = \sum_{i \in A_2(s)} \frac{R_i}{R} w_2(\delta_2^i - S + s) + \sum_{i \in A_3(s)} \frac{R_i}{R} w_1(\delta_1^i - s) +$$

$$\sum_{i \in A_4(s)} \frac{R_i}{R} ((w_1 + w_2)\delta_1^i + w_2\delta_2^i - w_1 s - w_2 S). \qquad (9)$$

As above, the average delay $D_R(s)$ is a continuous and convex piecewise linear function in s. A similar argument as in the previous section can be used to show that, if the average delay $D_R(s)$ has a unique optimal running time supplement s_R^*, then this optimal value s_R^* is such that the expression

$$\sum_{i \in A_2(s)} \frac{R_i}{R} w_2 - \sum_{i \in A_3(s) \cup A_4(s)} \frac{R_i}{R} w_1$$

is negative for $s = s_R^* - \Delta s$ and is positive for $s = s_R^* + \Delta s$ for a sufficiently small value of Δs. Note that this optimal value s_R^* is the value that is obtained by applying the stochastic optimization model. Fig. 3 represents parts of the graphs of the functions $D(s)$ and $D_R(s)$.

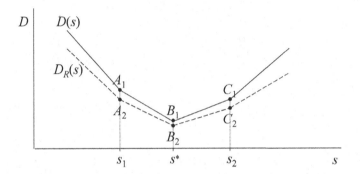

Fig. 3. The convex piecewise linear functions $D(s)$ and $D_R(s)$

4.3 Proof of Convergence

Theorem 1. *If the minimization problem has a unique optimal solution s^* with $0 < s^* < S$, then $\lim_{R \to \infty} P(s_R^* = s^*) = 1$.*

Proof. Let $0 \le s_1 < s^*$ be such that the interval (s_1, s^*) does not contain any value δ_1^i and such that the interval $(S - s^*, S - s_1)$ does not contain any value δ_2^i. Similarly, let $s^* < s_2 \le S$ be such that the interval (s^*, s_2) does not contain any value δ_1^i and such that the interval $(S - s_2, S - s^*)$ does not contain any value δ_2^i. Next, let Δ_1 and Δ_2 be defined by

$$\Delta_1 := \sum_{i \in A_2(s_1)} p_i w_2 - \sum_{i \in A_3(s_1) \cup A_4(s_1)} p_i w_1,$$

$$\Delta_2 := \sum_{i \in A_2(s_2)} p_i w_2 - \sum_{i \in A_3(s_2) \cup A_4(s_2)} p_i w_1.$$

Since s^* is a unique minimum of the average delay $D(s)$, $\Delta_1 < 0$ and $\Delta_2 > 0$. Note that Δ_1 and Δ_2 are represented in Fig. 3 by the differences $D(s^*) - D(s_1)$

and $D(s_2) - D(s^*)$. In other words, the slopes of the solid lines A_1B_1 and B_1C_1 are negative and positive, respectively.

Next we will show that, if R tends to infinity, then the probability that the differences $D_R(s^*) - D_R(s_1)$ and $D_R(s_2) - D_R(s^*)$ are also negative and positive tends to 1. In other words, if R tends to infinity, then the probability that the slopes of the dashed lines A_2B_2 and B_2C_2 in Fig. 3 are negative and positive, respectively, tends to 1. A consequence is that, if R tends to infinity, then the probability that $s_R^* = s^*$ tends to 1, as is to be proved.

To that end, first choose $\varepsilon > 0$ and let W be defined by $W := \max\{ w_1, w_2 \}$. Because of the Law of the Large Numbers, we know that for all $i \in I$ there exists an integer N_i such that for all $R > N_i$ the following holds: $P\left(\left|\frac{R_i}{R} - p_i\right| \geq \frac{-\Delta_1}{W|I|}\right) < \frac{\varepsilon}{2|I|}$. It follows that for all $R > \tilde{R}_1 := \max\{ N_i \mid i \in I \}$:

$$P \text{ (the slope of } A_2B_2 < 0) =$$

$$P\left(\sum_{i \in A_2(s_1)} \frac{R_i}{R} w_2 - \sum_{i \in A_3(s_1) \cup A_4(s_1)} \frac{R_i}{R} w_1 < 0 \right) =$$

$$P\left(\sum_{i \in A_2(s_1)} \left(\frac{R_i}{R} - p_i\right) w_2 - \sum_{i \in A_3(s_1) \cup A_4(s_1)} \left(\frac{R_i}{R} - p_i\right) w_1 < -\Delta_1 \right) \geq$$

$$P\left(\sum_{i \in I} \left|\frac{R_i}{R} - p_i\right| W < -\Delta_1 \right) \geq P\left(\bigcap_{i \in I} \left\{ \left|\frac{R_i}{R} - p_i\right| < -\frac{\Delta_1}{W|I|} \right\} \right) =$$

$$1 - P\left(\bigcup_{i \in I} \left\{ \left|\frac{R_i}{R} - p_i\right| \geq -\frac{\Delta_1}{W|I|} \right\} \right) \geq 1 - \sum_{i \in I} P\left(\left|\frac{R_i}{R} - p_i\right| \geq -\frac{\Delta_1}{W|I|} \right) > 1 - \frac{\varepsilon}{2}.$$

Similarly, there exists an integer $\tilde{R}_2$ such that for all $R > \tilde{R}_2$

$$P \text{ (the slope of } B_2C_2 > 0) =$$

$$P\left(\sum_{i \in A_2(s_2)} \frac{R_i}{R} w_2 - \sum_{i \in A_3(s_2) \cup A_4(s_2)} \frac{R_i}{R} w_1 > 0 \right) > 1 - \frac{\varepsilon}{2}.$$

As a consequence, for all $R > \max\{ \tilde{R}_1, \tilde{R}_2 \}$ the minimum of $D_R(s)$ is obtained for $s_R^* = s^*$ with probability at least $1 - 2(\frac{\varepsilon}{2}) = 1 - \varepsilon$. $\diamond$

Theorem 2. *If the minimization problem has a unique optimal solution s^* with $0 < s^* < S$, then for all $\delta > 0$ $\lim_{R \to \infty} P(|D_R(s_R^*) - D(s^*)| < \delta) = 1$.*

Proof. First, choose $\delta > 0$ and $\varepsilon > 0$, and let the positive number $\tilde{R}_0$ be such that $P(s_R^* = s^*) > 1 - \frac{\varepsilon}{2}$ for all $R > \tilde{R}_0$. According to the proof of Theorem 1, such a number $\tilde{R}_0$ exists. Next, we have the following (in)equalities:

$$|D_R(s^*) - D(s^*)| = \qquad (10)$$

$$\left| \sum_{i \in A_2(s^*)} \left(\frac{R_i}{R} - p_i \right) w_2(\delta_2^i - (S - s^*)) + \sum_{i \in A_3(s^*)} \left(\frac{R_i}{R} - p_i \right) w_1(\delta_1^i - s^*) + \right.$$

$$\left. \sum_{i \in A_4(s^*)} \left(\frac{R_i}{R} - p_i \right) ((w_1 + w_2)\delta_1^i + w_2\delta_2^i - w_1 s^* - w_2 S) \right| \leq M \times \sum_{i \in I} \left| \frac{R_i}{R} - p_i \right|,$$

where M is an appropriately chosen positive number. Again, because of the Law of the Large Numbers, we know that for all $i \in I$ there exists an integer N_i such that for all integers $R > N_i$ the following holds: $P\left(\left| \frac{R_i}{R} - p_i \right| \geq \frac{\delta}{M|I|} \right) < \frac{\varepsilon}{2|I|}$. Then, it follows that for all integers $R > \max\{ N_i \mid i \in I \}$ the following holds:

$$P\left(\sum_{i \in I} \left| \frac{R_i}{R} - p_i \right| < \frac{\delta}{M} \right) \geq P\left(\bigcap_{i \in I} \left\{ \left| \frac{R_i}{R} - p_i \right| < \frac{\delta}{M|I|} \right\} \right) = \qquad (11)$$

$$1 - P\left(\bigcup_{i \in I} \left\{ \left| \frac{R_i}{R} - p_i \right| \geq \frac{\delta}{M|I|} \right\} \right) \geq 1 - \sum_{i \in I} P\left(\left| \frac{R_i}{R} - p_i \right| \geq \frac{\delta}{M|I|} \right) > 1 - \frac{\varepsilon}{2}.$$

Combining the results in (10) and (11) gives that the following holds for all integers $R > \max\{ \{ N_i \mid i \in I \} \cup \{ \tilde{R}_0 \} \}$:

$$P(|D_R(s_R^*) - D(s^*)| < \delta) \geq P((|D_R(s_R^*) - D(s^*)| < \delta) \cap (s_R^* = s^*)) =$$

$$P(|D_R(s_R^*) - D(s^*)| < \delta \mid s_R^* = s^*) \times P(s_R^* = s^*) \geq$$

$$P\left(\sum_{i \in I} \left| \frac{R_i}{R} - p_i \right| < \frac{\delta}{M} \right) \times P(s_R^* = s^*) \geq \left(1 - \frac{\varepsilon}{2} \right)\left(1 - \frac{\varepsilon}{2} \right) > 1 - \varepsilon. \qquad \diamond$$

Note that the results of Theorems 1 and 2 also hold if the optimal solution s^* is not unique or if s^* equals 0 or S. However, slight modifications of the proofs are required then.

5   Computational Results

In this section, we describe the computational results that were obtained by applying the model described in Sect. 3 to a number of theoretical cases. Computational results for a real-life case are presented in Sect. 7.

All results in this section are based on equally and exponentially distributed disturbances and on equally weighted delays. However, as was noted earlier already, the latter is certainly *not* essential. It is also possible to use other probability distributions and different weights, including empirical ones.

The results described in this section were obtained by implementing the model in the modeling system OPL Studio and by solving it with the corresponding solver CPLEX 9.0 on an Intel Pentium IV PC with 3.0 GHz processor speed and 512 MB internal memory.

In all cases described in this section, the number of realizations R has been set to 1000. In our experiments, this number turned out to be large enough to generate stable results that were more or less independent of the detailed values of the disturbances. On the other hand, this number of realizations also led to acceptable running times of at most a couple of minutes.

5.1   Optimal Allocation of Running Time Supplements

In the experiments described in this section, the running times are subject to exponentially distributed disturbances with an average value of 1 minute. Furthermore, in this section the total amount of running time supplement S equals the total number of trips. That is, the total amount of running time supplement equals the average total disturbances. In Sect. 5.2, different amounts of running time supplement are applied. In all cases, the objective is to allocate the total amount of running time supplement to the trips in such a way that the average delay is minimal. Note that in the proportional allocation, each trip gets a running time supplement of 1 minute.

The results of the case with 10 trips are shown in Fig. 4. The horizontal axis shows the 10 trips, and the vertical axis shows the optimal amounts of running time supplement to be allocated to the trips. The vertical line in the figure represents the weighted average distance of the running time supplements from the starting point. This weighted average distance, WAD, is defined by:

$$WAD = \sum_{t=1}^{N} \frac{2t-1}{2N} \times s_t. \tag{12}$$

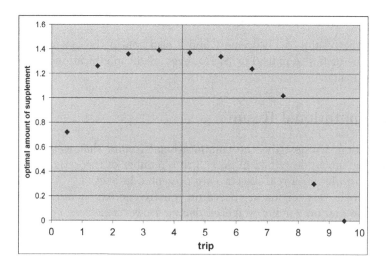

Fig. 4. The optimal allocation of the running time supplements for 10 trips

In Fig. 4, the value of the *WAD* equals about 0.425. For the proportional allocation of the running time supplements, the *WAD* equals exactly 0.5.

Note that, in comparison with the proportional allocation of the running time supplements, the allocation of the running time supplements according to Fig. 4 has a negative effect on the average *planned* travel times of the passengers. Indeed, if the numbers of travelers between all O/D-pairs of stations are more or less the same, then minimal average planned travel times of the passengers are obtained either by skipping the running time supplements completely, or by allocating the running time supplements as much as possible to the first or to the last trips along the line. However, it is likely that this allocation of the running time supplements leads to an unreliable timetable.

In Fig. 5, the upward bending line shows the average delay by the end of each trip for the optimal allocation of the running time supplements. The nearly diagonal line in this figure shows the average delay by the end of the trips for the proportional allocation of the running time supplements. Obviously, the optimal allocation performs better on almost all trips. Only on the last trips, the average delay increases quickly for the optimal allocation. This is due to the fact that the supplements have been shifted from the last trips towards the earlier trips.

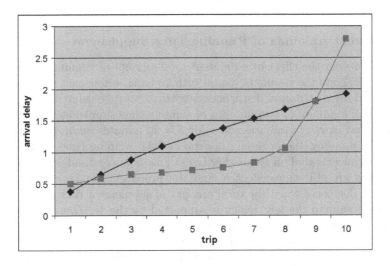

Fig. 5. The average delay by the end of the trips for 10 trips

Fig. 6 presents results that have been obtained by applying the model to 2 to 25 consecutive trips. The average delays of the optimal solutions are compared with the average delays of the proportional allocation of the running time supplements. The figure shows that the average delay decrease is only 1.2% for 2 trips, but the decrease is already 9.5% for 5 trips and 20.1% for 15 trips. Although the shapes of the supplement allocations and the average locations of the supplements are quite similar in all cases, the relative decreases in the average delay are far from equal for the different cases.

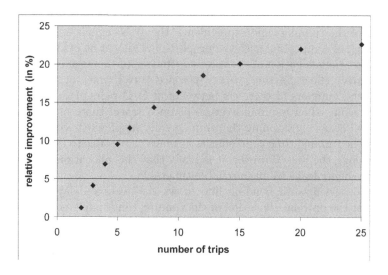

Fig. 6. The decrease in average delay for the optimal supplement allocation in comparison with the average delay for the proportional supplement allocation

5.2 Different Amounts of Running Time Supplement

Next, we consider the effect of a different total amount of running time supplement. Fig. 7 shows the results of a case with 10 trips, where each trip is subject to exponentially distributed disturbances with an average value of 1 minute. In this case half a minute or two minutes of running time supplement are available per trip. That is, $S = 5$ minutes (dots) or $S = 20$ minutes (diamonds).

If $S = 5$ minutes, the optimal allocation of the running time supplement is even more concentrated on the earlier trips. This is understandable, since early supplements are still more effective and the probability of an excessive early supplement decreases when the total amount of supplement decreases. There is an apparent shift to the left, which is supported by the WAD, which is 0.32 in this case. The decrease in the average delay goes up from about 16.3% for the case with $S = 10$ minutes to about 17.8% for the case with $S = 5$ minutes.

In the opposite situation with $S = 20$ minutes, where the total amount of running time supplement is twice as large as the total average disturbance, the allocation of the running time supplement shifts into the opposite direction. Now the WAD equals 0.492. The decrease in the average delay goes down to about 2.9% for the case with $S = 20$ minutes. In this case, the difference with the proportional allocation is small.

6   Several Trains on a Common Railway Infrastructure

In this section, we describe an extension of the stochastic optimization model presented in Sect. 3. Here a given cyclic timetable for a number of trains over

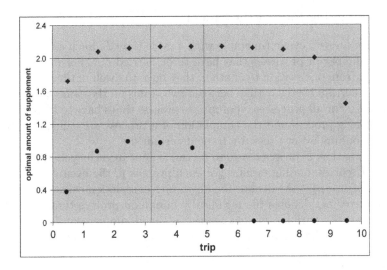

Fig. 7. The optimal allocation for half and double the total amount of supplement

a common part of the railway infrastructure is improved with respect to the average weighted delay of the trains. The underlying timetabling model shows some similarity with the well-known Periodic Event Scheduling Model (PESP, see Serafini and Ukovich [20]). Several researchers have studied the application of PESP for railway timetabling, see e.g. Nachtigall [15] and Peeters [16].

The extended model is again a stochastic optimization model. That is, in order to evaluate and optimize the timetable under construction, R realizations of the processes in each cycle time are operated subject to stochastic disturbances. The R realizations are operated one after another. As a consequence, the delayed trains in realization $r - 1$ may influence the trains in realization r if they use the same parts of the infrastructure. Thus, in contrast with the model described in Sect. 3, the R realizations are not independent of each other. These interactions between successive realizations are enabled by the cyclic nature of the timetable. They imply that the model is a non-standard variant of a stochastic optimization model (Birge and Louveaux [2]).

The model aims at improving a given cyclic timetable with respect to the average delay of the trains. The main purpose of the model is to optimally re-allocate buffer times and time supplements to the various process times in the timetable, thereby leaving the orders of the trains on the tracks unchanged. Furthermore, for ease of presentation, we assume that the given timetable does not contain complex cycles, for example caused by the rolling stock circulation or by circular chains of passenger connections. In general, this assumption is valid if the railway network has a tree-like structure and the railway traffic is considered in just one direction. It should be noted that this assumption can be relaxed, but then a more complex model results.

6.1   Timetabling Part of the Model

We consider a given cyclic timetable with a cycle time T. Such a timetable consists of a number of processes that have to be carried out. For example, trains have to run from one station to another, they have to dwell in the stations, there has to be a certain headway time between two trains on the same railway infrastructure, etc. For all processes, appropriate process times have to be determined. The begin of a process and the completion of a process are called *events*. Also the corresponding event times are to be determined.

We assume that P processes are to be carried out in each cycle time and that there are E corresponding events. For each process p, the events $b(p)$ and $c(p)$ (with $1 \leq b(p), c(p) \leq E$) denote the begin and completion events of process p. The parameter m_p denotes the technically minimum process time of process p. Furthermore, the variable s_p denotes the planned supplement for the process time of process p. The planned event time of event e is denoted by the variable v_e. We use a linear time axis. This implies that

$$v_{c(p)} - v_{b(p)} = m_p + s_p \text{ for } p = 1, \ldots, P.$$

Thus the planned process time of process p equals the difference between the planned completion time and the planned begin time of process p. Most of the constraints to be satisfied in a cyclic timetabling model can be expressed in terms of constraints on the process times (see e.g. Peeters [16]). In our model, each constraint therefore has the following form:

$$l_p \leq v_{c(p)} - v_{b(p)} \leq u_p \text{ for } p = 1, \ldots, P.$$

Here l_p and u_p are appropriate lower and upper bounds. For example, processes such as running along a track, dwelling in a station, and passenger connections between trains can be modeled in this way. Note that for modeling the headway processes, the assumption that the aim of the model is to improve a given cyclic timetable, thereby leaving the orders of the trains on the tracks unchanged, is essential. Indeed, in a cyclic timetable minimum headway times between trains depend on the orders of the trains on the tracks. If the latter are not known a priori, then additional binary variables are required to model these. This highly complicates the model. However, given the orders of the trains on the tracks, a minimum headway time of at least h minutes between two consecutive trains departing from the same station and entering the same track can be enforced by the following constraint:

$$v_{c(p)} - v_{b(p)} \geq h.$$

Here the events $b(p)$ and $c(p)$ denote the departures of the first and the second train, respectively (that is, the begin and completion events of the corresponding headway process p). Note that the headway times do not have to be bounded from above, since these have only positive effects. This is in contrast with running time supplements, which also have a negative effect, e.g. on the travel times.

In order to allocate a certain amount of time supplement to the process times, Q subsets of processes are selected. Each subset q of processes is connected with a certain amount of time supplement S_q to be allocated to the processes in the set q. Thus the following constraints are to be satisfied:

$$\sum_{p \in q} s_p \leq S_q \text{ for } q = 1, \ldots, Q.$$

For example, such a constraint may indicate that a certain total amount of running time supplement is to be allocated to the consecutive running times of a single train. This corresponds with the problem described in Sect. 3. However, a certain amount of time supplement may also have to be allocated to the trips of a *number* of trains together.

Final relevant constraints specify that, at each part of the infrastructure, the difference between the last and the first planned event time in each cycle time should not exceed the cycle time minus the minimum headway time $T - h$. Moreover, non-negativity constraints have to be imposed on the variables s_p, and if one wants to obtain a timetable that is specified in integer minutes, then integrality constraints have to be imposed on the corresponding event times.

6.2   Evaluation Part of the Model

In the same way as in the model described in Sect. 3, the timetable is evaluated during its modification by operating R realizations of the trains in each cycle time subject to predetermined stochastic disturbances. To that end, we use R realizations of the processes and the events in each cycle time.

The stochastic disturbance of realization r of process p is denoted by the parameter $\delta_{p,r}$ for $p = 1, \ldots, P$ and $r = 1, \ldots, R$. Furthermore, the realized event time of realization r of event e is denoted by the variable $\tilde{v}_{e,r}$ for $e = 1, \ldots, E$ and $r = 1, \ldots, R$.

Mainly the delays of the events corresponding to arrivals of trains are measured, but also other delays can be taken into account. The measured events are called the *relevant* events. The set of relevant events is denoted by $\tilde{E}$. The delay of realization r of relevant event e is denoted by the variable $D_{e,r}$. The average weighted delay over all processes is denoted by D.

The variables $D_{e,r}$ and D are assumed to be non-negative. The objective is to minimize the average weighted delay of the trains. Thus the objective is to

$$\text{minimize } D = \sum_{r=1}^{R} \sum_{e \in \tilde{E}} w_e D_{e,r} / R. \tag{13}$$

Here the weights w_e indicate the weights of the different delays. The constraints linking the event times of the processes to the technically minimum process times and the disturbances are the following:

$$\tilde{v}_{c(p),r} - \tilde{v}_{b(p),r} \geq m_p + \delta_{p,r} \text{ for } p = 1, \ldots, P; \ r = 1, \ldots, R. \tag{14}$$

Furthermore, processes do not begin too early, and a delay occurs if a relevant process ends too late. This results in the following constraints:

$$v_{b(p)} + rT \leq \tilde{v}_{b(p),r} \text{ for } p = 1, \ldots, P; \ r = 1, \ldots, R, \tag{15}$$

$$\tilde{v}_{e,r} - v_e - rT \leq D_{e,r} \text{ for } e \in \tilde{E}; \ r = 1, \ldots, R. \tag{16}$$

Note that here we use the cyclic character of the timetable, since the planned event time of realization r of event e equals $v_e + rT$.

As was mentioned earlier, the R realizations of the cycle times are operated one after another. As a consequence, the delayed trains in realization $r - 1$ may influence the trains in realization r if they use the same parts of the infrastructure. To that end, let e_1 be the *first* planned event in a cycle time on a certain part of the infrastructure, and let e_2 be the *last* planned event in a cycle time on the same infrastructure. Then the following constraint is to be satisfied:

$$\tilde{v}_{e_1,r} - \tilde{v}_{e_2,r-1} \geq h. \tag{17}$$

In other words, realization r of event e_1 cannot be carried out earlier than the headway time h after realization $r - 1$ of event e_2 has taken place. Also other interactions between successive realizations can be modeled. For example, realization $r - 1$ of a train that is one of the last trains in each cycle time may have a passenger connection with realization r of a train that is one of the first trains in each cycle time. As was noted earlier, these interactions imply that the model is a rather non-standard stochastic optimization model.

7   Computational Results

In this section, we describe the computational results that were obtained by applying the model described in Sect. 6 to a real-life case of NS Reizigers, the main Dutch operator of passenger trains. This analysis was carried out for study purposes. The results have not yet been implemented in the real timetable.

For solving the model to optimality, we again used the modeling system OPL Studio and CPLEX 9.0 on an Intel Pentium IV PC with 3.0 GHz processor speed and 512 MB internal memory. For this case we always used 500 realizations per run, because this gave sufficiently stable results and the computation times remained at an acceptable level of about one hour on the indicated hardware.

7.1   Case: Haarlem–Maastricht/Heerlen

The model was applied to improve the 2005 timetable of NS Reizigers on the corridor from Haarlem (Hlm) in the western part of the Netherlands to Maastricht (Mt) and Heerlen (Hrl) in the southern part. Throughout this section, this original timetable is called the *reference* timetable. All passenger trains on this corridor were included in the model. Cargo trains were left out, both from the reference timetable and from the improved one. As a consequence, the results for the two timetables are still comparable.

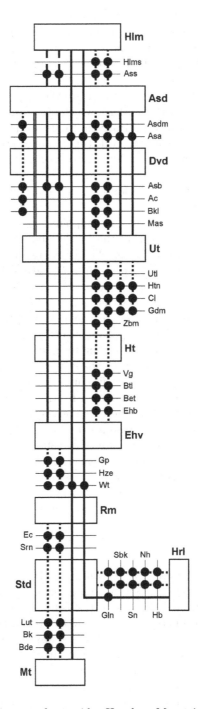

Fig. 8. The lines on the corridor Haarlem–Maastricht/Heerlen

The train lines on the studied corridor are shown in Fig. 8. In this figure, the dotted lines indicate the stoptrains, the other lines indicate the intercity lines. The main stations, represented by boxes, are the stations where the delays are measured. All trains dwell at these stations. The black dots indicate the other dwell stations of the trains.

The main lines on the studied corridor are the intercity lines 800 (Haarlem-Maastricht) and 900 (Haarlem-Heerlen), which are both operated once per hour. These lines have the corridor from Haarlem to Sittard (Std) in common. On this corridor, there is nearly a 30 minute cycle time, since almost all lines are operated there twice per hour with an exact 30 minute cycle time. The order of the trains on the different trips follows from the reference timetable. The overtakings of the stoptrains in Abcoude (Ac), Geldermalsen (Gdm) and 's-Hertogenbosch (Ht) remain unchanged.

Turn-around constraints at the line endpoints have not been taken into account in the model. This means that the southbound and the northbound trains are almost independent of each other. Therefore, two nearly independent problems are created: the southbound problem and the northbound problem. Only the southbound problem is described here.

The planned running times in the reference timetable include on average 7.92% of running time supplement on each trip. The only exceptions can be found on the line 3500, which bears additional supplements of 1 minute between Duivendrecht (Dvd) and Abcoude, and 4 additional minutes between Abcoude and Breukelen (Bkl). Only the trains that are overtaken have dwell time supplements. The other planned dwell times are equal to the minimum dwell times.

For the lines that are not fully covered by the corridor, the departure and arrival times at the stations where these lines enter or leave the model have been fixed to the event times in the reference timetable. In the realizations, these trains are assumed to enter the model at these stations without a delay.

In this case study, we mainly used exponential distributions for generating the disturbances to the process times. However, as was noted earlier, also other probability distributions could have been used. Anyway, in the sensitivity analysis, other probability distributions were used as well.

7.2   Results

First, the reference timetable was *evaluated* with the evaluation part of the model. That is, after fixing all event times according to this timetable, the stochastic optimization model was applied to operate 500 realizations of the trains according to this timetable but under stochastic disturbances.

Each trip between two measuring stations was disturbed with an average disturbance of 7.92% of the minimum running time. This average is the same as the planned running time supplement in the reference timetable. This percentage was also chosen in such a way that the evaluation led to a punctuality of 83.7%, which is comparable to the punctuality observed in practice. The evaluation led to an average delay of 1.38 minutes.

Table 1. The running time supplements in minutes for the lines 800 and 900 obtained by single line optimization and by corridor optimization

trip	avg. disturbance	running time supplements		
		per line 800	per line 900	corridor 800 & 900
Hlm-Asd	1.03	0.85	0.87	0.90
Asd-Dvd	0.81	1.01	1.02	1.16
Dvd-Ut	1.25	1.43	1.44	1.96
Ut-Ht	2.05	2.63	2.67	2.51
Ht-Ehv	1.32	1.71	1.72	1.55
Ehv-Rm	2.27	2.57	2.64	2.18
Rm-Std	1.10	0.72	0.78	0.67
Std-Mt	1.10	0.00	-	0.00
Std-Hrl	1.32	-	0.00	0.22

Next, the timetable was *optimized* by unfixing most of the event times and by operating 500 realizations of the trains under the *same* stochastic disturbances as in the evaluation of the reference timetable. The objective was to modify the timetable by re-allocating the running time supplements and the buffer times in such a way that the average delay at the ten measuring stations was minimal. Other relevant measures, such as the punctuality, were determined afterwards.

The optimization led to a model with 160,000+ variables and 300,000+ constraints. Because of the size of the model, the event times were allowed to be fractional, so that the model could be solved by Linear Programming. By the optimization, the average delay was reduced to 0.947 minutes, which is 31.4% less than the average delay in the reference timetable. The 3-minute punctuality increased from 83.7% for the reference timetable to 89.5%: this is a reduction of the number of late trains by 35.2%.

The optimal running time supplements for the lines 800 and 900 obtained by the optimization are shown in the last column of Table 1. The third and fourth column ("per line") show the results that were obtained by applying the single line model described in Sect. 3 to the lines 800 and 900 separately.

As in the reference timetable, an exact 30-minute cycle time was enforced for most lines on the corridor from Haarlem to Sittard, leading to identical supplements there for the lines 800 and 900 up to Sittard. Because of the longer running time between Sittard and Heerlen as compared to the running time between Sittard and Maastricht, there was 0.22 minute more running time supplement available for the line 900 than for the line 800. The latter could only be allocated to the trip Sittard-Heerlen, due to the 30-minute cycle time between Haarlem and Sittard.

For the rest of the corridor, the supplement allocation is very similar to the one found by the single line model described in Sect. 3. The only remark that can be made in this respect is that slightly larger supplements are found for the most busy parts of the infrastructure between Amsterdam and Utrecht, and smaller supplements for the somewhat quieter parts south of 's-Hertogenbosch.

7.3   Sensitivity Analysis

The timetable found by the stochastic optimization model is only optimal with respect to the *applied* disturbances. Therefore we carried out a sensitivity analysis in order to investigate the behavior of this timetable under other disturbances from the same distribution and under disturbances from other distributions. For this analysis, again only the southbound timetable was considered. In the following description, the *preferred* timetable is the timetable which is optimal with respect to exponentially disturbed running times with an average disturbance of 7.92% of the respective minimum running time.

First, we analyzed the consequences of other sets of random disturbances from the same disturbance distribution. The timetable was not optimized again for these other sets of disturbances, but both for the reference timetable and for the preferred timetable the delay propagation resulting from these disturbances was evaluated. Ten random sets of disturbances from the same distribution were used, leading to ten evaluations of both timetables.

This led to the results shown in Table 2. The range of the average delay and the unpunctuality has a width of at most 10%. This is relatively small in comparison with the differences between the reference and the preferred timetable.

Table 2. The influence of randomness on the punctuality measures

measure	timetable	avg.	min.	max.	range	σ
average delay	reference	1.34	1.30	1.38	6.0%	0.03
average delay	preferred	0.92	0.89	0.95	5.8%	0.02
unpunctuality	reference	15.6%	15.0%	16.3%	8.0%	0.5%
unpunctuality	preferred	10.0%	9.5%	10.5%	10.1%	0.4%

Next, we evaluated the preferred timetable, but now for sets of disturbances from other probability distributions. All distributions described here were multiplied by 0.0792 times the technically minimum running time. In this way, the original disturbance distribution could be described as an exponential distribution with an average value of 1 minute. First, the timetable was evaluated for the situation where a large part of the running times was not disturbed, and the rest was disturbed again by exponentially distributed disturbances. Furthermore, a uniform distribution and a triangular distribution were applied.

The results of this sensitivity analysis are summarized in Table 3. This table shows that the preferred timetable outperformed the reference timetable for all disturbance distributions that were applied. The worst results were obtained for the relatively large disturbances that occur with a low probability (80% 0 and

Table 3. Punctuality gain for different disturbance distributions. The parameter for the exponential distributions is the average (not the reciprocal of the average).

disturbance distribution	reference timetable avg.delay	punct.	preferred timetable avg.delay	punct.	improvement avg.delay	punct.
exp. 1	**1.38**	**83.7%**	**0.95**	**89.5%**	**31.4%**	**35.2%**
50% 0 and 50% exp. 1.5	1.43	83.6%	1.05	87.9%	26.4%	26.2%
80% 0 and 20% exp. 6	1.50	89.9%	1.24	91.4%	17.0%	14.9%
uniform (0,2.5)	1.66	79.8%	1.01	91.1%	39.1%	55.7%
triangular (0,0,4)	2.02	75.4%	1.33	85.3%	34.4%	40.2%

20% exp. 6). This was to be expected, since the running time supplements are intended for handling small disturbances only.

Finally, we compared the preferred timetable with the timetables that were obtained by *optimizing* the timetable under different disturbance distributions. Again, the optimization was carried out with respect to the average delay, and the punctuality was determined afterwards. The obtained results are shown in Table 4. In this table, the column "optimal timetable" represents the optimal timetable for our samples of the corresponding disturbance distribution. Hence, each row in this column corresponds to a different optimal timetable. Table 4 shows that the preferred timetable is close to the optimal timetable for each of the applied distributions. It can be concluded that the preferred timetable has a relatively high quality under a range of disturbance distributions. For further details of the experiments that were described here and for a description of further experiments that have been carried out we refer to Vromans [23].

Table 4. Optimality gap of the preferred timetable

disturbance distribution	preferred timetable avg. delay	punct.	optimal timetable avg. delay	punct.	improvement avg. delay	punct.
exp. 1	**0.95**	**89.5%**	**0.95**	**89.5%**	**0.0%**	**0.0%**
50% 0 and 50% exp. 1.5	1.05	87.9%	1.04	88.0%	1.1%	1.1%
80% 0 and 20% exp. 6	1.24	91.4%	1.17	92.2%	6.1%	10.0%
uniform (0,2.5)	1.01	91.1%	1.00	91.2%	1.6%	1.5%
triangular (0,0,4)	1.33	85.3%	1.31	85.3%	1.2%	0.2%

8   Final Remarks and Further Research

In this paper, we showed that stochastic optimization is a useful approach for locally improving a cyclic railway timetable. We first described a simple model for allocating a fixed amount of running time supplement to the consecutive trips of a single train. Thereafter, we extended the model such that it can be applied for optimally allocating time supplements and buffer times if several trains are operated according to a given cyclic timetable on a common railway infrastructure. This stochastic optimization model is a *symbiosis* of a *timetabling* model and a *simulation* model. Indeed, during the modification of the timetable, a number of realizations of the timetable under construction is operated under stochastic disturbances. The time supplements and buffer times are selected by the model such that the resulting total average delay is minimal.

The results obtained by the first model indicate that a proportional distribution of the running time supplements does not lead to a minimum average delay. A relatively large part of the running time supplements has to be shifted towards the earlier trips of a train. The motivation is that delay reductions early on are counted on all subsequent trips. A consequence is that running time supplements on the last trips are relatively low, since these decrease the delay on the last trips only. The difference between the optimal allocation and the proportional allocation of the running time supplements is largest when the average amount of running time supplement per trip is relatively small in comparison with the average size of the disturbances.

The results obtained by the second model indicate that a significant reduction of the average delay can be achieved by re-allocating time supplements and buffer times in a given timetable. We focused on the timetable on the corridor between Haarlem and Maastricht/Heerlen in the Netherlands. On this corridor, the average delay could be reduced within the model by about 30%. Further experiments, for example including both directions of the studied corridor and integer departure times, will be carried out to further validate the model.

So far, we mainly experimented with disturbances from exponential probability distributions. However, the latter is not at all essential. Any probability distributions, including empirical ones, can be used. Moreover, each process can be subject to disturbances from its own probability distribution. Obviously, the applied probability distributions should be such that the disturbances are as much as possible realistic in practice. This may require a lot of field research in order to determine the appropriate probability distributions.

As was noted earlier, the allocation of the time supplements that leads to a minimal average delay of the trains need not lead to minimal average *planned* or *realized* travel times of the passengers. Indeed, if the objective is to minimize these, then in principle it is optimal to skip all time supplements: each minute of time supplement brings the risk that it is redundant in certain realizations, since there are no disturbances. Moreover, if the objective is to minimize the average travel times of the passengers, given the allocation of a certain *fixed* amount of running time supplements, then these running time supplements should mainly be allocated to the first or the last trips of the trains: these are usually the parts of a line with the lowest numbers of passengers. However, it is likely that a timetable with such an allocation of the running time supplements or without any time supplements is quite unreliable. Therefore, a trade-off has to be made between the improved reliability that is caused by the time supplements and several other criteria, such as the travel times of the passengers and the efficiency of rolling stock and crew. This is a subject for further research.

In our further research, we will also experiment with relaxations of the assumptions described in Sect. 6. That is, we will first assume that the given timetable may contain cycles that are caused by the rolling stock circulation or by chains of passenger connections. Under certain conditions, this seems to be a relatively easy extension. Next, we will assume that the orders of the trains on the tracks have not been fixed a priori. Since this extension requires many additional binary variables, it strongly reduces the computability of the timetable. However, the gain will be that the model will not only be able to *improve* a given timetable, but even to generate *from scratch* a timetable that is optimal with respect to the average delay of the trains.

References

1. Bergmark, R.: Railroad capacity and traffic analysis using SIMON. In: Allan, J., Brebbia, C.A., Hill, R.J., Sciutto, G. (eds.) Computers in Railways V, pp. 183–191. WIT Press, Ashurst (1996)
2. Birge, J.R., Louveaux, F.: Introduction to Stochastic Programming, Springer, New York (1997)
3. Carey, M.: Ex ante heuristic measures of schedule reliability. Transportation Research B 33(7), 473–494 (1999)
4. Goverde, R.M.P.: The Max-Plus Algebra approach to railway timetable design. In: Mellitt, B., Hill, R.J., Allan, J., Sciutto, G., Brebbia, C.A. (eds.) Computers in Railways VI, pp. 339–350. WIT Press, Ashurst (1998)
5. Haldeman, L.: Automatische Analyse von IST-Fahrplen, Master's thesis (in German), ETH Zürich, Switzerland (2003)
6. Hallowell, S.F., Harker, P.T.: Predictuing on-time performance in scheduled railroad operations: methodology and application to train scheduling. Transportation Research A 32(6), 279–295 (1998)
7. Higgins, A., Kozan, E.: Modelling train delays in urban networks. Transportation Science 32(4), 346–357 (1998)
8. Huisman, T., Boucherie, R.J.: Running times on railway sections with heterogeneous train traffic. Transportation Research Part B 35, 271–292 (2001)
9. Jochim, H.E.: Verspätung auf Stammstrecken: wie man ihrer nicht Herr wird. Der Nahverkehr (in German), 3 (2004)
10. König, H.: VirtuOS: Simulieren von Bahnbetrieb. Betrieb und Verkehr (in German) 50(1-2), 44–47 (2001)
11. de Kort, A.F.: Advanced railway planning using Max-Plus algebra. In: Allan, J., Brebbia, C.A., Hill, R.J., Sciutto, G. (eds.) Computers in Railways VII, pp. 257–266. WIT Press, Ashurst (2000)
12. Linderoth, J.T., Shapiro, A., Wright, S.J.: The emprirical behavior of sampling methods for stochastic programming. Optimization Technical Report 02-01. University of Wisconsin-Madison (2002)
13. Middelkoop, D., Bouwman, M.: Train network simulator for support of network wide planning of infrastructure and timetables. In: Allan, J., Brebbia, C.A., Hill, R.J., Sciutto, G. (eds.) Computers in Railways VII, pp. 267–276. WIT Press, Ashurst (2000)

14. Mühlhans, E.: Berechnung der Verspätungsentwicklung bei Zugfahrten. Eisenbahntechnische Rundschau 39, 465–468 (1990)
15. Nachtigall, K.: Periodic network optimization with different arc frequencies. Discrete Applied Mathematics 69, 1–17 (1996)
16. Peeters, L.W.P.: Cyclic railway timetable optimization. Ph.D. thesis, Erasmus University Rotterdam, Rotterdam School of Management (2003)
17. Petersen, E.R., Taylor, A.J.: A structured model for rail line simulation and optimization. Transportation Science 18, 192–206 (1982)
18. Rudolph, R.: Allowances and margins in railway scheduling. In: Proceedings of WCRR 2003, Edinburgh, pp. 230–238 (2003)
19. Schwanhäußer, W.: The status of German railway operations management in research and practice. Transportation Research A 28(A), 495–500 (1994)
20. Serafini, P., Ukovich, W.: A mathematical model for Periodic Event Scheduling Problems. SIAM Journal on Discrete Mathematics 2, 550–581 (1989)
21. Soto y Koelemeijer, G., Iounoussov, A.R., Goverde, R.M.P., van Egmond, R.J.: PETER, a performance evaluator for railway timetables. In: Allan, J., Brebbia, C.A., Hill, R.J., Sciutto, G. (eds.) Computers in Railways VII, pp. 405–414. WIT Press, Ashurst (2000)
22. U.I.C.: Timetable recovery margins to guarantee timekeeping - Recovery margins, Leaflet 451-1, U.I.C. Paris, France (2000)
23. Vromans, M.J.C.M.: Reliability of railway services. Ph.D. thesis, Erasmus University Rotterdam, Rotterdam School of Management (2005)
24. Wahlborg, M.: Simulation models: important aids for Banverket's planning process. In: Allan, J., Brebbia, C.A., Hill, R.J., Sciutto, G. (eds.) Computers in Railways V, pp. 175–181. WIT Press, Ashurst (1996)

Timetable Information: Models and Algorithms*

Matthias Müller-Hannemann[1], Frank Schulz[2], Dorothea Wagner[2], and Christos Zaroliagis[3]

[1] Darmstadt University of Technology, Department of Computer Science,
Algorithmics Group, Hochschulstr. 10, 64289 Darmstadt, Germany
`muellerh@algo.informatik.tu-darmstadt.de`
[2] University of Karlsruhe, Department of Computer Science, P.O. Box 6980, 76128
Karlsruhe, Germany
`{fschulz,dwagner}@ira.uka.de`
[3] Computer Technology Institute, P.O. Box 1122, 26110 Patras, Greece, and
Department of Computer Engineering and Informatics, University of Patras,
26500 Patras, Greece
`zaro@ceid.upatras.gr`

Abstract. We give an overview of models and efficient algorithms for optimally solving timetable information problems like "given a departure and an arrival station as well as a departure time, which is the connection that arrives as early as possible at the arrival station?" Two main approaches that transform the problems into shortest path problems are reviewed, including issues like the modeling of realistic details (e.g., train transfers) and further optimization criteria (e.g., the number of transfers). An important topic is also multi-criteria optimization, where in general all attractive connections with respect to several criteria shall be determined. Finally, we discuss the performance of the described algorithms, which is crucial for their application in a real system.

1 Introduction

The first electronic timetable information systems were established in the late eighties of the last century. Current systems are for example HAFAS [13], which is used by many European railway companies, or EFA [8], which is mainly used for local traffic limited to smaller regions in Europe. Empirically, the resulting connections are satisfying in the majority of cases. There are cases, however, for which the suggested itineraries are clearly not optimal (given some optimization criterion). The main reason for such non-optimal connections is that the algorithms behind the systems employ heuristic methods to reduce the search space (in order to achieve an acceptable response time) that do not always guarantee optimal solutions. Such heuristic approaches often work in two phases as described below.

In the last few years the question arose whether models and algorithms for optimally solving timetable information problems are feasible. In this work we

---

* Partially supported by the Future and Emerging Technologies Unit of EC (IST priority - 6th FP), under contract no. FP6-021235-2 (project ARRIVAL).

F. Geraets et al. (Eds.): Railway Optimization 2004, LNCS 4359, pp. 67–90, 2007.

want to give an overview of such approaches, which solve timetable information queries by finding a shortest path in an appropriately defined graph. Hence, the problems are directly transformed into shortest path problems.

In the remainder of the introduction we give a brief overview of heuristic two-phase approaches and direct shortest-path approaches. In Section 2 the timetable information problems are formally specified. Two main approaches for modeling timetable information directly as shortest paths are described in detail in Section 3, where first a simplified problem specification is considered. Later on, in Section 4, extensions of the approaches that cover realistic details are outlined. Multi-criteria optimization is discussed in Section 5, and studies investigating the performance of the algorithms described in this paper are summarized in Section 6. We conclude the survey with some final remarks in Section 7.

1.1   Two-Phase Approaches

We want to mention two predecessors of "real" timetable information systems. Around the year 1988, the Dutch and German train companies started to use electronic timetable information systems. Heuristics that usually yield good solutions, but cannot always guarantee an optimal solution, are used to keep the search spaces small enough. The two systems we describe work both in two phases, where the first phase heuristically restricts the search space.

TRAINS

Tulp and Siklòssy [37] describe the TRAINS system, which was used by the Dutch railways (NS) at that time as a prototype: It is based on a graph where nodes represent cities. They distinguish two levels of the network, a "static" level which consists of arcs between nodes representing distances, and a "dynamic" level where the arcs include information about the departure and arrival times of trains. The algorithm uses the static level to cut out the "interesting" part of the network, without considering any information about time. Note that this cutting is heuristic in the sense that optimal connections may be lost by that step, which is not permitted in the models investigated later in this paper. Then, a train connection is calculated by a modification of Dijkstra's algorithm [7] trying to incorporate time for train changes at stations. Once a connection to the destination station is found, a backward search tries to improve the result (e.g., to find a connection that departs later and has the same arrival time).

ARIADNE

Baumann and Schmidt [2] outline an algorithm called ARIADNE, which can be regarded as ancestor of HAFAS [13], the timetable information system that is nowadays used by the German railway company Deutsche Bahn AG and many other railway companies worldwide. As in TRAINS, the algorithm considers two different networks: a static graph representing the topographic railway network, and a dynamic network including time, traffic days, train classes etc. The ARIADNE algorithm works in two phases: The first phase ("Wegesuche") searches feasible paths in the static network by a bidirectional version of Dijkstra's algorithm and outputs a subgraph of the network to be considered in the second phase. Note that again—as in the TRAINS algorithm—optimal solutions may be lost by this step. The second phase ("Zeitsuche") computes on the dynamic, time-dependent version of the network, limited by the subgraph computed in the first phase, several feasible train connections. These are rated according to measures like travel time, quality of trains, direct connection, etc.

1.2   Direct Shortest Path Approaches

Two main approaches have been proposed for modeling timetable information as shortest path problem: the *time-expanded* [19,20,21,22,26,30,28,32,33,34], and the *time-dependent* approach [4,5,16,23,24,25,30,28,29,32]. The common characteristic of both approaches is that a query is answered by applying some shortest path algorithm to a suitably constructed graph.

The Time-Expanded Approach

A time-expanded graph is constructed in which every node corresponds to a specific time event (departure or arrival) at a station and edges between nodes represent either elementary connections between the two events (i.e., served by a train that does not stop in-between), or waiting within a station. Depending on the optimization criterion, the construction assigns specific fixed lengths to the edges. This naturally results in the construction of a very large (but usually sparse) graph. The simplified version of the earliest arrival problem—where details like transfer rules and traffic days are neglected—has been extensively studied:

In [33], Schulz, Wagner and Weihe explicitly use the time-expanded approach to model a simplified version of the earliest arrival problem as a shortest path problem in a static graph, and solve the problem optimally. An extensive experimental study has been conducted and—at least in the simplified scenario—it could be demonstrated that the running time of the time-expanded approach on state-of-the-art computers is acceptable. To achieve this result, several speed-up techniques, which guarantee optimal solutions, were applied to Dijkstra's algorithm for computing the shortest path. More details on the speed-up techniques are provided in Section 6.

An extension of the time-expanded approach incorporating train transfers and an extensive experimental study focused on multi-criteria problems is presented by Müller-Hannemann and Weihe in [22]. The results of this study are quite promising: in practice (among other data also the time-expanded graph was considered) the number of Pareto-optimal paths is often very small, and labeling approaches are feasible. In [21], Müller-Hannemann, Schnee and Weihe focus on more realistic and complex real-world scenarios for timetable information, in particular with respect to space limitations. Further extensions towards realistic models and also further optimization criteria as well as bicriteria problems are presented by Pyrga, Schulz, Wagner and Zaroliagis in [30,28] (see also [32]), where the authors also conducted an experimental comparison with the time-dependent approach (see below). Multi-criteria optimization in the time-expanded graph by a labeling approach is extensively investigated by Müller-Hannemann and Schnee [20]; the notion of Pareto-optimal connections is relaxed (cf. 5.2). Möhring suggests the time-expanded model as a graph-theoretic concept for timetable information in [19]. He further discusses algorithms for solving multi-criteria problems, and focuses on a distributed approach for timetable information, which is also the topic of the recent projects DELFI [6] and EU-Spirit [11]: the railway network is considered as consisting of several (overlapping) subnetworks (e.g., each subnetwork is operated by a different company or institution), and a global solution is constructed from several subqueries to the conventional timetable information systems operated on the respective subnetworks. In a sense such new systems operate like meta search engines for the web.

The Time-Dependent Approach

The idea is to avoid the maintenance of a node per event. Instead, the time-dependent graph is used in which every node represents a station, and two nodes are connected by an edge if the corresponding stations are connected by an elementary connection. The lengths on the edges are assigned "on-the-fly": the length of an edge depends on the time in which the particular edge will be used by the shortest path algorithm to answer the query. Dynamic programming approaches for a time-dependent shortest-path problem have first been studied by Cooke and Halsey [5]. Later, Kostreva and Wiecek [16] generalized this approach towards multiple criteria. However, no performance guarantees are given for these dynamic programming approaches. Orda and Rom [24,25] thoroughly investigated the complexity of time-dependent shortest path problems and give efficient algorithms for special cases. Brodal and Jacob [3,4] argued that in the simplified case of the earliest arrival problem, Dijkstra's algorithm considers many redundant edges in the time-expanded approach. They suggest to use a time-dependent network instead and proved by a detailed theoretical analysis of operation counts in both approaches that a variant of a time-dependent shortest-path algorithm introduced by Orda and Rom is more efficient than the time-expanded approach. Pyrga, Schulz, Wagner and Zaroliagis extended the time-dependent model to cope with realistic problem specifications [29]. A subsequent study [30,28,32] compares these models experimentally with the time-expanded models, where also bicriteria problems are considered.

The work of Nachtigal [23] can also be classified as a time-dependent approach to timetable information. The problem specification he uses is different to most other approaches: given a source station, for all other stations arrival functions depending on the departure time shall be computed. Hence, the departure time is not part of the query, and solutions are computed for all possible departure times.

2 Problem Specification

2.1 Data

A *timetable* consists of data concerning: stations (or bus stops, ports, etc), trains (or busses, ferries, etc), connecting stations, departure and arrival times of trains at stations, and traffic days. More formally, we are given a set of *trains* $\mathcal{Z}$, a set of stations $\mathcal{B}$, and a set of *elementary connections* $\mathcal{C}$ whose elements c are 5-tuples of the form $c = (Z, S_1, S_2, t_d, t_a)$. Such a tuple (elementary connection) is interpreted as train Z leaves station S_1 at time t_d, and the next stop of train Z is station S_2 at time t_a. If x denotes a tuple's field, then the notation $x(c)$ specifies the value of x in the elementary connection c.

The *departure* and *arrival times* $t_d(c)$ and $t_a(c)$ of an elementary connection $c \in \mathcal{C}$ within a day are integers in the interval $[0, 1439]$ representing time in minutes after midnight. The *length* of an elementary connection c, denoted by *length*(c), is the time that passes between the departure and the arrival of c.

A timetable is valid for a number of N *traffic days*, and every train is assigned a bit-field of N bits determining on which traffic day the train operates (for overnight trains the departure of the first elementary connection counts).

At a station $S \in \mathcal{B}$ it is possible to transfer from one train to another only if the time between the arrival and the departure at that station S is larger than or equal to a given, station-specific, *minimum transfer time*, denoted by *transfer*(S). There may also be more complicated transfer rules, for example the transfer time can be smaller for trains that depart from the same platform. The most general notion is to specify a station-specific minimum transfer time, and exceptions in the form of a set of feasible and a set of forbidden transfer trains for each arrival of a train at a station.

Between stations that are located close to each other it is possible to walk by foot. Such data is available through so-called *foot-edges* between stations. Each foot-edge is associated with a natural number representing the time in minutes needed for the walk. Formally, we treat a foot-edge like an elementary connection c, where the train Z and the departure and arrival times t_d and t_a are invalid, and *length*(c) is the associated walking time. Foot-edges are independent of traffic days.

2.2 Connections

Let $P = (c_1, \ldots, c_k)$ be a sequence of elementary connections (and foot-edges) together with departure times $dep_i(P)$ and arrival times $arr_i(P)$ for each elementary connection c_i, $1 \leq i \leq k$. We assume that the times $dep_i(P)$ and $arr_i(P)$ include data regarding also the departure/arrival day by counting time in minutes from the first day of the timetable. A time value t is of the form $t = a \cdot 1440 + b$, where $a \in [0, 364]$ and $b \in [0, 1439]$. Hence, the actual time within a day is $t \pmod{1440}$ and the actual day is $\lfloor t/1440 \rfloor$.

Such a sequence P is called a *consistent connection* from station $A = S_1(c_1)$ to station $B = S_2(c_k)$ if it fulfills some consistency conditions: the departure station of c_{i+1} is the arrival station of c_i, and the time values $dep_i(P)$ and $arr_i(P)$ correspond to the time values t_d and t_a, resp., of the elementary connections (modulo 1440) and respect the transfer times at stations. More formally, P is a *consistent connection* if the following conditions are satisfied:

$$c_i \quad \text{is valid on day } \lfloor dep_i(P)/1440 \rfloor,$$
$$S_2(c_i) = S_1(c_{i+1}),$$
$$dep_i(P) \equiv t_d(c_i) \pmod{1440},$$
$$arr_i(P) = dep_i(P) + length(c_i),$$
$$dep_{i+1}(P) - arr_i(P) \geq \begin{cases} 0 & \text{if } Z(c_{i+1}) = Z(c_i) \text{ or} \\ & c_i \text{ is a foot-edge, and} \\ transfer(S_2(c_i)) & \text{otherwise.} \end{cases}$$

2.3   Criteria and Queries

For the timetable information problems we are additionally given a large, on-line sequence of *queries*. A query generally defines a set of valid connections, and an optimization criterion (or criteria) on that set of connections. The problem is to find the optimal connection (or a set of optimal connections) w.r.t. the specific criterion or criteria.

The Basic Query

The most fundamental query is also referred to as the *earliest arrival problem*. A query (A, B, t_0) consists of a departure station A, an arrival station B, and a departure time t_0. Connections are *valid* if they do not depart before the given departure time t_0, and the optimization criterion is to minimize the difference between the arrival time and the given departure time. Additionally, one may ask among all connections that are solutions to such a query for the connection that departs as late as possible.

Extended Queries

Other important optimization criteria involve the *number of transfers* and the *price* of a connection. In the *minimum number of transfers problem*, the query is to ask, given two stations A and B, for a connection with as few transfers as possible, which doesn't involve a departure or arrival time at all. Similarly, one can ask for a connection with the lowest price.

A query can also contain a sequence of via stations together with the duration of the stays at the respective stations. Further, certain trains or train classes can be excluded from the set of trains, e.g., if one intends to use a ticket that is valid only for local trains the intercity trains should be excluded. Also, instead of specifying the departure time, as in the earliest arrival problem, the aspired arrival time may be given. Alternatively, such time specifications can be given as time intervals.

3   Basic Modeling: The Earliest Arrival Problem

In this section, we review models for solving the first and most fundamental basic query, namely the earliest arrival problem, in both the time-expanded and the time-dependent approach. We consider throughout this section a simplified specification of train connections: We assume that (i) a transfer between trains at a station takes negligible time, i.e., $transfer(S) = 0$ for each station S, (ii) every train is operated daily, i.e., every day is the same in the timetable, and (iii) there are no foot-edges. In the following section we show how the models can be extended to comply with the realistic specification, and also consider the extended types of queries.

3.1   Time-Expanded Model

The time-expanded model [33] is based on the directed *time-expanded graph* which is constructed as follows. There is a node for every time *event* (departure or arrival) at a station, and there are two types of edges. For every elementary connection (Z, S_1, S_2, t_d, t_a) in the timetable, there is a *train-edge* in the graph connecting a *departure node*, belonging to station S_1 and associated with time t_d, with an *arrival node*, belonging to station S_2 and associated with time t_a. In other words, the endpoints of the train-edges induce the set of nodes of the graph. For each station S, all nodes belonging to S are ordered according to their time values. Let $v_1, \ldots, v_k$ be the nodes of S in that order. Then, there is a set of *stay-edges* (v_i, v_{i+1}), $1 \leq i \leq k - 1$, and (v_k, v_1) connecting the time events within a station and representing waiting within that station. The length of an edge (u, v) is $t_v - t_u$ (for edges over midnight the length is $1440 + t_v - t_u$, respectively), where t_u and t_v are the time values associated with u and v, respectively. Figure 1 illustrates this definition.

A shortest path in the time-expanded graph from the first departure node s at the departure station A with departure time later than or equal to the given start time t_0 to one of the arrival nodes of the destination station B constitutes a solution to the earliest arrival problem in the time-expanded model. The actual path can be found by Dijkstra's algorithm [7].

3.2   Time-Dependent Model

The time-dependent model [4] is also based on a digraph, called *time-dependent graph*. In this graph there is only one node per station, and there is an edge e from station A to station B if there is an elementary connection from A to B. The set of elementary connections from A to B is denoted by $\mathcal{C}(e)$. The definition is illustrated in Figure 1. The length of an edge $e = (v, w)$ depends on the time at which this particular edge will be used during the algorithm. In other words, if T is a set denoting time, then the length of an edge (v, w) is given by $f_{(v,w)}(t) - t$, where t is the departure time at v, $f_{(v,w)} : T \to T$ is a function such that $f_{(v,w)}(t) = t'$, and $t' \geq t$ is the earliest possible arrival time at w. The time-dependent model is based on the assumption that overtaking of trains on

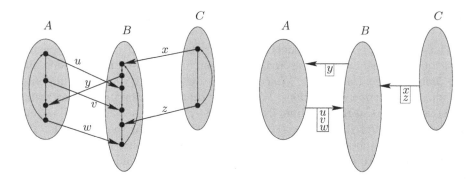

Fig. 1. The time-expanded graph (left) and the time-dependent graph (right) of a timetable with three stations A, B, C. There are three trains that connect A with B (elementary connections u,v,w), one train from C via B to A (x,y) and one train from C to B (z).

an edge is not allowed, i.e., for any two given stations A and B, there are no two trains leaving A and arriving to B such that the train that leaves A second arrives first at B.

A modification of Dijkstra's algorithm [7] can be used to solve the earliest arrival problem in the time-dependent model [4]. Let D denote the departure station and t_0 the earliest departure time. The differences, w.r.t. Dijkstra's algorithm, are: set the distance label $\delta(D)$ of the starting node corresponding to the departure station D to t_0 (and not to 0), and calculate the edge lengths "on-the-fly". The edge lengths (and implicitly the time-dependent function f) are calculated as follows. Since Dijkstra's algorithm is a label-setting shortest-path algorithm, whenever an edge $e = (A, B)$ is considered the distance label $\delta(A)$ of node A is optimal. In the time-dependent model, $\delta(A)$ denotes the earliest arrival time at station A. In other words, we indeed know the earliest arrival time at station A whenever the edge $e = (A, B)$ is considered, and therefore we know at that stage of the algorithm which train has to be taken to reach station B via A as early as possible: the first train that departs later than or equal to the earliest arrival time at A. The particular connection $c \in C(e)$ can be easily found by binary search if the elementary connections $\mathcal{C}(e)$ are maintained in a sorted array (or with more sophisticated techniques in constant time). The edge length of e, $\ell_e(t)$, is then defined to be the time to wait for the departure of c plus $length(c)$. Consequently, $f_e(t) = t + \ell_e(t)$. The correctness of the algorithm is based on the fact that f is *non-decreasing* ($t \leq t' \Rightarrow f(t) \leq f(t')$) and has *non-negative delay* ($\forall t, f(t) \geq t$).

3.3   Comparison of Models

In the simplified scenario we are investigating in this section, the graphs that are used in the two approaches are strongly related: Contracting all nodes that belong to the same station in the time-expanded graph and deleting parallel edges afterwards yields the time-dependent graph. Further, the algorithm used in the time-dependent approach can be viewed as an improved implementation of the simple shortest-path search by Dijkstra's algorithm in the time-expanded approach: If the first edge from some station A to another station B has already been processed by Dijkstra's algorithm in the time-expanded graph, all other edges e'_{AB} from station A to station B do not have to be considered anymore. The reason is that such an edge doesn't provide an improvement since the path through the first edge extended by some stay-edges to the head of the edge e'_{AB} has the same length. In a sense, the time-dependent algorithm implements this observation.

Note, however, that on the one hand the edge lengths have still to be computed in the time-dependent algorithm, which consumes running time as well, so that it is not immediately clear which algorithm is faster. We will discuss this question in Section 6. On the other hand, the similarity of the graphs and the algorithms is disturbed when the realistic specifications are incorporated into the models in the following section.

4   Realistic Modeling

In this section, we explain how the approaches introduced so far for the simplified earliest arrival problem can be extended towards realistic problem specifications and other optimization criteria.

4.1   Transfers Rules

We summarize first how transfer times at stations can be incorporated in the time-expanded and the time-dependent models, and after that discuss the case of extended transfer rules.

Time-Expanded Model

To incorporate transfer times in the time-expanded model the *realistic time-expanded graph* is constructed as follows (cf. [22,30,28]). Based on the time-expanded graph, for each station, a copy of all departure and arrival nodes in the station is maintained which we call *transfer-nodes*; see Figure 2. The stay-edges are now introduced between the transfer-nodes. For every arrival node there are two additional outgoing edges: one edge to the departure of the same train, and a second edge to the transfer-node with time value greater than or equal to the sum of the time of the arrival node and the minimum time needed to change trains at the given station. If the earliest arrival problem shall be solved, the edge lengths are defined as in the definition of the original model (see Section 3.1).

Time-Dependent Model

The original time-dependent model is extended in [29] (cf. also [4]) using information on the routes that trains may follow. Hence, we assume that we are given a set

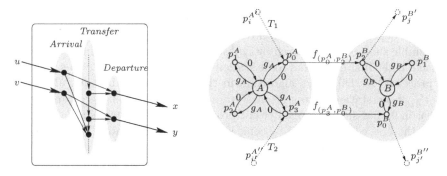

Fig. 2. Modeling transfer times in the time-expanded approach using the realistic time-expanded graph (left) and in the time-dependent approach using the train-route graph (right)

of train routes and their respective time schedules. In the following, we describe the construction of the *train-route graph*. We say that stations $S_0, S_1, \ldots, S_{k-1}$, $k > 0$, form a *train route* if there is some train starting its journey from S_0 and visiting consecutively $S_1, \ldots, S_{k-1}$ in turn. If there are more than one trains following the same schedule (with respect to the order in which they visit the above nodes), then we say that they all belong to the same train route.

The node-set of the train-route graph consists of the station-nodes $\mathcal{S}$ representing the stations, and for each station S of one additional route-node per route that passes through the station S, denoted by p_i^S, where i is an index of the specific route passing through station S. There are three types of edges: (i) edges from each station-node to the route-nodes belonging to the same station model boarding a train belonging to the specific route; (ii) edges from each route-node to the station-node model getting off a train at that station; (iii) for each train route $S_0, \ldots S_{k-q}$ edges that connect the corresponding route-nodes model the actual train trips.

To solve the earliest arrival problem with transfer times, edge lengths are defined as follows. The edges modeling boarding a train at a station S are assigned the transfer time $g_S = \mathit{transfer}(S)$, edges modeling getting off a train are assigned zero length, and the edges representing the train routes have time-dependent lengths as in the basic modeling described in Section 3.2. Given the query to solve, all internal edges are assigned zero length, and the modified version of Dijkstra's algorithm (cf. Section 3.2) is applied.

Extended Transfer Rules

Additionally to the transfer times at stations, exceptions which explicitly allow specific transfers can be modeled in the realistic time-expanded graph as additional edges connecting the arrival with the departure node of the corresponding elementary connections. Concerning the time-dependent approach, in [29] a graph similar to the train-route graph is constructed that allows to model also variable transfer times.

4.2   Foot-Edges

In the time-dependent approach a foot-edge from station A to station B can be directly modeled as an edge in the train-route graph between the two nodes representing the stations A and B. For the earliest arrival problem, such an edge is assigned a constant edge length: the walking time.

The straight-forward modeling of foot-edges in the time-expanded case is of course done by time expansion. For each transfer node of A in the (realistic) time-expanded graph an additional edge is maintained to the first possible transfer node at B. Another solution for the time-expanded approach is to apply the time-dependent idea and compute the additional edges during the algorithm (the node at B has to be calculated depending on the arrival time at B) instead of explicitly constructing them.

4.3   Traffic Days

Edges representing elementary connections of trains that are not valid can be simply ignored during Dijkstra's algorithm in the time-expanded approach, and the test whether an elementary connection is valid or not can be done by a lookup in the traffic day bit-mask of the corresponding train, if the day of departure is known. However, the algorithm has to be slightly modified because it may happen that an optimal connection stays more than a day at a station, and such connections would not be found otherwise. See [30] for details concerning the modification of the algorithm. In the time-dependent approach, the traffic days have to be considered in the calculation of the time-dependent edge lengths.

Problems with traffic days occur when speed-up techniques for Dijkstra's algorithm are applied that require preprocessing (cf. Section 6), because then every day is different and the preprocessing basically has to be done separately for each day.

4.4   The Minimum Number of Transfers Problem

The realistic time-expanded graph as well as the train-route graph (cf. Section 4.1) can be used to solve a minimum number of transfers query with a similar method (cf. [20,22,30,28]): edges that model transfers are assigned a length of one, and all the other edges are assigned length zero. In the time-expanded case all incoming edges of transfer nodes have length one, whereas in the time-dependent case the edges that represent getting off a train, except those belonging to the departure station, are assigned length one, and all other edges have length zero. Note that the edge lengths in the time-dependent train-route graph are all static here.

A shortest path in one of the graphs from a node belonging to (resp. representing) the departure station to a node belonging to (resp. representing) the arrival station provides a solution to the minimum number of transfers problem.

4.5   Extended Queries

Latest Departure

Determining a connection optimized for the latest departure combined with the earliest arrival can be done in the time-expanded case by introducing the latest departure as second criterion and determining the lexicographically first solution. In the time-dependent model the standard approach is to carry out a backward search from the destination station to the arrival station once the earliest arrival at the destination station is known.

Time Intervals

In pre-trip planning one often seeks the fastest connection which starts within a certain departure interval $[t_0, t_1]$ (or arrives within a certain arrival time interval). This variant can also be solved by Dijkstra's algorithm. With a single starting point, Dijkstra's algorithm starts by labeling the corresponding event node with distance 0 and puts it into the priority queue. With a time interval the only difference is that one initially inserts a label for each departure event between t_0 and t_1 into the priority queue and marks all corresponding nodes with a distance label of 0.

Excluding Trains

The exclusion of specific trains or of train classes, the exclusion or required inclusion of train attributes with respect to a given query can be handled like traffic days: We simply mark train edges as *invisible* for the search if they do not meet all requirements of the given query.

Via Stations

A query may contain one (or more) so called *vias*, i.e., stations the connection has to visit and where at least a specified amount of time can be spent. Assume that we have a query $(A = S_0, B = S_k, t_0)$ with via stations S_1, S_2, S_{k-1} and corresponding stay durations $d_1, d_2, \ldots d_k$. To answer such a query, one can simply split the query into basic queries without vias. More precisely, we first answer the query (A, S_1, t_0) and may find out that the earliest arrival time at S_1 is t_1. Then we answer the query $(S_1, S_2, t_1 + d_1)$. If t_i denotes the earliest arrival time at S_i (provided that we have visited $S_1, S_2, \ldots, S_{i-1}$ before), we continue with basic queries of the form $(S_i, S_{i+1}, t_i + d_i)$ for $i = 1, \ldots, k - 1$. Finally, we concatenate the connections found in each of the basic queries to get the connection which solves the earliest arrival problem.

Cheapest Connections

Many customers are interested in finding a cheapest connection from station A to B within a certain time interval. Unfortunately, given the complicated fare regulations in most countries, this goal seems to be intractable. Recall that a shortest path problem on a given digraph usually assumes that the length of a path can be calculated as the sum over the edge lengths which constitute the path. Given nonnegative lengths, the separability of the objective function then suffices to apply Dijkstra's algorithm.

Even in the standard tariff, the fare of a connection is usually not additive based on its elementary connections. The situation becomes substantially worse if one would also like to consider the many exceptions and special offers which exist and frequently change. Hence, there is no hope to solve the cheapest connection problem exactly and simultaneously efficiently.

However, what one can do is to use fare estimations based on a simpler model. Müller-Hannemann and Weihe [22] and Müller-Hannemann and Schnee [20] use a simplified fare model which assumes that the basic fare is proportional to the distance traveled. Depending on the train class an extra supplementary fare is charged. For the fastest trains like German ICE and French TGV, this supplement is assumed to be proportional to the speed of the train, whereas certain trains like Eurocity and Intercity trains have a constant surcharge which has to be paid at most once. With such a simplified fare model we can again use Dijkstra's algorithm if we store in our distance labels also flags indicating which train classes have been used.

Müller-Hannemann and Schnee [20] use these fare estimates to find low cost connections in a framework which does not only look for a single best connection but for several attractive connections.

5 Multi-criteria Optimization

In the previous sections we focused on single-criterion optimization, and in particular on finding fastest connections. In practice, however, one wishes to find optimal connections under several criteria. For instance, a customer may want to ask for a connection with a small number of transfers that departs later than a given time and does not arrive at the destination too late.

Other additional criteria of interest are fares, convenience (for example, measured by the used train classes in a connection), stability of a connection in case of delays (for example, measured by the minimum buffer time of a transfer within a connection, where the *buffer time* of a transfer is the difference between the waiting time and the minimum transfer time), seat reservability (is it possible to get a seat reservation for all those parts of a connection where the used trains do in principle allow a seat reservation), etc.

Computing optimal connections under multiple criteria reduces (in a completely analogous to the single-criterion case) to the *multi-criteria or multi-objective shortest path* (MOSP) problem – a fundamental problem in the area of multi-criteria or multi-objective optimization [10]. An instance of a multi-criteria optimization problem is associated with a set of feasible solutions Q and a d-vector function $\mathbf{f} = [f_1, \ldots, f_d]^T$ (d is typically a constant) associating each feasible solution $q \in Q$ with a d-vector $\mathbf{f}(q)$ (w.l.o.g we assume that all objectives f_i, $1 \leq i \leq d$, are to be minimized). In multi-criteria optimization we are interested not in finding a single optimal solution, but in computing the *trade-off* among the different objective functions, called the *Pareto set (or curve)* $\mathcal{P}$, which is the set of all feasible solutions in Q whose vector of the various objectives is *not* dominated by any other solution (a solution p *dominates* another solution q iff $f_i(p) \leq f_i(q)$, for all $1 \leq i \leq d$, and $f_j(p) < f_j(q)$, for at least one j, $1 \leq j \leq d$).

Multi-objective optimization problems are usually NP-hard (as indeed is the case for MOSP), since the Pareto set is typically exponential in size (even in the case of two objectives). Hence, exact methods (i.e., methods that find all Pareto optimal solutions) may be efficient under certain circumstances that depend on the particular instance of a given problem. On the other hand, even if a decision maker is armed with the entire Pareto set, s/he is still left with the problem of which is the "best" solution for the application at hand. Consequently, three natural approaches to deal with multiobjective optimization problems are to: (i) study approximate versions of the Pareto set that result in (guaranteed) near optimal but smaller Pareto sets; (ii) optimize one objective while bounding the rest (*constrained approach*); and (iii) proceed in a normative way and choose the "best" solution by introducing a utility (often non-linear) function on the objectives (*normalization or decision maker's approach*).

In the following, we will discuss the above exact and non-exact approaches in computing optimal connections under multiple criteria.

5.1 Exact Approaches

The straightforward approach in dealing with multiple criteria is to find the entire Pareto set, that is, all Pareto-optimal (i.e., feasible and undominated) solutions (connections). As mentioned above, this may be a hard problem. Even very simple instances of graphs – actually chains of parallel arcs with just two criteria – may have exponentially many different Pareto optima [14]. In certain cases, however, the cardinality of the Pareto set may not be too large and the Pareto set can be computed efficiently. In the following, we discuss such cases.

Size of the Pareto Set

If the range of valid values of some criterion is restricted to a small discrete set, adding such a criterion cannot lead to an exponential blow-up. For example, for the number of transfers we can safely assume a small constant k as an upper bound on its number in practice. Hence, if we add the number of transfers as an additional criterion to one or more other criteria, the size of the Pareto set can only increase by a factor of k.

Empirical results indicate that one may have small Pareto sets in timetable information. In a recent study with 25000 queries to the server of Deutsche Bahn AG, Müller-Hannemann and Schnee observed only 3.6 Pareto optimal connections on average if the three criteria travel time, number of transfers, and fares are used. The observed maximum was 19 Pareto-optimal connections.

Müller-Hannemann and Weihe [22] studied the size of the Pareto set also from a theoretical point of view. They considered certain characteristics found in the application scenario in an attempt to explain the huge gap between the potentially exponentially sized Pareto set and the small sizes observed in practice.

An important characteristic of our application is that we can partition the edge set of our graph models into a small number of different *edge classes* such that each edge class has a certain semantics. Naturally, we can take the different types of edges as individual classes. Moreover, we can refine the class of those edges modeling elementary train connections into further classes derived from the type of train (*train classes*). A similar way to partition the edge set is to group them by average speed.

Average speed as the ratio of travel distance and travel time relates two criteria together. If we now assume that there are only k different average speeds (and therefore only k different edge classes), we arrive at the *ratio-restricted lengths model*. More generally than just considering speed, but still with two criteria, we assume in this model that every edge class is equipped with a value r which denotes the ratio between the length values of the first and second criterion, respectively. But even in a bicriteria model with at most $k > 1$ different ratios, the number of Pareto optima can still be exponentially large (Lemma 2.1 in [22]).

Another important characteristic of our application is that Pareto-optimal connections typically show a certain pattern with respect to the order of used train classes. Namely, if we order the train classes by their maximum speed, from slowest to fastest means of transportation, then most connections turn out to be *bitonic*: they consist of one acceleration and one deceleration phase. In the acceleration phase, we start with a slower train and from transfer to transfer we monotonically use trains with higher average speed until we reach a maximum. Then the deceleration phase starts and we use again slower and slower trains. Of course, there are exceptions: for example, in cities like Paris or London one may have to use the subway when changing between two high-speed trains. But it seems reasonable to assume that the number of changes between acceleration and deceleration phases is rather limited for Pareto-optimal connections. This model has been called *restricted non-monotonical*.

Combining the ratio-restricted lengths model with the restricted non-monotonical case it is possible to give tight polynomial worst case bounds on the size of the Pareto set (Lemma 2.3 and Lemma 2.7. in [22]).

Finding the Pareto Set

The standard approaches to the case that all Pareto optima have to be computed are generalizations of the standard algorithms for the single-criterion case (for a survey see Ehrgott and Gandibleux [10]).

The main difference is that instead of one scalar distance label, each node v maintains a list of d-dimensional distance labels assuming that we work with d criteria. Such a list contains a set of Pareto-optimal paths from the start node to v.

This immediately leads to a "Pareto version" of Dijkstra's algorithm (first described by Hansen [14] and Martins [18]). Instead of storing temporarily labeled

nodes, the priority queue now maintains d-dimensional labels. In each iteration
of the main loop, we extract the lexicographic smallest label from the priority
queue instead of choosing the node with smallest distance label. If v is the corresponding node to the extracted label one updates for all feasible edges of type
(v, w) the list of Pareto optima stored at the head node w. More precisely, a
tentative new d-dimensional label is created and compared to all labels in the
list of Pareto-optima held at node w. It is only inserted into that list if it is not
dominated by any other label in the list. Moreover, labels dominated by the new
label are removed. For a more detailed description of the generalized Dijkstra
algorithm and a correctness proof we refer to [19] and [35].

An adaption of this algorithm has been used by Müller-Hannemann and
Schnee to build the timetable information server PARETO [20] which relies
on the time-expanded graph model.

In [30,28], Pareto-optimal connections concerning the earliest arrival and minimum number of transfers have been considered for the time-dependent approach.
In this work, it is further shown that the Pareto-set in the special case of a bicriteria problem involving the earliest arrival as one criterion can be determined
in the time-expanded approach by running Dijkstra's algorithm on the realistic time-expanded graph with lexicographically ordered distance labels. The
Pareto-optimal solutions are enumerated by the solutions that Dijkstra's algorithm reports at the destination station (i.e., the algorithm is not terminated
until all Pareto-optimal solutions have been found).

5.2   Approximation Approaches

The ultimate goal of a traffic information system is to offer a small set of highly
attractive connections as an answer to a customer query. In that respect finding
the whole set of Pareto-optimal solutions bears two problems:

1. Not every Pareto-optimal solution is really noteworthy for a customer.
2. Many attractive connections are dominated only slightly.

The first of these two problems can be solved easily by filtering out unattrac-
tive connections. To tackle the second problem, we need a proper notion of
approximate Pareto optimal solutions. Current research concentrates along two
directions: the (recently re-investigated) concept of $(1 + \varepsilon)$-Pareto set [27], and
the concept of relaxed Pareto dominance, usually called ε-efficiency [17,42].

Approximate Pareto Sets

Despite so much research in multiobjective optimization [9,10], only recently a
systematic study of the complexity issues regarding the construction of approx-
imate Pareto sets has been initiated [27]. Informally, an $(1 + \varepsilon)$-*Pareto set* $\mathcal{P}_\varepsilon$
is a subset of feasible solutions such that for any Pareto optimal solution and
any $\varepsilon > 0$, there exists a solution in $\mathcal{P}_\varepsilon$ that is no more than $(1 + \varepsilon)$ away in
all objectives. Although this concept is not new (it has been previously used in

the context of bicriteria and multiobjective shortest paths [14,41]), Papadimitriou and Yannakakis in a seminal work [27] show that for *any* multiobjective optimization problem there exists a $(1 + \varepsilon)$-Pareto set $\mathcal{P}_\varepsilon$ of (polynomial) size $|\mathcal{P}_\varepsilon| = O((4B/\varepsilon)^{d-1})$, where B is the number of bits required to represent the values in the objective functions (bounded by some polynomial in the size of the input). They also provide a necessary and sufficient condition for its efficient (polynomial in the size of the input and $1/\varepsilon$) construction. In particular, $\mathcal{P}_\varepsilon$ can be constructed by $O((4B/\varepsilon)^d)$ calls to a GAP routine that solves (in time polynomial in the size of the input and $1/\varepsilon$) the following problem: given a vector of values **a**, either compute a solution that dominates **a**, or report that there is no solution better than **a** by at least a factor of $1 + \varepsilon$ in all objectives. Extensions to that method to produce a constant approximation to the smallest possible $(1 + \varepsilon)$-Pareto set for the cases of 2 and 3 objectives are presented in [38], while for $d > 3$ objectives inapproximability results are shown for such a constant approximation.

Apart from the above general results, there has been very recent work on improved approximation algorithms (FPTAS) for multiobjective shortest paths by Tsaggouris and Zaroliagis in [36]. In that paper, a new and remarkably simple algorithm is given that constructs $(1 + \varepsilon)$-Pareto sets for the single-source multiobjective shortest path problem, which improves considerably upon previous approaches. The algorithm can be viewed as a generalization of the Bellman-Ford algorithm. It proceeds in rounds. In each round i and for each node v, the algorithm maintains a d-dimensional label representing an approximate Pareto set to all Pareto optimal s-v paths with no more than i edges (s is the source node). When an edge (u, v) is considered during round i, the algorithm performs (instead of a relaxation) an extend-&-merge operation. This operation extends the node label of u in round $i-1$ with the edge (u, v) and merges the resulting set with the label associated with v by keeping the solution that approximately dominates all other solutions. This keeps the size of all labels polynomially bounded, contrary to previous label correcting or setting approaches which used to keep all undominated solutions and thus resulting in exponentially large sets of labels.

Relaxed Pareto Dominance

Müller-Hannemann and Schnee [20] recently generalized the concept of *relaxed Pareto dominance* (also known as ε-efficiency [17,42]) and applied it to traffic information. In relaxed Pareto dominance, a solution p dominates (in the relaxed-Pareto sense) another solution q iff $f_i(p) + h_i(p,q) \leq f_i(q)$, for all $1 \leq i \leq d$, and $f_i(p) + h_i(p,q) < f_i(q)$, for at least one j, $1 \leq j \leq d$, where $h_i(p,q)$ is an appropriately chosen relaxation function. The idea is to make more pairs of connections mutually incomparable by redefining the dominance relation for certain criteria. For example, if we do not want to suppress a connection with a slightly longer travel time, say of less than 5 minutes, then we would define that connection A will dominate connection B with respect to travel time only if the travel time of A plus these 5 minutes are less or equal to the travel time of B. For more examples how to apply relaxed dominance, see [20].

5.3   Normalization Approaches

In this approach, a utility function is introduced that translates (in a linear or non-linear way) the different criteria into a common cost (utility) measure. For instance, when traveling in a traffic network one typically wishes to minimize travel distance and time; both criteria can be translated into a common cost measure (e.g., money), where the former is linearly translated, while the latter non-linearly (small amounts of time have relatively low value, while large amounts of time are very valuable). Under the normalization approach, we seek for a single optimum in the Pareto set (a feasible solution that optimizes the utility function). We distinguish between the case where all criteria are linearly translated to the common cost measure and to the case where some (or all) of the criteria are non-linearly translated.

The Linear Case – Weighted Sum of Criteria

The straightforward (and simplest) approach could be to express the relative importance of optimization criteria by weights and then to optimize a weighted sum of the criteria. This approach reduces the multi-criteria problem to a single-criterion optimization which can be solved by the standard Dijkstra algorithm provided we use a graph model where each criterion is non-negative and additive on the edges. Setting all but one weight to zero, we get the single-criterion optimization as a special case.

Such an approach has two serious drawbacks. First, it will inevitably miss many attractive connections as it will find just one single solution (and not all Pareto optima). The second drawback of such an approach lies in the choice of suitable weight parameters. Each potential customer has its own preference system, but typically this preference system is not given explicitly in terms of weight parameters. The user (customer or salesperson of a train company) of a traffic information system and/or the system itself might set the parameters incorrectly as neither of them will typically know the customer's preference system to its full extent.

The Non-linear Case

The case of non-linear utility function is the most interesting one, since it reflects realistic scenaria in traffic optimization. Experience shows that users of traffic networks value certain attributes (e.g., time) non-linearly [15]: small amounts have relatively low value, while large amounts are very valuable. Also, the vast majority of transit systems have a non-additive (non-linear) fare structure [12]. Consequently, the most interesting theoretical models for traffic equilibria involve minimizing a monotonic non-linear utility function. In this case, the problem of computing optimal connections reduces to the so-called *non-additive short-est path* (NASP) problem: given a digraph whose edges are associated with d-dimensional cost vectors, the task is to find a path that minimizes a certain d-attribute non-linear cost function.

Very recently, Tsaggouris and Zaroliagis [36] presented the first FPTAS for NASP. In particular, they showed how the FPTAS for multiobjective shortest paths can provide a FPTAS for NASP for any number of objectives and for a rather general form of a utility function that includes all polynomials of bounded degree with non-negative coefficients. For the bicriteria case, a FPTAS for NASP was independently presented in [1].

5.4 Lexicographical Ordering

One other possibility is to settle for only one specific Pareto-optimal solution: the *lexicographically first* one. Dijkstra's algorithm works not only for non-negative real edge weights, but in general for semi-rings [31]. In our case, edge weights and node labels are d-tuples, with lexicographical ordering and element-wise addition.

With the simplified version of the time-expanded approach the lexicographically first solution can be computed for any d-tuples as edge weights. For example, if $d = 2$, the first element being the travel time and the second one the number of transfers, among all fastest connections the one with the minimum number of transfers is computed. With the realistic version of the time-expanded approach only tuples can be used where the first criterion is travel time. This restriction is due to the 24-hour cycles induced by the stay-edges belonging to each station. A special case are pairs as edge weights with travel time as first criterion. In this case all Pareto-optimal solutions can be computed by Dijkstra's algorithm (cf. the last paragraph of Section 5.1).

In the time-dependent approach the edge weights are required to be non-decreasing (cf. Section 3.2). This is not necessarily true for arbitrary d-tuples as edge weights, but it can be shown that for the case $d = 2$, where the first element is the number of transfers and the second one is the travel time, the time-dependent version of Dijkstra's algorithm can be extended to find the lexicographically first solution. See [28,29,32] for further details.

6 Performance

As mentioned in the introduction, the performance of the core algorithms is crucial for a timetable information system. The average performance is particularly important in a scenario of a central server that has to answer several hundreds of (on-line) queries which are issued, for example, through the Internet or through terminals at train stations. We review the results of experimental studies involving the approaches introduced in the previous sections.

6.1 Simplified Earliest Arrival Problem

The approaches introduced in Section 3 for solving the simplified earliest arrival problem have been extensively studied, both in the time-expanded and the time-dependent approach.

Time-Expanded Approach

Schulz, Wagner, and Weihe [33] conducted an experimental study based on the time-expanded graph (cf. Section 3.1) with realistic timetable data of the German railway company Deutsche Bahn and a sample of half a million of real-world customer queries. Using a single 336 MHz Ultra-SPARC-II processor, the average running time per query of Dijkstra's algorithm was 0.103 seconds. The main contribution of the study is that it demonstrates that the average running time of Dijkstra's algorithm can be drastically improved by applying distance-preserving speed-up techniques (which still guarantee optimal solutions): a speed-up of a factor of 34 could be observed, yielding an average running time of 0.003 seconds. More precisely, they used a geometric speed-up technique based on angular sectors limiting the reachable nodes through an edge, and a graph decomposition technique based on a small "backbone graph" for finding the shortest path. Both techniques reduce the search space of the algorithm and rely on a preprocessing step in which the additional information is pre-computed. Wagner and Willhalm [40] have shown that other geometric containers are better suited and yield even higher speed-up factors, in particular bounding boxes around the reachable nodes through an edge show good results.

The second technique has been generalized by Schulz, Wagner, and Zaroliagis in [34]. They demonstrated, also conducting experiments with the same time-expanded graph as in [33], that several hierarchical levels (3 or 4 levels for the data used) of backbone graphs yield better running times than only one additional level (by a factor of roughly 3). See also [39] for a survey on speed-up techniques for shortest path algorithms.

Time-Dependent Approach

Brodal and Jacob proved in [4] by a detailed theoretical analysis of operation counts in both approaches that the time-dependent approach is more efficient than the time-expanded approach. This was also the starting point of an experimental comparison of the two approaches conducted by Pyrga, Schulz, Wagner, and Zaroliagis [30,28]. They revealed that indeed the time-dependent approach is faster than the time-expanded approach by factors in the range 12 to 40 depending on the data set (timetables consisting of French and German long-distance traffic as well as two timetables consisting of local traffic have been used).

Basically, the preprocessing speed-up techniques mentioned above for the time-expanded case can also be applied in the time-dependent approach; however, we are not aware of experimental studies dealing with this issue.

6.2   Realistic Single-Criterion Problems

In the experiments mentioned above [28], also the realistic specifications have been considered. Solving the minimum number of transfers problem is clearly faster (by a factor of roughly 4) in the train-route graph than in the realistic time-expanded graph. This is due to the fact that in this case the train-route graph is also static and smaller than the time-expanded graph.

Concerning the realistic earliest arrival problem, the picture looks different: The average running times of the time-expanded and the time-dependent approach are almost equal, only a speed-up factor of 1.5 was observed. Comparing the average running time for solving the simplified earliest arrival problem to the realistic earliest arrival problem, the time-expanded implementations solved the simplified version only slightly faster (by a factor of less than 2), while the simplified time-dependent implementation was faster by a factor of 5.

6.3 Multi-criteria Optimization

Finding all attractive connections with respect to travel time, fare, and number of interchanges is of course more expensive than just searching for a fastest connection. In the implementation of [20], such a search needs about 10 times as long as the search with a single criterion. For the multi-criteria case, we still need more effective speed-up techniques.

7 Conclusion

We have discussed time-expanded and time-dependent models for several kinds of single- and multi-criteria optimization problems for timetable information systems that provide optimal solutions via shortest paths. Extensions that model realistic requirements (like train transfers) can be integrated in both approaches.

The time-dependent approach is clearly superior with respect to performance when the simplified earliest arrival problem is considered, and speed-up factors in the range from 10 to 40 were observed. When considering extensions of the models for the solution of realistic versions of optimization problems in the single-criterion case, the performance of the two approaches is almost equal. Speed-up techniques yield running times indicating that these approaches are applicable in practice. The main open question is how these speed-up techniques—relying on additional information computed beforehand—can be extended to deal with dynamic changes of the timetable; such a change of the timetable invalidates the preprocessed information. Possibly, the additional information can be adapted by small updates to cope with both "off-line" changes like the treatment of different traffic days and "on-line" changes caused for example by accidents.

For other optimization criteria, it is more likely that the integration can be modeled directly by edge lengths in the time-expanded model than in the time-dependent model: In case the criterion can be expressed as additive costs for elementary connections, these costs induce edge lengths in the time-expanded graph. In contrast, in the time-dependent approach it is not clear if the costs can be mapped to feasible edge lengths, since only the first elementary connection per edge is considered. Because of this most studies concerning multi-criteria optimization have focused on the time-expanded approach. In that, (relaxed) Pareto-optimal solutions are desirable, and it turns out that in practice the size of the Pareto frontier is quite small, such that labeling approaches are feasible. For practical application, the multi-criteria optimization techniques provide the most satisfactory solutions. However, further speed-up techniques are required (the techniques for the single-criterion problems cannot be directly applied for the general multi-criteria algorithms) in order to yield a performance that is acceptable for a real-world timetable information system.

References

1. Ackermann, H., Newman, A., Röglin, H., Vöcking, B.: Decision making based on approximate and smoothed pareto curves. In: Deng, X., Du, D.-Z. (eds.) ISAAC 2005. LNCS, vol. 3827, pp. 675–684. Springer, Heidelberg (2005)
2. Baumann, N., Schmidt, R.: Buxtehude–Garmisch in 6 Sekunden. Die elektronische Fahrplanauskunft (EFA) der Deutschen Bundesbahn. Die Bundesbahn. Zeitschrift für aktuelle Verkehrsfragen, 10, 929–931 (1988)
3. Brodal, G.S., Jacob, R.: Time-dependent networks as models to achieve fast exact time-table queries. Technical Report ALCOMFT-TR-01-176, BRICS, University of Aarhus, Denmark (2001),
   http://www.brics.dk/ALCOM-FT/TR/ALCOMFT-TR-01-176.html
4. Brodal, G.S., Jacob, R.: Time-dependent networks as models to achieve fast exact time-table queries. In: Proceedings of the 3rd Workshop on Algorithmic Methods and Models for Optimization of Railways (ATMOS 2003). Electronic Notes in Theoretical Computer Science, vol. 92, Elsevier, Amsterdam (2004) (A previous version appeared as [3])
5. Cooke, K.L., Halsey, E.: The shortest route through a network with time-dependent internodal transit times. Journal of Mathematical Analysis and Applications 14, 493–498 (1966)
6. DELFI. Durchgängige elektronische Fahrplaninformation, http://www.delfi.de/
7. Dijkstra, E.W.: A note on two problems in connexion with graphs. Numerische Mathematik 1, 269–271 (1959)
8. EFA. A timetable information system by Mentz Datenverarbeitung GmbH, München, Germany, http://www.mentzdv.de/
9. Ehrgott, M.: Multicriteria Optimization. Springer, Heidelberg (2000)
10. Ehrgott, M., Gandibleux, X.: Multiobjective combinatorial optimization. In: Multiple Criteria Optimization — State of the Art Annotated Bibliographic Surveys, pp. 369–444. Kluwer Academic Publishers, Boston, MA (2002)
11. EUSpirit. European travel information system, http://www.eu-spirit.com/
12. Gabriel, S., Bernstein, D.: The traffic equilibrium problem with nonadditive path costs. Transportation Science 31(4), 337–348 (1997)
13. HAFAS. A timetable information system by HaCon Ingenieurgesellschaft mbH, Hannover, Germany, http://www.hacon.de/hafas/
14. Hansen, P.: Bicriteria path problems. In: Fandel, G., Gal, T. (eds.) Multiple Criteria Decision Making Theory and Applications. Lecture Notes in Economics and Mathematical Systems, vol. 177, pp. 109–127. Springer, Berlin (1979)
15. Hensen, D., Truong, T.: Valuation of travel times savings. Journal of Transport Economics and Policy, 237–260 (1985)
16. Kostreva, M.M., Wiecek, M.M.: Time dependency in multiple objective dynamic programming. Journal of Mathematical Analysis and Applications 173, 289–307 (1993)
17. Loridan, P.: ε-solutions in vector minimization problems. Journal of Optimization Theory and Applications 43, 265–276 (1984)

18. Martins, E.Q.V.: On a multicriteria shortest path problem. European Journal of Operations Research 16, 236–245 (1984)
19. Möhring, R.: Verteilte Verbindungssuche im öffentlichen Personenverkehr: Graphentheoretische Modelle und Algorithmen. In: Angewandte Mathematik – insbesondere Informatik, Vieweg, pp. 192–220 (1999)
20. Müller-Hannemann, M., Schnee, M.: Finding all attractive train connections by multi-criteria Pareto search. In: Proceedings of the 4th Workshop in Algorithmic Methods and Models for Optimization of Railways (ATMOS 2004), To appear in the same volume (2004)
21. Müller-Hannemann, M., Schnee, M., Weihe, K.: Getting train timetables into the main storage. In: Proceedings of the 2nd Workshop on Algorithmic Methods and Models for Optimization of Railways (ATMOS 2002). Electronic Notes in Theoretical Computer Science, vol. 66, Elsevier, Amsterdam (2002)
22. Müller-Hannemann, M., Weihe, K.: Pareto shortest paths is often feasible in practice. In: Brodal, G.S., Frigioni, D., Marchetti-Spaccamela, A. (eds.) WAE 2001. LNCS, vol. 2141, pp. 185–198. Springer, Heidelberg (2001)
23. Nachtigal, K.: Time depending shortest-path problems with applications to railway networks. European Journal of Operations Research 83, 154–166 (1995)
24. Orda, A., Rom, R.: Shortest-path and minimum-delay algorithms in networks with time-dependent edge-length. Journal of the ACM, 37(3) (1990)
25. Orda, A., Rom, R.: Minimum weight paths in time-dependent networks. Networks, 21 (1991)
26. Pallottino, S., Scutellà, M.G.: Shortest path algorithms in transportation models: Classical and innovative aspects. In: Equilibrium and Advanced Transportation Modelling, ch. 11, Kluwer Academic Publishers, Dordrecht (1998)
27. Papadimitriou, C., Yannakakis, M.: On the approximability of trade-offs and optimal access of web sources. In: Proc. 41st IEEE Symp. on Foundations of Computer Science – FOCS 2000, pp. 86–92 (2000)
28. Pyrga, E., Schulz, F., Wagner, D., Zaroliagis, C.: Experimental comparison of shortest path approaches for timetable information. In: Proceedings of the Sixth Workshop on Algorithm Engineering and Experiments, SIAM, pp. 88–99 (2004)
29. Pyrga, E., Schulz, F., Wagner, D., Zaroliagis, C.: Towards realistic modeling of time-table information through the time-dependent approach. In: Proceedings of the 3rd Workshop on Algorithmic Methods and Models for Optimization of Railways (ATMOS 2003). Electronic Notes in Theoretical Computer Science, vol. 92, pp. 85–103. Elsevier, Amsterdam (2004)
30. Pyrga, E., Schulz, F., Wagner, D., Zaroliagis, C.: Efficient Models for Timetable Information in Public Transportation Systems. ACM Journal of Experimental Algorithmics, 12(2.4) (2007)
31. Rote, G.: Path problems in graphs. In: Tinhofer, G., Mayr, E., Noltemeier, H., Syslo, M. (eds.) Computational Graph Theory, pp. 155–190. Springer, Heidelberg (1990)
32. Schulz, F.: Timetable Information and Shortest Paths. PhD thesis, Universität Karlsruhe (TH), Fakultät Informatik (2005)
33. Schulz, F., Wagner, D., Weihe, K.: Dijkstra's algorithm on-line: An empirical case study from public railroad transport. Journal of Experimental Algorithmics, 5(12) (2000)
34. Schulz, F., Wagner, D., Zaroliagis, C.: Using multi-level graphs for timetable information in railway systems. In: Mount, D.M., Stein, C. (eds.) ALENEX 2002. LNCS, vol. 2409, pp. 43–59. Springer, Heidelberg (2002)

35. Theune, D.: Robuste und effiziente Methoden zur Lösung von Wegproblemen. Teubner Verlag, Stuttgart (1995)
36. Tsaggouris, G., Zaroliagis, C.: Multiobjective optimization: Improved FPTAS for shortest paths and non-linear objectives with applications. Theory of Computing Systems (to appear, 2007)
37. Tulp, E., Siklóssy, L.: TRAINS, an active time-table searcher. In: Eighth European Conf. on AI, pp. 170–175 (1988)
38. Vassilvitskii, S., Yannakakis, M.: Efficiently computing succinct trade-off curves. In: Díaz, J., Karhumäki, J., Lepistö, A., Sannella, D. (eds.) ICALP 2004. LNCS, vol. 3142, pp. 1201–1213. Springer, Heidelberg (2004)
39. Wagner, D., Willhalm, T.: Speed-up techniques for shortest-path computations. In: Thomas, W., Weil, P. (eds.) STACS 2007. LNCS, vol. 4393, pp. 23–36. Springer, Heidelberg (2007)
40. Wagner, D., Willhalm, T., Zaroliagis, C.: Geometric containers for efficient shortest-path computation. ACM Journal of Experimental Algorithmics, 10 (2005)
41. Warburton, A.: Approximation of pareto optima in multiple-objective shortest path problems. Operations Research 35, 70–79 (1987)
42. White, D.J.: Epsilon efficiency. Jorunal of Optimization Theory and Applications 49, 319–337 (1986)

Estimates on Rolling Stock and Crew in DSB S-tog Based on Timetables

Michael Folkmann, Julie Jespersen, and Morten N. Nielsen

Danish State Railways (DSB)
S-tog a/s, Production Planning
Kalvebod Brygge 32
1560 Copenhagen V, Denmark
{mfolkmann,jjespersen,monnielsen}@s-tog.dsb.dk

Abstract. Reliable evaluation of the efficiency of timetables is very important for transportation companies today, because of increased competition on the market in Europe. Two major contributions to the cost of a timetable are the rolling stock (RS) and crew. Especially when the service level is increased, it is necessary to evaluate the need for new RS and hiring of crew in advance. In this paper we describe two models, one for evaluating the amount of RS, and one for evaluating the crew needed for a given timetable. The decision makers use the models as tools to choose between different timetables, to dimension the workforce, and to decide on the optimal fleet composition before buying new RS.

1 Introduction

DSB S-tog a/s (S-tog) is the Danish train company responsible for the trains in the greater Copenhagen area. S-tog is owned by DSB which runs most of the trains in Denmark. The railway was established in Denmark in 1847, and the S-tog part was founded in 1934. Since then the network has grown and the complexity of the planning process has grown as well. The timetable for 2007 is currently under preparation in S-tog. The discussion is based on a questionnaire among the customers to evaluate the trade-off between decreased travel time and more frequent trains. Based on the answers, a number of possible timetables have been produced and these must be evaluated regarding functionality and cost. Since the major part of the cost of a timetable comes from investments and daily use of RS and crew, it is important to have reliable models for evaluating the cost of these two factors at the tactical and strategic levels. The present flow in this process is that, based on the timetable, an RS plan is made. Based on the RS plan, a crew plan for drivers is constructed. The process is lengthy and currently performed manually.

This paper describes two models based on Integer Programming (IP) to evaluate the need for RS and crew. Both models are based entirely on data from the timetable and experiences from the past, i.e. the models can be run in parallel without the need of the usual workflow where the crew planning cannot start before the RS plan is finished.

F. Geraets et al. (Eds.): Railway Optimization 2004, LNCS 4359, pp. 91–107, 2007.

These models solve a number of problems common to many companies: Comparing the budget for different suggested brands (e.g. timetables), increasing service level without increasing the current budget, and increasing satisfaction among employees (i.e. drivers) by evaluating possible changes in labor rules, still within a reasonable budget.

The RS model for the Capacity Assignment Problem (CAP) has been used both at the strategic level to decide on the composition of the fleet and at the tactical level to decide on the distribution of the fleet during rush hours. The Crew model solving the Manpower Planning Problem (MPP) has been used at the tactical level to estimate the number of drivers necessary for three day types. Previously, the *cost per train kilometer* was calculated based on timetables, and this factor was used to evaluate the total cost for drivers in new timetables. Recently it was discussed in S-tog to change the "shape" of the production from one peak (morning rush hour) to two peaks (morning and evening rush hour). Here the old method is expected to be not sufficiently precise. In relation to special plans, the model is also used to estimate the number of drivers as a goal for the daily crew planners and as a quality measure for their work.

Finally, both models have been used to compare timetables where different objectives have been incorporated, e.g. the trade-off between higher frequency and faster travel times.

The paper is organized as follows: Sect. 2 gives background information about S-tog and terminology. Sect. 3 describes our solution for the CAP, and Sect. 4 presents our solution for the MPP. Finally, Sect. 5 contains conclusions.

2   Background and Terminology

2.1   The S-tog System

The timetable for S-tog is cyclic with a twenty minutes period. Departures are divided into *lines*. Each line has a fixed stopping pattern and two termini. The network consists of a central segment where most lines stop at every station and six rays emerging from the central segment. A timetable has a certain format. That is, it has predetermined cyclic periods and its lines cover predetermined rays. An example of a timetable format is high frequency. A line typically runs on two rays, but a special set of lines, called "The Ring", connects several rays. Each ray is covered by more than one line, implying that large stations have a higher frequency than twenty minutes. Since most passengers and trains pass through the central segment, this is often the main focus for strategic and tactical questions. The lines running along the rays are either *stop lines*, stopping at all stations, or *quick lines* stopping only at a subset of stations. A *train set* is a train following the timetable and consists of *train units*. A *round* for a train set is the time it takes to run from one terminus to the other terminus and back again. A *closed circuit* is the set of train sets required to cover a round. That is, the first train set in a round succeeds the last train set in the previous round.

The frequency is decisive to the number of train sets necessary for making an entire closed circuit on each line, e.g. line H driving between the stations of Farum and Frederikssund needs ten train sets to ensure the steady frequency of twenty minutes as it, including turn around at the termini, takes three hours and twenty minutes to drive from Farum to Frederikssund and back again. Fig. 1 illustrates the infrastructure for S-tog, i.e. the network composed of stations, lines, and rays. The RS *types* at S-tog is named 2g and 4g, where *g* is an abbreviation for *generation*. Among the 4g units both small and large units exist. For 4g units, two small units are almost equal to one large unit. The large unit has a slightly larger capacity than two small units. The current RS quantity consists of 40 units of 2g, 93 units of large 4g, and no small 4g units. In the beginning of 2005, the first small units of 4g were introduced and in the beginning of 2006 S-tog expects that 31 units of small 4g are available. The feasible compositions are as follows: 2g can be combined and 4g can be combined. It is not possible to mix 2g with 4g. At most two 2g units can be combined. Up to four 4g small units can be combined and two small units can be exchanged for one large.

S-tog has approximately 530 drivers including standby personnel. As in many other companies, a complex set of rules governs the daily work for drivers. The daily activities of drivers include driving, having a break, and deadheading, but no local shunting tasks.

2.2 The Current Planning Process

The first step of the current planning process in S-tog is to determine stopping patterns of the lines. Then, the timetable is constructed using these lines. The main objective is to minimize the total number of train sets needed to run the timetable subject to restrictions on connection times between lines at certain stations. Given the timetable, the RS is planned. First, waiting times at the termini are determined. Most often, each line has it own set of RS during the day, but some times two lines are *merged* to share RS. This is also determined in this step. Second, the number of units for each train is determined along with the shunting moves at the termini. To balance the number of RS at the depots, empty RS movements are added to the plan. After the RS plan is finished, the crew plan is worked out. The normal plan for RS and crew is usually valid for one year, but almost every week special plans are made due to maintenance of the tracks. In a special plan both the plan for RS and the plan for crew can be altered.

2.3 The Problems

Input to the CAP model are the trains in the timetable and the passenger demands for each train. The CAP model assigns train units to each train set on each line according to the best fit to the demand size, given constraints on the train composition and capacity demands. All demands must be covered with at least one train unit. To ensure feasibility, the model allows standing passengers. Standing passengers must not exceed limits of maximum 35 passengers and maximum 5% standing passengers in a train set. The objective is to minimize the total number of train kilometers, minimize the (percent wise) number of standing passengers, or maximizing either the reserve of RS (available to the repair

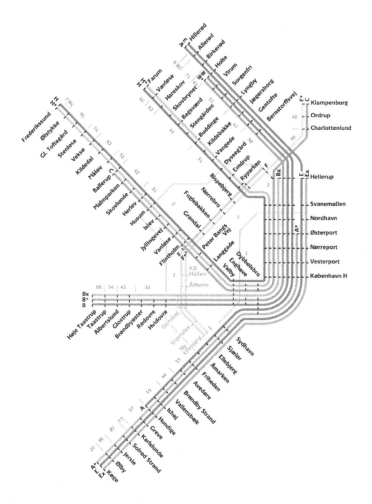

Fig. 1. The S-tog network. 'The Ring' corresponds to the two lines F and F+. The main crew centre is København H.

shop), or to maximize the capacity allocated to peak hours. The CAP focuses on a closed circuit for all lines at a time, i.e. benefits of coupling and decoupling are not evaluated. Instead the model calculates the RS need for the different periods of the day. We run several scenarios of the CAP and compare them, thereby obtaining information for making strategic decisions.

For the MPP model, the day is divided into time intervals. The inputs to the model are the *RS workload*, i.e. the number of train sets running in each time interval, and a set of *duty templates*. A duty template is a schedule for a working day for a driver. A schedule is composed of values between 0 and 1 for a number of time intervals. These values indicate the fractional amount of the time interval that a driver following the schedule is able to drive a train.

Neither the RS workload nor the duty templates say anything about the particular lines to be driven in the time interval. Each train set requires only one driver regardless of the number of units. The model covers the RS workload by the schedules for the duty templates in the solution, while minimizing the total number of duty templates used.

The main application of a solution to the MPP is to compare timetables considered by management for future operation. The reason for introducing a mathematical model for the MPP is to enable a comparison of timetables with different shapes of the RS workload. Using the previous method, different profiles with the same total RS workload would require the same number of drivers, but using the new model we are able to distinguish different profiles by the number of drivers needed.

2.4 Literature Review

A very similar model for evaluating the need for RS has been suggested by Abbink et al. [1]. They present an IP model used to assign material types and subtypes to the eight o'clock cross section (the busiest time). Otherwise, the literature has mainly focused on the operational problems. The RS problem has been investigated in a number of papers. Schrijver [14] developed a network model for minimizing the number of train units during the week for Dutch Railways on the Amsterdam-Vlissingen line. CPLEX is used to solve the instances and speedup is suggested by finding convex hulls of constraints and first branching on variables corresponding to rush hours. The model does not include maintenance and shunting, and no standing passengers are allowed in the trains. Brucker et al. [3] investigated a local search approach (simulated annealing). The model takes empty RS movements into account, but no maintenance. Shunting moves are partly solved by allowing a minimum time for shunting operations in the model. The model is tested on data from Württenberg (Germany). Ben-Khedher et al. [2] study the problem of allocating train units to the French High Speed Trains. Cordeau et al. [4] present a Benders decomposition approach for the locomotive and car assignment problem. The model is extended to include maintenance constraints in Cordeau et al. [5]. Lingaya et al. [11] study the problem of allocating carriages to trains at VIA Rail in Canada. Coupling and decoupling of carriages and the order of the carriages are considered. Finally, Peeters et al. [12] present a branch and price approach outperforming a CPLEX run of the same model with similar branching.

The literature on crew planning is very large and the following list of references is not complete. The problem has been studied in the context of bus drivers [6,13], airline crew [7,15] and railway crew [8,9,10].

3 The Capacity Assignment Problem

The CAP is the problem of allocating train units to train sets. In this section the model is described in the three different versions mentioned above; *Minimizing driven kilometers*, *Maximizing reserves*, and *Maximizing capacity in peak hours*.

The CAP is, when taken out of a context, an allocation of RS units to trains. At S-tog we have used the CAP for especially the morning peak, which is the most constrained time of the day with respect to the number of RS units in use. Using a forecast of the demand of the morning peak estimating two, five or ten years ahead enables us to predict the maximum load in the future and hence enables strategic choices with respect to, for example, the timetable format. The results of the CAP are integrated with other informations as e.g. the expected possibility of resizing trains during the day and expected transports of trains to the workshop and cleaning facilities to give the total cost of driving a timetable of a certain format. Often, if the results of the CAP for a certain timetable are poor with respect to utilizing RS units expediently, the evaluation process is aborted.

3.1   Describing the CAP Model

The model is described by the lines, $l \in L$, where $L = \{1, \ldots, |L|\}$ and $|L|$ is the number of lines, and the train sets on each line, $s \in S_l$, where $S_l = \{1, \ldots, |S_l|\}$ and $|S_l|$ is the number of train sets in each line l. Also described by indices are the train type, $t \in \{2g, 4g\}$, and the number of train units, $a_n = n \in N_t$, where $N_t = \{1, \ldots, |N_t|\}$ and $|N_t|$ is the maximum number of units coupled in a train set for train type t. The length of train units is in the presented model 2 for both types of train units. It is denoted $length_t = \{2, 2\}$. The maximum length of a train is $MaxLength = 4$.

The decision variables of the model are binary and describe whether a given number of train units of a certain type are put on a train set on a given line. The variables can be described by the expression below:

$$x_{l,s,t,n} = \begin{cases} 1 \text{ if there are } n \text{ trains of type } t \text{ on train set } s \text{ of line } l \\ 0 \text{ otherwise.} \end{cases}$$

For every train set on each line we also include a decision variable giving the number of standing passengers, $v_{l,s}$. Also decision variables, $p_{l,s}$, of percent wise standing passengers are included in the model. The mathematical model for the CAP has been set up with three different objective functions, namely *Minimizing driven kilometers*, *Maximizing reserves*, and *Maximizing capacity in peak hours*. The constraints are the same for the three versions except for a few adjustments or additional constraints.

The objective functions consist of two terms. One term referring to $x_{l,s,t,n}$ and one referring to $v_{l,s}$ and $p_{l,s}$. All solutions are evaluated subjectively and this creates the need for being able to tune the weights in order to affect the variables. The weights in the last term, W_2 and W_3, are besides being weighted according to W_1, also normalized in relation to each other. That is, the weight W_3 reflects that $p_{l,s}$ is measured in percent compared to $v_{l,s}$.

To keep the sum of standing and percent wise standing passengers down, both sets of variables are included in all three objectives.

Each ray in the timetable is assigned its length in kilometers and the number of trips the train makes during the considered time period of the run. The objective when *Minimizing driven kilometers* is to cover the demand on trains by assigning train units to trains such that the sum of driven kilometers is minimized while satisfying the constraints.

$$\min \ [W_1 \cdot \sum_l km_l + \sum_{l,s}(W_2 \cdot v_{l,s} + W_3 \cdot p_{l,s})] \tag{1}$$

The number of kilometers driven on each ray is calculated from the binary variables:

$$km_l = \sum_{s,t,n} trainKM_l \cdot a_n \cdot length_t \cdot x_{l,s,t,n} \quad \forall l \in L \tag{2}$$

In the version *Maximizing reserves*, the train units are allocated to the train sets only according to the size of demand. The model still allows standing passengers but the driven kilometers are not included in the objective. Since the number of train units are minimized, the reserves are maximized.

$$\min \ [W_1 \cdot \sum_{l,s,t,n} (Cap_{t,n} \cdot x_{l,s,t,n}) + \sum_{l,s}(W_2 \cdot v_{l,s} + W_3 \cdot p_{l,s})] \tag{3}$$

The version *Maximizing capacity in peak hours* differs from the other versions as it seeks the best use of the entire fleet in the peak hours, i.e. all the fleet is allocated in the best possible way according to demand. This is interesting when evaluating the ability to cover future demands in the different timetables. The capacity allocated to predefined peak hour trains is maximized, i.e. the model maximizes the total number of seats in peak hours, P.

$$\max \ [W_1 \cdot \sum_{l,t,n} \sum_{s \in P}(Cap_{t,n} \cdot x_{l,s,t,n}) - \sum_{l,s}(W_2 \cdot v_{l,s} + W_3 \cdot p_{l,s})] \tag{4}$$

For all feasible solutions the total sum of used train units for each train type is bounded by the maximum capacity of units of each type.

$$\sum_{l,s,n} a_n \cdot x_{l,s,t,n} \leq TrainCapacity_t \quad \forall t \in \{2g, 4g\} \tag{5}$$

Each train must be covered by at least one train unit.

$$\sum_{l,t,n} x_{l,s,t,n} \geq 1 \quad \forall s \in S_l \tag{6}$$

The upper bound on the length of each train must be less than two, which is the maximum train length, cf. Sect. 2.1.

$$\sum_{t,n} length_t \cdot a_n \cdot x_{l,s,t,n} \leq MaxLength \quad \forall l \in L, s \in S_l \tag{7}$$

The seat capacity allocated to a train plus the amount of standing passengers must cover the demand for that specific train.

$$\sum_{t,n} (seats_{t,n} \cdot x_{l,s,t,n}) + v_{l,s} \geq Demand_{l,s} \quad \forall l \in L, s \in S_l \tag{8}$$

Politically it has been decided that S-tog, if possible, should offer seats to all passengers. This restriction is formulated in the model partly by bounds on the standing passengers and partly by bounds on the percent wise number of standing passengers. Both sets of variables are included as both are formal requirements in the company defining the softly formulated political demand. The number of standing passengers on each train must be less than a minimized percentage of the capacity on a train. As this is non-linear, we allow the approximation below, i.e. the number of standing passengers on a train must be less than a minimized percentage of the demand. This is acceptable as the capacity often lies very close to the demand.

$$v_{l,s} \leq Demand_{l,s} \cdot p_{l,s} \quad \forall l \in L, s \in S_l \tag{9}$$

The model also contains restrictions on preventing the allocation of certain train type units to certain rays. These restrictions are kept by fixing a subset of the decision variables $x_{l,s,t,n}$ at zero, where l is a given ray in the network and t is a given type.

3.2   Complexity

Even though the model is a mixed integer programming model, it is in its present form easy to solve – CPLEX 8.0 has a running time of a fraction of seconds. Adding one or more train types will increase the complexity of the problem. Also the S-tog network contains only a limited number of rays of small length. Therefore, the number of RS to which units are assigned is small. It is important to mention that the standing passengers are characterized by real variables instead of integers. Due to the structure of the model, the solutions will be integer regarding the standing passenger variables.

3.3   Calculating Demands

The CAP model has three major inputs; *a timetable*, a set of *preference-constants* and a set of *passenger demands*. First the information from the *timetable* is used to determine the number of lines and the operating time of each of these lines. The operating time varies across lines. As the model considers a limited period of a day, the lines included are those with operating hours in the time interval of the model. Passengers have certain preferences for stop versus quick lines and these vary over the rays in the network. In the CAP we assume these preferences constant for each ray and we will refer to them as *preference-constants*. The preference-constants depend on the structure of the timetable and they should be adjusted for each timetable evaluated in the CAP. The preference-constants are used when the *passenger demands* are calculated for each train set in the model. They are multiplied with the demands for an entire ray giving demands for each line on that ray. For each line there are again demands for each train set on that line. It is adequate to model the CAP in this way without stations because we are able to characterize all train sets by defining a closed circuit. All train sets are included exactly once in the model.

3.4   The Problem Core of the CAP

The CAP model is used to evaluate timetables against each other. The CAP is used in three different versions to evaluate the effect of changing to other timetables than the current. Each version evaluates the timetables individually, i.e. for each timetable the versions are solved and the results are compared to corresponding results on other timetables. The CAP model is run for different periods during the day and comparisons are then made within each period. The time periods considered are "morning peak", "day", "afternoon peak", and "late evening". Due to the construction of our demand during the day, we assume that "early morning" resembles "late evening" and "day" resembles "late afternoon"/"early evening". Therefore, the analysis covers a complete day. The over-all effect for a timetable is calculated as the sum of all effects on the versions calculated over all periods of a day. Both weekdays, Saturdays and Sundays have been considered.

3.5   Results

We have generated a substantial amount of results using the CAP. However, most of these are confidential and cannot be published in detail. Below, we give a flavor of the results on a more general level of detail.

We have compared a 10 minutes frequency timetable (10TT) with a 20 minutes frequency timetable (20TT). The comparison was made as a part of a larger series of tests aiming to clarify which of two train types, small or large, are preferable in future production. The solution for the 10TT showed a distribution between allocated small and large units of 37.5%/62.5% whereas the solution for the 20TT showed a distribution of 4.2%/95.8%. The results confirmed very clearly that smaller train units are far more important in the 10TT.

When comparing the two timetables, 10TT and 20TT, with respect to the number of train units used and the higher number of departures offered for the passengers, the 10TT is clearly preferable as it employs a relatively lower number of train units. The 20TT uses 21.75% of the total fleet. Using this allocation of units to trains on a timetable of 10 minutes frequency would mean doubling the use of train units to 43.5% of the fleet. Compared to this the CAP run of 10TT gives a solution using 36.1% of the fleet. The 10TT is, as expected, preferable with respect to offering an expanded product to the customer and minimizing the train units allocated in the solution. The relatively low utilization of the RS is a result of the less strained time period considered. The number of drivers necessary for driving the 10TT is much higher than the need for drivers in the 20TT as the number of departures is doubled. The exact number of necessary drivers depends on the solution in the MPP as this is where the driver efficiency is found. Note that the cost of RS per unit is much higher than the cost of a driver. Therefore the result of CAP will often be decisive with respect to the expansion of the production.

In the transition phase of exchanging all RS with new ones, the aim is to offer the new 4g train units to as many passengers as possible. This objective must be combined with the objective of covering the demand with as few train units as possible. A solution was found with the CAP fulfilling both objectives. It showed a distribution of passengers on 4g and 2g of 68.3%/31.7%. The present solution, at that time, had a distribution of 20.3%/79.7%. Also the CAP solution decreased the utilized number of train units in use by 22.6%. Part of this high decrease must be seen relative to the fact that the seat capacity on 4g is higher than on 2g. That is, a passenger demand may be covered by two 2g train units or alternatively a single 4g train unit. In practice, the results initiated a two step manual process of first changing the distribution of 4g and 2g to offer 4g to more passengers and secondly reducing the number of train units used to cover the demand. This manual improvement did not reach the level of the optimal solution found with the CAP, mostly due to the manual planner's distrust in the level of estimated demand.

3.6   Related and Future Work

We have focused on comparing proposed timetables and on the economical evaluation of new timetables regarding RS and drivers. It is however possible to use the CAP in other contexts. Several of the problems mentioned below have already been addressed and one of them is in progress.

Capability to Handle Increase in Demand

Given the present timetable, it is interesting to see how large an increase in demand we can handle. The timetable is fixed and different levels of increase in demand are investigated. It is relevant to investigate each of the versions of minimizing kilometers, maximizing reserves and maximizing capacity allocated to peak hours.

Locating Potential Areas of Savings

The CAP has been used to evaluate the saving potentials in different areas in the network. Presently the RS plans are constructed manually and therefore the automated search for potential saving areas becomes an important tool. The model allocates train units to lines while minimizing the driven kilometres.

Evaluating a Possible Other Formation of the Fleet

Another core problem considered is the prospect of having more of the shorter units of 4g material. Given a possibility to change the composition of the order-contract of short and long train units of the new RS type, it is interesting to compare different possible scenarios of different fleet formations.

Assuring the Best Possible Phase Out of the 2g RS

New material will soon eliminate the need for the old type of RS. The new train units are received over a long time period and it is necessary to start using the new RSs before all units have been delivered. This implies a need for a phase out plan for the old RS type. S-tog is receiving one type of new RS, which can have two different lengths. There have been considerations of how to exploit the shorter one of the new RS type. Because of difficulties in shunting and the possible need for more shunting personnel, the savings of letting the new train units drive in combinations of a short and a long unit had to be clarified. That is, if the savings of combining the short and long units are high, the increased complexity and time needed for shunting is ignored. Also earlier experiences on driving short units alone have not been good for other types of RS. Therefore there has been a need to document any saved cost of driving trains with only one short unit assigned. The estimate of the savings was found by looking for the solution with the fewest possible driven kilometres and allowing all possible combinations of the new RS type.

4   The Manpower Planning Problem

The MPP is the combinatorial problem of covering an RS workload profile for a day with duty templates of certain structures. An optimal solution to an instance of MPP is a schedule of a minimum number of duty templates located during the day. In other words the model seeks to minimize the number of drivers by finding the best pattern of check-ins. The detail of the time intervals in the model implies a solution to be too crude for the actual allocation of the drivers but useful for planning the number of drivers needed for the timetable related to the instance.

4.1   Workload Profiles and Duty Templates

The input to our model is the workload profile for the RS and a set of duty templates resembling real duties.

It is possible to determine the exact start- and end times for each train set in the RS plan. Subsequently, the number of running train sets in the network is determined, where each of these needs a driver. We have chosen to consider time intervals of fifteen minutes where the driver can check-in, cf. Fig. 2(a). Thereby, the coarse intervals make the location of the exact time for the workload for each train set uncertain within the particular interval.

According to collective agreements a duty template must be constructed following specific rules. Considering the structure of a duty, the following rules must be respected:

1. The duty time must be between six and nine hours;
2. There may be one break with a minimum length of 30 minutes, or two breaks with one break at least 20 minutes and the sum of two breaks at least 45 minutes;
3. The working day starts with fifteen minutes check-in and ends with ten minutes check-out.

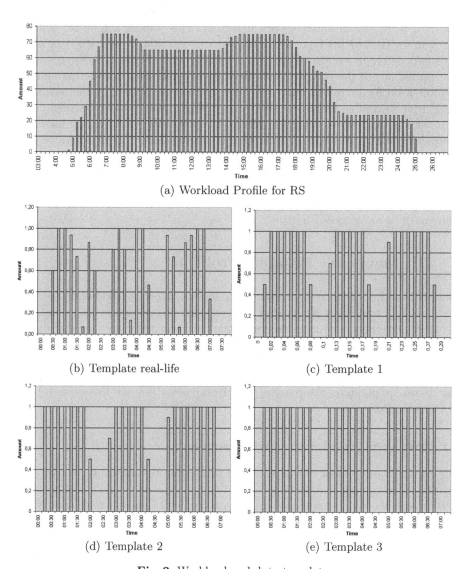

(a) Workload Profile for RS

(b) Template real-life

(c) Template 1

(d) Template 2

(e) Template 3

Fig. 2. Workload and duty templates

The templates in the model are based upon a historical set of duties. It is assumed that these duties satisfy union rules. Distinction will be made between *driving* time and *other* time. Driving time is characterized by the time the driver is at hand for driving or handling the RS and with exclusion of e.g. waiting time at the termini where no option for a break is possible. A graphical representation of a template is shown in Fig. 2(b). The x-axis shows time during the duty in fifteen minutes' intervals and the y-axis shows the percentage of driving time in each quarter. For each template the first and last quarters are always check-in and check-out time, hence having zero driving time.

We include waiting time at the termini as driving time in the template, i.e. time associated with RS and round up the driving, cf. Fig. 2(e). Not all minutes according to the rules are multiples of fifteen minutes and rounding up the driving time implies that the efficiency of Fig. 2(e) is larger than in practice. A template where no rounding up the driving time in the time intervals is done gives a more detailed template. Caused by the mentioned rules with different types of time off we constructed two templates showing these details; Fig. 2(d) illustrates details handled for the breaks only, and Fig. 2(c) also considers details in relation to check-in and check-out.

Along with the rules for each duty, the following rules across duties need to be fulfilled in order to make a set of duties acceptable to the collective agreement:

4. Average duty time for a person is at most seven hours, and therefore we restrict the entire plan to have an average duty time of at most seven hours;
5. The number of duties with one break must be less than the number of duties with two breaks. This rule is managed by the average number of breaks over all duties which is a value between 1.5 and 2.

Rule 4 clearly affects the solution and the feasibility, given the different duty templates. If none of the possible duty templates considered has a length below seven hours, the problem is infeasible.

Some rules, even though demanded by the collective agreement, are not considered in the model. One example is the 11-hours rule, stating that each driver should have at least 11 hours off in between two consecutive duties. Experience shows that this is well handled during the roster planning.

4.2 Description of the Model

As mentioned previously, the model considers a period of one day in time intervals of fifteen minutes: $Time = \{1, \ldots, T\}$; where T is 96. The set $Time_{cin}$ is the set of possible check-in times of a driver, where $Time_{cin} \subseteq Time$. Let $D = \{1, \ldots, |D|\}$ be a set of duty templates. We define T_d to be the length of duty template $d \in D$ measured in time units.

With reference to Sect. 4.1, a duty template is characterized for every time unit in the template, denoted $d(l)$ with $l \in \{1, \ldots, T_d\}$. The value for $d(l)$ is the availability for driving in the time slot, that is $d(l) \in [0; 1]$.

The number of drivers to be checked in at cin with a duty template d is described by $x_{cin,d}$. Let $N = Time_{cin} \times D$ be the set of all combinations of check-in times and all duty templates. The constraint of covering the workload b_i in time interval i is expressed by the multi cover inequalities:

$$y_i \geq b_i \text{ for } i \in Time, \quad \text{with} \tag{10}$$

$$y_i = \sum_{d \in D} \sum_{cin \in [i - T_d + 1; i]} x_{cin,d} \cdot d(l) \tag{11}$$

for $l = i - cin + 1$, i.e. the corresponding time interval in the duty template. In other words, y_i denotes the number of drivers that are available in the ith time slot.

The rule of a maximum average time of $maxTime$ over all duties is enforced by:

$$\sum_{d \in D} \sum_{cin \in Time_{cin}} x_{cin,d} \cdot (T_d - maxTime) \leq 0 \tag{12}$$

In order to satisfy the rule of an average for the breaks we implement this with bounds by:

$$LB \cdot \left(\sum_{d \in D} \sum_{cin \in Time_{cin}} x_{cin,d} \right) \leq \sum_{d \in D} \sum_{cin \in Time_{cin}} x_{cin,d} \cdot breaks(d)$$

$$\tag{13}$$

$$UB \cdot \left(\sum_{d \in D} \sum_{cin \in Time_{cin}} x_{cin,d} \right) \geq \sum_{d \in D} \sum_{cin \in Time_{cin}} x_{cin,d} \cdot breaks(d)$$

where, $breaks(d)$ equals the number of breaks in the duty template d; LB/UB define the lower/upper bound for the average number of breaks over all duties, according to rule 5 in Sect. 4.1.

The objective is to minimize the total number of selected duty templates:

$$\min \sum_{d \in D} \sum_{cin \in Time_{cin}} w_{cin,d} \cdot x_{cin,d} \tag{14}$$

Adding a specific weight to each duty template according to e.g. the time of the day allows us to implement cost for early versus late working hours.

The complete model then consists of the objective function in (14), the bookkeeping variable y_i in (11), and the inequalities in (10), (12) and (13).

4.3 The Test Setup

To validate the model, we created instances of the MPP for which we have a manually produced feasible solution. The data are from a day in the spring of 2004.

Demand for drivers must include the need for deadheading trips necessary at any given time interval. This is included in the model simply by adding the need for deadheading trips at any given time interval to the workload.

Considering the workload illustrated in Fig. 2(a), only 82 of the 96 quarters of a day actually refer to operating RS. From approximately 1:30 to 5:00 there is no demand for drivers. Because of the structure in the workload, the combination of the breaks has significant impact on the solution. To ensure the most general model, we allow a duty to start at any point in time.

We define three sets P1, P2, and P3 of duty templates containing six templates with the same level of detail. The three different levels of detail are illustrated by the three templates, Fig. 2(c)– 2(e). The six templates in each set are derived from six duties used on the particular day in 2004. The average efficiency of the sets of duties is respectively 71%, 73% and 77%, the difference implied by the level of detail regarding the utilization of each quarter.

The average duty time for the manually produced solution was 6:55, which is the value for $maxTime$ in equation (12). The solution for the problem instance has 67 duties with one break and 174 duties with two breaks, a total of 241. This gives an average number of breaks of 1.722. The bounds for the average number of breaks are chosen as ±1%, ±5% and ±10%. We have chosen the most tight as 1% due to the fact that the break average is an approximation itself. The ±10% value seems to be too large, but it is interesting to see the effect of this. For a solution to instances with ±10% considered, the number of duties with one break lies in between 26 and 108.

The runs have been made on a DELL Latitude D600 with a GenuineIntel Pentium M 1700 MHz processor with 2GB RAM. The software was GAMS IDE 2.0.23.10 on a Windows 2000 SP4 environment using ILOG CPLEX 8.1.0. All instances have been run for six hours with zero tolerance for the gap.

4.4 Results

We ran two types of experiments, one minimizing the number of duties and one minimizing the total duty time. This is implemented by the weights in the objective function, cf. (14). For the first type, the weights are ones and the results are presented in Table 1(a). For the latter, the weights equal the duty times corresponding to the duty templates and Table 1(b) contains the results.

The LB column is the value of the LB in the last iteration before CPLEX termination; UB is the best solution found, the gap % column is the gap from the LB to the UB, but the absolute gap is practically constant through each profile. For P3 in Table 1(b) the solutions found were optimal. The templates show the importance in the degree of realism in the duties, where the most

Table 1. Different bounds on average breaks with different sets of duty templates

(a) Minimizing the number of duties

Sets	±%	LB	UB	gap%	Duty time
P1	1	238	241	1.24	1665
P1	5	236	239	1.26	1652
P1	10	233	236	1.27	1631
P1	∞	228	231	1.30	1596
P2	1	229	232	1.29	1603
P2	5	227	230	1.30	1588
P2	10	225	228	1.32	1574
P2	∞	223	225	0.89	1554
P3	1	219	220	0.45	1521
P3	5	219	220	0.45	1520
P3	10	218	219	0.46	1513
P3	∞	217	218	0.46	1508

(b) Minimizing the total duty time

Sets	±%	LB	UB	gap%	Duties
P1	1	1647	1664	1.02	241
P1	5	1632	1648	0.97	239
P1	10	1614	1631	1.04	236
P1	∞	1581	1596	0.94	231
P2	1	1583	1598	0.94	233
P2	5	1571	1583	0.76	231
P2	10	1558	1572	0.89	229
P2	∞	1544	1555	0.71	226
P3	1	OPT	1518	0.00	221
P3	5	OPT	1513	0.00	220
P3	10	OPT	1508	0.00	220
P3	∞	OPT	1503	0.00	220

realistic templates (P1) have a smaller percentage of work and more duties in the solution. The results for P1 and P3 for the experiments without restriction on the average number of breaks (∞) increase the solution by 6.0% from 218 to 231, while the efficiency decreases by 7.8% from 77% to 71%. This indicate that the more realistic and detailed P1 templates combine the duties in the solution in a better way than the solutions found using the P3 templates. Therefore we expect that using smaller time intervals will increase the reliability of the solution significantly.

As one might expect, the bounds for the average number of breaks is important for the result but, as the tables show, the bounds are not important for the gap. The two tables show no significant difference in the solution depending on types of the objective function.

We see that the result found for P1 with bound $\pm 1\%$ in the average number of breaks, 241, equals the value found by the manual solution, cf. Sect. 4.3. The tests for the MPP have showed that the selection of templates and the average number of breaks is crucial for the quality of the solution. The MPP provides a good tool to investigate the effect of the additional rules and the benefits of adjusting them.

4.5   Related and Future Work

To make the model more usable to management, the weight in the objective function needs further exploration to estimate the total cost of a solution more precisely. The labor rules often use small time units, e.g. two or five minutes, and in order to make the estimation more precise we need to refine the time units for the workload. This will increase the number of variables and thereby increase the complexity of the model.

5   Conclusion

We have presented two IP models for evaluating the need for RS and crew used when comparing timetables. Both models have been used for strategic decisions in S-tog. The RS model was used to evaluate the need for small vs. large 4g units on different timetables. The crew model was used as a daily tool for crew planners when changing the plans due to for instance track works and to evaluate the size of the work force needed on different timetables in the year 2006.

References

1. Abbink, E.J.W., Van den Berg, B.W.V., Kroon, L.G., Salomon, M.: Allocation of railway rolling stock for passenger trains. Transportation Science 38, 33–41 (2004)
2. Ben-Khedher, N., Kintanar, J., Queille, C., Stripling, W.: Schedule optimization at SNCF: From conception to day of departure. Interfaces 28, 6–23 (1998)
3. Brucker, P., Hurink, J., Rolfes, T.: Routing of railway carriages: A case study. Osnabrucker Geschriften zur Mathematik, Heft 205 (1998)

4. Cordeau, J.F., Desrosiers, J., Soumis, F.: A Benders decomposition approach for the locomotive and car assignment problem. Transportation Science 34, 133–149 (2000)
5. Cordeau, J.F., Desrosiers, J., Soumis, F.: Simultaneous assignment of locomotives and cars to passenger trains. Operations Research 49, 531–548 (2001)
6. Fores, S., Proll, L., Wren, A.: A column generation approach to bus driver scheduling. In: Bell, M. (ed.) Transportation Networks: Recent Methodological Advances. Pergamon, pp. 195–208 (1998)
7. Hoffman, K., Padberg, M.: Solving airline crew scheduling problems by branch and cut. Management Science 39, 657–682 (1993)
8. Kroon, L.G., Fischetti, M.: Scheduling train drivers and guards: The Dutch "Noord-Oost" case. In: Proceedings of the 34th HICSS conference, Hawaii (2000)
9. Kroon, L.G., Fischetti, M.: Crew scheduling for Netherlands Railways - Destination: Customer. In: Daduna, J., Voss, S. (eds.) Computer Aided Scheduling in Public Transport, Springer, Heidelberg (2001)
10. Kwan, A., Kwan, R., Parker, M., Wren, A.: Producing train driver schedules under differing operating strategies. In: Wilson, N.H.M. (ed.) Computer-Aided Scheduling in Public Transport, Springer, Heidelberg (1999)
11. Lingaya, N., Cordeau, J.F., Desaulniers, G., Desrosiers, J., Soumis, F.: Operational car assignment at VIA Rail Canada. Transportation Research B 36(9), 755–778 (2002)
12. Peeters, M., Kroon, L.G.: Circulation of railway rolling stock: a branch-and-price approach. Technical report, Rotterdam School of Management, Erasmus University Rotterdam (2003)
13. Rousseau, J., Wren, A.: Bus driver scheduling – an overview. In: Daduna, J., Branco, I., Paixao, J.M.P. (eds.) Computer-Aided Scheduling in Public Transport, Springer, Heidelberg (1995)
14. Schrijver, A.: Minimum circulation of railway stock. CWI Quarterly 6(3), 205–217 (1993)
15. Wedelin, D.: An algorithm for large scale 0-1 integer programming with application to airline crew scheduling. Annals of Operations Research 57, 283–301 (1995)

A Capacity Test for Shunting Movements

John van den Broek[1] and Leo Kroon[2]

[1] Dept. of Mathematics and Computer Science,
Eindhoven University of Technology,
NS Reizigers, Utrecht,
The Netherlands
`j.j.j.v.d.broek@tue.nl`
[2] Rotterdam School of Management,
Erasmus University Rotterdam,
NS Reizigers, Utrecht,
The Netherlands
`L.Kroon@rsm.nl`

Abstract. One of the bottlenecks in the logistic planning process at Netherlands Railways is the capacity of the infrastructure at the larger railway stations. To provide passenger trains with the right composition of rolling stock, many shunting movements between platform tracks and shunting areas are necessary, especially just before and after the peak hours. These shunting movements use the same infrastructure as the timetabled passenger and cargo trains.

In this paper we describe a capacity test that has been developed to test at any moment during the planning process, whether the capacity of the infrastructure between the platform tracks and the shunting areas is sufficient for facilitating all the shunting movements that have to be planned in between the already timetabled train movements. With this test it is not necessary anymore to plan every detail of the shunting movements far before the actual operations.

The capacity test is based on a mixed integer programming model. The running time of the Branch-and-Bound algorithm of CPLEX 9.0 is sufficiently small, as was observed in computational experiments related to three stations in the Netherlands.

1 Introduction

In this paper we focus on shunting processes related to passenger trains. Shunting processes belong to the backstage processes in a railway system. They are carried out in and around the large stations in a railway network in order to provide passenger trains with the right composition of rolling stock, and to facilitate the inside and outside cleaning and the short term maintenance of the rolling stock. During the rush hours, passenger trains are usually operated at full capacity. However, outside the rush hours, an operator of passenger trains usually has a surplus of rolling stock. This surplus has to be parked at a shunting area in order to be able to fully exploit the main railway infrastructure.

F. Geraets et al. (Eds.): Railway Optimization 2004, LNCS 4359, pp. 108–125, 2007.

At the larger railway stations, timetable related shunting movements between platform tracks and shunting areas are necessary in the following cases:

1. Extending a train. Rolling stock is added to a passenger train in order to increase the train's capacity. A shunting movement is necessary to bring the rolling stock from a shunting area to a platform track.
2. Shortening a train. Rolling stock is uncoupled from an arriving passenger train. A shunting movement is necessary to bring the uncoupled rolling stock from a platform track to a shunting area.
3. Starting trains. These are the first trains in the morning that have to depart from a station. To provide these trains with rolling stock, a shunting movement from a shunting area to a platform track is necessary.
4. Ending trains. These are trains which arrive at a station and, after arrival, the rolling stock of these trains is not used anymore on the same day. The rolling stock has to be brought from a platform track to a shunting area.

Besides the timetable related shunting movements, also many shunting movements related to the cleaning and maintenance of the rolling stock have to be carried out. However, these shunting movements usually take place at the shunting areas themselves, without too much interaction with the timetabled trains. We only have to consider these shunting movements if they use infrastructure outside the shunting areas.

Shunting processes involve highly complex routing and scheduling problems with capacity restrictions on time, space, and personnel. Especially the limited capacity of time and space (routing and storage) leads to several bottlenecks in the railway process. The routing and scheduling aspects of the shunting processes are strongly interrelated. Usually, the capacity of the required shunting crew is less a bottleneck, since personnel is a relatively cheap resource.

For each shunting movement, an appropriate route over the railway infrastructure and an appropriate time instant have to be determined. This is particularly relevant for the shunting movements between the shunting area and the platform area of a station. These shunting movements have to take place between the timetabled passenger and cargo trains. They should not disturb these trains. From a robustness point of view, it is desirable that the shunting movements are scheduled as far (in time and space) as possible from the train movements of the passenger and cargo trains.

Shunting processes are highly dependent on the timetable and on the rolling stock circulation of a railway operator. First, as was indicated already, the shunting movements share the capacity of the railway infrastructure with the timetabled passenger and cargo trains. Moreover, as soon as the planned timetable or the planned rolling stock circulation is modified, the number, the order, and the compositions of the shunting movements usually change as well. Therefore also the shunting plans have to be modified in such cases. Currently this requires a lot of manual planning work.

In the current planning process, every detail of the shunting movements is planned as soon as possible, sometimes months before the actual operations. Adding one train to a plan or a small change in the rolling stock circulation may result in many changes in the original shunting plans of a number of railway stations, which means a lot of replanning. The time spent on detail planning has to be reduced by postponing the detail planning. The only reason to plan every detail of the shunting movements far before actual operations, is to be certain that the capacity of the infrastructure is sufficient.

Our first contribution is the recognition that in practice it is useful to make a distinction between a capacity test to be carried out a relatively long time before the operations, and a planning tool to be used for finalizing the detailed plans briefly before the operations. The main contribution is the capacity test. It verifies whether the capacity of the infrastructure between the shunting areas and platform tracks is sufficient. Two mixed integer programming models have been developed which both minimize the number of shunting movements that can not be planned because of lack of capacity of the infrastructure. In the first model the routes of the trains are fixed beforehand and in the second model the routes are determined by the model. The disadvantage of the last model is the larger computation time. If our models indicate that there is sufficient capacity, then it is rather sure that there is sufficient capacity in practice. If our models indicate that the capacity is insufficient for some shunting movements, a detailed plan has to be made at that moment in time and it could be that the rolling stock circulation has to be adjusted.

The rest of the paper is structured in the following way. In Section 2 we give a literature review of research related to shunting processes. Section 3 contains a detailed description of the problem and the goal is explained in more detail. The problem is formulated as a mixed integer program which is described in Section 4. Section 5 contains computational results for a few railway stations in the Netherlands. Some conclusions are given in Section 6.

2   Literature Review

A prototype model for a capacity test as described in the previous section has been developed in earlier research by Van den Broek [1]. This model assumes that the routes for all the shunting movements are fixed beforehand and verifies that each shunting movement can be scheduled at such a time instant that each element of the infrastructure is occupied by at most one movement at the same time. The model is a mixed integer program that is solved by CPLEX.

Research that was carried out by Duinkerken [8] deals with the storage capacity of a shunting area. This research provided a prototype of a tool to determine whether the storage capacity of a shunting area is sufficient for storing a certain set of train units. Duinkerken describes an integer linear program that is solved by CPLEX. This model does not only take into account the total number of rolling stock units that have to be stored concurrently at the shunting area, but also the arrival and departure times of these rolling stock units.

Other research related to shunting processes was carried out by Freling et al. [10]. Their research aims at the development of planning tools that support planners to generate detailed shunting plans from scratch. This is in contrast with our model, which focuses on the development of a global capacity test for the mid term planning. Freling et al. take into account many small details. They split the shunting problem into a matching problem for arriving and departing train units, a routing problem for routing train units to the shunting tracks, and a parking problem for storing the train units at the shunting tracks. The first step is solved by CPLEX, the second step by applying column generation to a set covering model, and the last step by applying A* search.

Also Di Stefano and Koci [9] did research related to shunting processes. They looked at how to order trains on the available shunting tracks in order to minimize the number of required shunting movements on the next morning. They assume that each track is long enough to host the trains assigned to it. Their main objective is minimizing the number of required shunting tracks. They consider several variants of their shunting problems, distinguished from each other by the ends of the shunting tracks that can be used for entering or leaving these tracks. For example, in the SISO-variant (Single Input Single Output), all trains enter the shunting area from one end of the tracks and all trains leave the shunting area into one end of the tracks. For several variants of their problem they provide computational complexity results.

Next, Tomii et al. [12] and Tomii and Zhou [13] describe a genetic algorithm that handles both storage of train units and several related processes, such as cleaning and maintenance. However, the shunting part of their problem is of a less complex nature than the general shunting problem, since in their context at most one train unit can be parked on each shunting track at the same time.

Papers on shunting trams and buses in their storage depots have been written by Winter and Zimmermann [14], Blasum et al. [3], Di Miele and Gallo [7], and Hamdouni et al. [11]. Winter and Zimmermann [14] focus on storage areas in which the trams are stored one behind the other in dead-end sidings. They assume that the earliest departure takes place after the last arrival. They also describe real-time dispatch strategies. Their model assigns trams to depot positions, thereby minimizing the number of necessary shunting movements. Blasum et al. [3] study similar problems, especially focusing on a smooth start-up process of the tram system in the early morning. Gallo and Di Miele [7] describe a model for parking buses in a storage area based on Minimal Non-Crossing Matching and Generalized Assignment. This model also takes into account the fact that the vehicles have different lengths. Moreover, they present an approach for dealing with mixed arrivals and departures. That is, the earliest departure takes place before the last arrival. Another application of bus dispatching is described by Hamdouni et al. [11]. Here robust solutions are emphasized by having as little different bus types as possible in each lane of the depot, and by grouping in each lane the buses of the same type as much as possible together.

The shunting problem has some similarity with the problem of routing trains through railway stations that is described by Zwaneveld et al. [15], [16]. In this problem, the arrival and departure times of a set of trains at a certain railway station are given, and the question is whether the trains can be routed through the railway station in such a way that trains do not conflict with each other.

Zwaneveld et al. try to assign trains to platforms and to minimize the number of necessary shunting movements. The routing problem is proved to be NP-hard and modeled as a weighted node packing problem. A Branch-and-Cut algorithm has been developed to solve the problem. Main difference with our capacity test is that they don't verify whether the capacity of the infrastructure between the shunting area and the platform area is sufficient to carry out the shunting movements. A similar problem is studied by Billionnet [2].

Cornelsen and Di Stefano [6] also look at assigning trains to platforms given a timetable. They model the platform assignment problem as a graph coloring problem on a conflict graph. The vertices of the graph represent the trains and two vertices are adjacent if the corresponding trains cannot be assigned to the same platform due to their arrival and departure times. Cornelsen and Di Stefano consider the platform assignment problem both on a linear time axis and on a cyclic time axis. The main difference with the capacity test and the model of Zwaneveld et al. [16] is that Cornelsen and Di Stefano don't take into account any shunting movement nor the capacity of the switch zone. They distinguish between variants with and without the so-called midnight constraint. The midnight constraint means that the earliest departure takes place after the last arrival. They present several complexity results and approximation methods.

Note that, apart from research on shunting processes for vehicles for passenger transport (trains, trams, and buses), there is a lot of research going on related to shunting processes for cargo trains. Since shunting processes for cargo trains are usually carried out at dedicated locations, they fall outside the scope of the current paper. A recent overview of the use of Operations Research in railway systems focusing on train routing and scheduling problems is provided by Cordeau et al [5], also focusing on shunting problems related to cargo trains. Bussieck, Winter and Zimmermann [4] give a survey of the application of discrete optimization techniques in public rail transport.

3   Problem Description

To verify whether the capacity of the infrastructure is sufficient to facilitate all the shunting movements between the platform area of a certain station and the corresponding shunting area, a global capacity test is needed. The capacity of the infrastructure between those areas has to be shared by passenger trains, cargo trains and shunting movements. The test has to check whether it is still possible to schedule and route the shunting movements between the passenger and cargo trains. This section gives a formal description of this capacity test.

In the *planning* process, each train movement has a unique time instant which corresponds with an event on the platform tracks, and which is called the *plan time* of the movement. The plan time of an arriving passenger train corresponds with the arrival time and the plan time of a departing passenger train is equal to the departure time of that train. For a cargo train, the plan time is the time instant at which the train passes a platform track. The plan time of a shunting movement corresponds with the arrival or departure on a platform track.

In the planning process, each shunting movement has a given departure and arrival track as input. These tracks can be a shunting area, a shunting track or a platform track. Depending on the infrastructure of a railway station, a shunting movement can be routed along several possible routes to get from its departure track to its arrival track. These routes differ in the used tracks, switches and crosses. The possible routes between a pair of tracks consist of one *priority route* and a maximum of nine possible *alternative routes*.

Moreover, each shunting movement has a feasible *time window* which contains the allowed plan times of the shunting movement. This time window is given by an earliest and a latest possible plan time which are based on the availability of railway tracks, platform tracks, and shunting tracks. For example, a shunting movement bringing empty rolling stock from a platform track to a shunting area cannot start before the passengers got out after the arrival of the rolling stock on the platform track and has to be completed before the next train arrives at the same platform track.

Passenger and cargo trains are *planned in detail* far before the operations. Planned in detail means that the arrival- and departure tracks, the route and the plan time of a movement are fixed. Shunting movements are train movements with a relatively low priority. Therefore they are preferably planned in detail only briefly before the operations. Indeed, otherwise there is the risk that they have to be replanned several times, e.g. due to a modified rolling stock circulation or due to the fact that additional passenger or cargo trains have to be facilitated on the infrastructure. However, the current practice is that the shunting movements are also planned far before the operations. In fact, the shunting plans themselves serve as a capacity check for the capacity of the infrastructure of the stations. This current practice is due to the fact that creating shunting plans is a difficult problem, and that intelligent support is currently lacking.

Planners build some robustness into the plans by taking into account a certain *headway time* between each pair of train movements. Therefore at least a certain minimum number of minutes has to be scheduled between the plan times of two consecutive movements that use a common element of the infrastructure. This minimum amount of time is given by the planning norms. These norms depend for example on whether trains are cargo trains or passenger trains, or whether trains are arriving or departing trains.

Saw movements are movements that arise when the arrival track of a shunting movement can not be reached from the departure track by one forward movement, see Figure 1. Most of these saw movements arise when the arrival and departure track are parallel to each other or when at least one of the two tracks is a track which can only be approached from one side. Rolling stock that carries out a saw movement has to change direction on a track in between. In Figure 1 such a track can be found at ②. Such a track is called a *saw track*. At a saw track the driver has to walk from one side of the train to the other side. This means that the plan times of the two parts of a saw movement have to be separated in time to give the driver the opportunity to walk to the other side of the train and results in the occupation of the saw track for a number of minutes. Usually, there are several saw tracks to choose from. Which saw track is chosen depends on the other train movements that use the possible saw tracks. However, in the current paper it is assumed that for every saw movement, the saw track is given a priori.

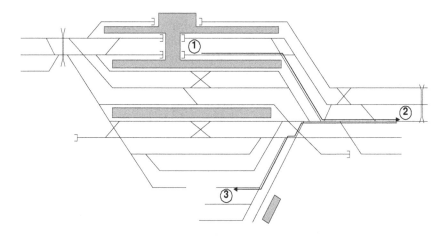

Fig. 1. Example of a saw movement

As was indicated before, shunting movements depend on the timetable and on the rolling stock circulation. So the timetable and the rolling stock circulation are assumed to be known when the capacity test is applied. This means also that the length of each train is known, which makes it possible to verify whether the rolling stock fits on a certain platform track. This is done before applying the capacity test. The minimum amount of time between the plan times of the two parts of a saw movement depends on the length of the train. Since the length of each train is assumed to be known, also this minimum amount of time between those plan times can be determined a priori.

The global capacity test has to indicate whether it is still possible to find for each shunting movement an appropriate route from its given departure track to its given arrival track within its given time window which does not conflict with the train movements which are already planned in detail nor with each other. The test is not allowed to change the arrival tracks, the departure tracks, and the plan times of the passenger and cargo trains. The test has to be used at the moment that the shunting movements between the platform area and the shunting area of the station still have to be scheduled and routed, which is *after* the timetabled trains have been scheduled and routed through the station.

During day time a large part of the rolling stock serves as a passenger train and will not be parked at a shunting area. This means that the capacity of the shunting areas will be sufficient during day time and only has to be verified during night. Therefore, it is assumed that the capacity and the detailed layout of a shunting area are not relevant for the capacity test. The capacity test focuses on the capacity of the infrastructure *between* the platform area of a station and the shunting area, not on the shunting area itself. Because the rolling stock circulation is known, it has been verified already whether the capacity of the shunting area is sufficient. This can be done e.g. by applying the model described by Duinkerken [8]. As a consequence, when our capacity test is applied, the shunting area can be seen as a set of tracks with sufficient capacity.

It is also not necessary to take into account the crew planning. In comparison with the infrastructure, crew is a relatively cheap resource and there is usually a sufficient number of train drivers available to carry out the shunting movements. Therefore, the details of the crew planning are skipped in this paper.

Note that the capacity test is *not* intended to be a detailed planning tool. As will be explained later, the constraints taken into account by the capacity test are somewhat stricter than the constraints taken into account in practice. As a consequence, if the result of the test is that the capacity of the infrastructure is sufficient, then it can be expected that the capacity is indeed sufficient in practice. If there is still a sufficient amount of time and space for each shunting movement, then no specific action of the planners is required: detailed planning of the shunting movements can be postponed. On the other hand, if the capacity test gives as a result that the capacity is not sufficient, then appropriate actions of the planners are required. In such cases, some shunting movements are critical and have to be planned in detail.

4   The Mathematical Programming Model

In this section we describe two mixed integer programming models that have been developed for the global capacity test. The models check whether it is possible to plan all the shunting movements, which have not been planned in detail yet, between the movements already planned in detail. In the first model, the route of each shunting movement is given and the plantime is to be selected. The second model is an extension of the first one. In the second model it is also allowed to select the route of each train movement from a pre-specified set.

4.1   Parameters of Model with Fixed Routes (MFR)

In both models, a train movement has to arrive at or depart from a platform track or a track parallel to the platform tracks. For example, a cargo train that doesn't stop at a station is split into two movements. The first movement arrives at a platform track and the second movement departs from this track and has an arrival track that is equal to the arrival track of the original movement. The set of movements is described by S.

At the moment of testing, some movements have been planned in detail and some have not been planned in detail. As a consequence, a set S_p of planned movements and a set S_n of not yet planned movements is introduced. The set S_p contains all the timetabled trains, the cargo trains and possibly the already planned shunting movements. The set S_n contains the shunting movements which have not been planned in detail yet.

Trains can arrive and depart from more than one track. For example: a long passenger train can arrive at platform tracks 7A, 7B and 7C. Also, several movements use more tracks than just their arrival and departure tracks. These other tracks are part of the route and have to be empty when the movement is carried out. The set T_j is defined as the set of tracks which are used by movement j along the given route of movement j. This set contains at least the arrival and departure tracks of movement j.

The plan time of a movement is defined as the number of minutes after the starting time of the planning horizon, thereby assuming that time zero corresponds with this starting time. For every movement j, the release date r_j is the first possible time instant that is allowed to be the plan time of movement j. The due date d_j of movement j is the last possible time instant that is allowed to be the plan time of movement j. For a movement $j \in S_p$, the release date and the due date are equal to the already determined plan time. Movement $j \in S_n$ has a time window $[r_j, d_j]$, which contains all the possible time instants that are allowed to be the plan time of movement j.

To check that all tracks on the route of a certain movement are not occupied, the concept of the *successor* of a movement is introduced as follows:

Definition 1. *The successor of movement j with respect to track t is the next movement after movement j that must use track t after movement j has arrived at track t or after it has left track t.*

Every other movement that uses track t has to be carried out before movement j or after the successor of movement j. Therefore no other movement is allowed to use track t after movement j as long as the successor of j has not been carried out. For example: if an ending train arrives at a platform, the arriving movement has as its successor the shunting movement that brings the rolling stock to a shunting area. As long as the shunting movement has not been carried out, no other movement is allowed to use the platform track. If a track is a shunting area or empty after movement j, then movement j has no successor with respect to this track. The set s_j of successors of movement j can be deduced from the input data. Set s_j^t is defined as the set of successors of movement j with respect to track t.

Trains that do not stop at a station are separated in an arriving movement and a departing movement. The departing movement is the successor of the arriving movement with respect to an appropriate track.

In the model, saw movements are split into two or even more movements. The first movement is the movement from the departure track to the saw track, and the second one departs from the saw track and arrives at the arrival track of the original movement. The second part is the successor of the first part.

Two movements are defined to be *route dependent* if they have an element of the infrastructure in common in their routes. Such an infra-element can be a track, a switch or a crossing. If two movements have no infra-element in common, then they do not directly influence each other's plan time. The plan times of two route dependent movements have to be separated by the headway time of which the value is given by the planning norms. The parameter b_{jk} is defined as the minimum amount of time between the plan times of the route dependent movements j and k if movement j is carried out before movement k. The transfer time of a driver and reversing the direction of the rolling stock are also covered by this parameter if movements j and k together form a saw movement.

Obviously, a movement which has a time window around 6:00 am does not influence a shunting movement which has a time window around 9:00 pm. If two movements have time windows that are in time far from each other, then the order of operation of the movements is known a priori. On the other hand, two movements are *time dependent* if their time windows differ less than the norm between those two movements. This means that movements j and k are time dependent if and only if $r_j < d_k + b_{kj}$ and $d_j > r_k - b_{jk}$. If two movements are route dependent and time dependent, then they are *dependent*. Summarizing, the definition of dependency of two movements is as follows:

Definition 2. *Two movements are dependent if they have an element of the infrastructure in common in their routes and the order of operation of the two movements is unknown a priori. Two movements are independent if one of the two conditions is not satisfied.*

Now assume that movement k is the successor of movement j and that movement m uses their common track. To avoid that movement m is scheduled over their common track between movements j and k, movement m is dependent of the movements j and k if $d_m + b_{mj} > r_j$ and $r_m < d_k + b_{km}$. If movements j, k and m are not fulfilling these constraints, then it is known a priori whether m is carried out before j or after k. Hence, the order of operation of the three movements is known a priori and hence the movements are independent.

In the model the following parameter is used to indicate whether two movements j and k are dependent:

$$a_{jk} := \begin{cases} 1 & \text{if movements } j \text{ and } k \text{ are dependent} \\ 0 & \text{if movements } j \text{ and } k \text{ are independent} \end{cases}$$

If two train units have to be combined, then they arrive at the track where they are combined from the same direction or from different directions. If they arrive from the same direction, then the order in which they enter the track is known and the train unit that arrives last is the successor of the one that arrives first. If the train units arrive from different directions and at least one of them has not yet been planned in detail, then they may arrive in either order. Because of technical and safety reasons they can not arrive at the same time, so the movements are dependent. By giving each movement the combined train as its successor, the order of operation with the combined train is known and hence both movements are independent of that train. Now each movement can arrive at the track after the other one has arrived.

If a train arrives on a track and is split into two parts, then we have a similar situation. If both parts leave the track into the same direction, then the order in which they leave the track is known and the train unit that leaves last is the successor of the other. If they leave the track in different directions and at least one of them has not yet been planned in detail, then they may leave the track in either order. Again, because of technical and safety reasons, they can not leave the track at the same time, so the departing movements are dependent. By giving the arriving train both departing movements as its successor with respect to the arrival track, the departing movements are independent of the arriving train. Now the departing movements can leave the track in either order.

4.2   Model with Fixed Routes (MFR)

The goal of the capacity test is to verify if it is possible to plan all the movements not yet planned in detail within their time windows. Therefore, the model minimizes the number of movements not yet planned in detail which can not be planned within their time window. If the objective value is zero, then it can be concluded that the capacity of the infrastructure is still sufficient. If the objective value is strictly positive, then not all the shunting movements of S_n can be planned within their time window.

The decision variables used in the model with fixed routes are the following:

- y_j = the plan time of movement j in minutes
- $U_j = \begin{cases} 1 & \text{if movement } j \text{ can not be planned within its time window} \\ 0 & \text{if movement } j \text{ can be planned within its time window} \end{cases}$
- $x_{jk} = \begin{cases} 1 & \text{if movement } j \text{ has to be operated before movement } k \\ 0 & \text{if movement } j \text{ has to be operated after movement } k \end{cases}$

The decision variable y_j is a real variable which gives the plan time of movement j as the number of minutes after the starting time of testing. The decision variable U_j can be derived from the decision variable y_j. If $r_j \leq y_j \leq d_j$, then the variable U_j is zero, else it is one. No conflicts between planned movements gives $y_j = r_j = d_j$ for every movement $j \in S_p$, which gives $U_j = 0$. The decision variable x_{jk} is only defined if movements j and k are dependent. The variable gives the order in which the movements j and k have to be operated. It would be sufficient to define x_{jk} only for movements $j < k$, but then the model becomes hard to read and the computation time is not influenced, because CPLEX eliminates the 'extra' variables.

The problem with fixed routes is described with the following mixed integer programming model:

minimize $\qquad\qquad\qquad \sum_{j \in S_n} U_j$

subject to:

$$y_j = r_j \qquad\qquad \forall j \in S_p \qquad\qquad (1)$$
$$U_j = 0 \qquad\qquad \forall j \in S_p \qquad\qquad (2)$$
$$y_j \geq r_j \qquad\qquad \forall j \in S_n \qquad\qquad (3)$$
$$y_j \leq d_j + U_j M \qquad \forall j \in S_n \qquad\qquad (4)$$
$$x_{jk} + x_{kj} = 1 \qquad \forall j, k \in S \text{ with } a_{jk} = 1 \qquad (5)$$

$$y_j + b_{jk} \leq y_k + (1 + U_j - x_{jk})M \; \forall j, k \in S \text{ with } a_{jk} = 1 \tag{6}$$

$$y_j + b_{jk} \leq y_k \qquad \forall j, k \in S_n \text{ with } k \in s_j \tag{7}$$

$$x_{jm} = x_{km} \qquad \forall j, k, m \in S, \text{ with } a_{jm} = a_{km} = 1, \tag{8}$$
$$\exists t \in T_m : s_j^t = k$$

$$x_{jk} \in \{0, 1\} \qquad \forall j, k \in S \text{ with } a_{jk} = 1 \tag{9}$$

$$y_j \in \mathbb{R}^+ \qquad \forall j \in S \tag{10}$$

$$U_j \in \{0, 1\} \qquad \forall j \in S \tag{11}$$

The meaning of the first three constraints is obvious. Constraint (4) handles the fact that the plan time of a movement not yet planned in detail preferably does not exceed the movement's due date. Constraints (5) and (6) take care that there is enough time between the plan times of two dependent movements. Constraints (4) and (6) contain a big-M, a large constant integer value. Due to these constraints, there always exists a feasible solution for the model. Constraint (6) is binding only if $U_j = 0$ and $x_{jk} = 1$. If movement k is a successor of movement j, then its plan time has to be larger than the plan time of movement j plus the required minimum amount of time between movements j and k. Constraint (7) handles this. If movement k is a successor of movement j with respect to track t, then constraint (8) takes care that movement m, which uses track t and is dependent of movements j and k, is operated before movement j or after movement k.

4.3   Model with Variable Routes (MVR)

In order to increase the flexibility of the model (MFR), an extension (MVR) has been developed in which it is possible to select the routes for all train movements, not only for the shunting movements, but also for the passenger and cargo trains. In order to facilitate the additional flexibility, a few parameters and decision variables have been added. The added parameters are the following:

- R_j = the set of possible routes of movement j.
- T_{jr} = is the set of tracks used by movement j if route $r \in R_j$ is chosen.
- $a_{jrks} = \begin{cases} 1 \text{ if movements } j \text{ by route } r \text{ and } k \text{ by route } s \text{ are dependent} \\ 0 \text{ if movements } j \text{ by route } r \text{ and } k \text{ by route } s \text{ are independent} \end{cases}$

The definition of dependency is the same as in the model (MFR). The route of every movement has to be determined by the model. So a new decision variable is introduced, namely:

$$z_{jr} = \begin{cases} 1 \text{ if movement } j \text{ has to be routed along route } r \\ 0 \text{ otherwise} \end{cases}$$

The model minimizes the weighted number of not yet planned movements which can not be planned within their time window. The parameter w_n represents the penalty if a not yet planned movement can not be planned.

A second term has been added to the objective function. The goal of this term is to prevent that a lot of alternative routes are chosen, especially for the already planned movements. Alternative routes are less comfortable for the passengers, since they usually use more switches. The parameter w_{jr} represents the penalty if alternative route $r \in R_j$ is chosen for movement j. These penalties have to be much smaller than the penalty w_n, because the most important objective is to find a solution such that all movements can be planned. Parameter w_{j0} is the weighting factor if the priority route of a movement is chosen. Now the mixed integer programming model (MVR) can be described as follows:

minimize $\qquad w_n \sum_{j \in S_n} U_j + \sum_{j \in S} \sum_{r \in R_j} w_{jr} z_{jr}$

subject to:

$$
\begin{array}{lll}
y_j = r_j & \forall j \in S_p & (12) \\
U_j = 0 & \forall j \in S_p & (13) \\
y_j \geq r_j & \forall j \in S_n & (14) \\
y_j \leq d_j + U_j M & \forall j \in S_n & (15) \\
y_j + b_{jk} \leq y_k & \forall j, k \in S_n \text{ with } k \in s_j & (16) \\
\sum_{r \in R_j} z_{jr} = 1 & \forall j \in S & (17) \\
x_{jk} + x_{kj} = 1 & \forall j, k \in S : \exists r \in R_j, & (18) \\
& \exists s \in R_k \text{ with } a_{jrks} = 1 \\
y_j + b_{jk} \leq y_k + M(3 + U_j - x_{jk} - z_{jr} - z_{ks}) & \forall j, k \in S, \forall r \in R_j, & (19) \\
& \forall s \in R_k \text{ with } a_{jrks} = 1 \\
z_{mq} - 1 \leq x_{jm} - x_{km} \leq 1 - z_{mq}, & \forall j, k, m \in S, \text{ with} & (20) \\
& d_m + b_{mj} > r_j, r_m < d_k + b_{km}, \\
& \forall q \in R_m \exists t \in T_{mq} : s_j^t = k \\
x_{jk} \in \{0,1\} & \forall j, k \in S \text{ with } \exists r \in R_j, & (21) \\
& \exists s \in R_k : a_{jrks} = 1 \\
z_{jr} \in \{0,1\} & \forall j \in S, \forall r \in R_j & (22) \\
y_j \in \mathbb{R}^+ & \forall j \in S & (23) \\
U_j \in \{0,1\} & \forall j \in S & (24)
\end{array}
$$

Also for this model defining x_{jk} only for $j < k$ would be sufficient, but for sake of readability $j > k$ is included into the model. This doesn't influence the computation time of CPLEX. Several constraints of (MVR) are the same as those in (MFR). Therefore, we only describe the differences. For every movement only one route can be chosen. Constraint (17) looks after this. Constraints (18) and (19) take care that, if there are routes r and s for movements j and k such that these movements along these routes are dependent, then there will be sufficient time between the plan times of these movements if routes r and s are selected.

Obviously, constraint (19) has an effect on the plan times of movements j and k only if $U_j = 0$ and $x_{jk} = z_{jr} = z_{ks} = 1$. If track t of route q for movement m is occupied by successive movements j and k during part of the time window of movement m, then m is not allowed to use track t as long as this track is occupied. If route q is chosen for movement m, then m has to be carried out before movement j or after movement k. Constraint (20) handle this.

5 Application to Railway Stations in the Netherlands

To check whether the models can be solved quickly and effectively, they have been tested for three railway stations in the Netherlands. These stations are Groningen, Utrecht and Zwolle. A description of the stations is given in Section 5.1. Thereafter the results are presented in Sections 5.2 and 5.3.

5.1 Introduction Railway Stations Groningen, Utrecht and Zwolle

Railway station Groningen is a relatively small station in the northern part of the Netherlands with one shunting area (see Figure 1). Because the shunting area is located parallel to the platform tracks, many saw movements are required. There are even many saw movements which have to be split into three parts, because the shunting area can only be reached from one side. A typical 24-hour weekday for station Groningen has about 575 train movements where the saw movements have been split already into separate movements. Approximately 175 of these 575 movements are shunting movements.

A second railway station which is used to test the model is station Utrecht. This is the largest railway station in the Netherlands. Trains from this station depart to all directions of the Netherlands. Utrecht has two shunting areas, a large one at the southern part of the station (OZ) and a small one at the northern part of the station (Landstraat). The shunting area Landstraat can only be reached from a few platform tracks and can only be entered from one direction. The shunting area OZ is a large shunting area which is always entered or left via the same track. To get to this shunting area, a shunting movement needs to use the same infrastructure as a lot of trains that arrive from or depart to the southern and eastern part of the Netherlands.

Not much trains start or end at railway station Utrecht, so there is a relatively small number of shunting movements. A typical weekday for station Utrecht has approximately 1800 movements including 150 shunting movements. Saw movements only take place twice or three times a day. But if a saw movement takes place, then planners take a transfer time for the driver of at least 10 minutes. As a consequence, each saw movement is a serious bottleneck.

The third railway station that is used for testing the model is station Zwolle in the north-eastern part of the Netherlands. This station is chosen because it is known as one of the hardest stations of the Netherlands with respect to shunting. This is caused by the fact that it has several smaller shunting areas and because many shunting movements are related to the internal and external cleaning of rolling stock. A typical weekday at station Zwolle has approximately 900 movements including 175 shunting movements.

5.2   Results of MFR

The model introduced in Section 4.2 is solved by the standard MIP solver CPLEX 9.0. The computations were carried out on an Intel Pentium M, 1.8 GHz processor with 512 MB internal memory. The model is tested on the stations introduced in the previous section and with a time interval of testing from Tuesday 2:00 am to Wednesday 2:00 am during a normal week. The CPU times CPLEX needs to solve the model for the different stations are given in Table 1. For the three railway stations the model is solved quickly.

Table 1. Running times of CPLEX on MFR

	All preferred routes		Real plan	
	Objective	CPU Time (sec)	Objective	CPU Time (sec)
Groningen	9	2.81	0	2.79
Utrecht	1	1.09	1	1.14
Zwolle	5	4.58	3	4.72

The used data were derived from real shunting plans. The plan times of the already planned passenger and cargo trains were kept the same as in the real plans and for the shunting movements a time window was derived. In the first data set of each station, it was assumed that all movements use their priority route and in the second data set their routes were taken equal to the ones in the real plan.

Because real data should be conflict free, the model should not find movements that could not be planned. But for railway stations Utrecht and Zwolle the model found shunting movements that could not be planned. This can be explained by the fact that planners have the opportunity to violate the norms.

A first impression may be that it is a bit strange that Utrecht has the smallest CPU time. This can be explained by the fact that Utrecht has less shunting movements in comparison with the number of passenger and cargo trains and by the fact that it only has two or three saw movements a day. Especially saw movements are responsible for many dependent movements. The latter require a lot of variables and constraints. That the CPU time for railway station Utrecht is small can also be explained by the fact that the capacity of the infrastructure is so scarce that there is not much to decide for the model.

5.3   Results of MVR

The model introduced in Section 4.3 is also solved with the standard MIP solver CPLEX 9.0 and the computations were also carried out on an Intel Pentium M, 1.8 GHz processor with 512 MB internal memory. The model is tested with the same data sets as used for testing the model MFR.

Solving the model MFR with the routes as given in the real plans results in plan times for all the movements not yet planned in detail. The resulting solution can be used as a starting solution for the model MVR. This results in a good upper bound and a smaller running time. The values of the weighting factors are taken 1000 for w_n and r for w_{jr}. The latter implies that lower numbered routes are preferred, which represents the current practice.

The CPU times CPLEX needs to solve the model for the different stations and for varying numbers of allowed alternative routes are given in Table 2. The CPU times are given in seconds. Table 3 shows the results for the case that an objective difference of five is used. This means that CPLEX terminates as soon as the absolute difference between the lower bound and the upper bound is less than five. Especially for station Zwolle this reduces the running time of CPLEX on most instances significantly.

Table 2. Running times of CPLEX without objective difference on MVR

Number of alternative routes	Groningen		Utrecht		Zwolle	
	Objective	CPU time	Objective	CPU time	Objective	CPU time
0	9000	0.54	1000	0.29	5000	4.55
1	1008	31.53	1	15.76	1004	142.70
2	10	49.74	1	43.29	6	280.20
3	10	29.14	1	59.48	6	323.19
4	10	31.34	1	105.85	6	492.26
5	10	33.42	1	184.56	6	647.00
6	10	33.39	1	237.71	6	646.00
7	10	33.60	1	154.12	6	957.68
8	10	33.36	1	200.93	6	1127.91
9	10	33.38	1	146.63	6	1357.99

If for all movements only the priority route is allowed, then the number of shunting movements that can not be planned is obviously the same as in the previous section. For station Groningen already eight of the nine infeasible shunting movements can be planned if one alternative route is allowed. Allowing two or more routes for every movement makes it possible to plan all the movements. The objective value of 10 can be explained by the fact that eight shunting movements get a first alternative route and one movement gets its second alternative route. The running times for station Groningen are very small.

If only the priority route of the movements is allowed, then station Utrecht has only one shunting movement that can not be planned. This movement is part of the only saw movement in the data set. If one alternative route is allowed, then all movements can be planned. For station Utrecht the running time for solving the model becomes larger than for station Groningen if two or more alternative routes are allowed. This can be explained by the fact that the number of possible alternative routes in Utrecht is much larger, which results in much more possible solutions for station Utrecht in comparison with station Groningen. But the running time of CPLEX is still very small for station Utrecht.

Table 3. Running times of CPLEX with objective difference 5 on MVR

Number of	Groningen			Utrecht			Zwolle		
alternative routes	LB	UB	CPU time	LB	UB	CPU time	LB	UB	CPU time
0	9000	9000	0.53	1000	1000	0.22	5000	5000	4.97
1	1008	1011	10.80	1	1	15.00	1003	1004	119.04
2	10	10	57.23	1	1	40.26	6	7	233.41
3	10	10	36.25	1	1	58.89	5	6	264.03
4	10	10	32.55	1	1	88.07	6	6	426.19
5	10	10	39.67	1	1	179.75	5	6	597.23
6	10	10	39.36	1	1	223.72	6	6	573.88
7	10	10	39.18	1	1	153.10	4	6	804.32
8	10	10	38.97	1	1	198.34	5	6	1022.56
9	10	10	37.83	1	1	133.50	4	6	1208.04

For station Zwolle there are five shunting movements which can not be planned if all movements have to take their priority route. Allowing one alternative route gives only one shunting movement which can not be planned. If two alternative routes are allowed, then all movements can be planned. Allowing an extra possible route for a movement gives a large increase in the running time for station Zwolle. But also in these cases, the running times of CPLEX are still less than half an hour.

6   Conclusions and Future Research

In this paper a global capacity test for the infrastructure between shunting areas and platform areas of railway stations is described. Two mixed integer programming models are introduced. Both models verify whether there is still enough capacity of the infrastructure to plan all the not yet planned shunting movements in between the already planned train movements. In the first model the routes of the movements are given and can not be changed. This model can be solved very quickly by the MIP solver CPLEX 9.0. The second model allows selecting the route of a movement, which results in much more feasible solutions. Nevertheless, for stations Groningen and Utrecht, the model can be solved very quickly. For station Zwolle, the solution process takes somewhat more time. It can be concluded that the computation times are small enough for practical use.

Currently, the capacity test is being implemented in practice. This will facilitate the postponement of planning the details of the non-critical shunting movements until briefly before the operations. The latter will result in a reduction of the amount of replanning of the shunting movements. In the end, this will also lead to a reduction of the throughput time of the complete logistic planning process.

In our future research we will focus on relaxing the assumption that the saw track of each saw movement is known a priori. This makes the problem more complex. However, we intend to enable this increased complexity by using the model on a rolling horizon with shorter time horizons of at most a couple of hours per run. Note that the experiments so far covered a time horizon of a complete day.

References

1. Van den Broek, J.J.J.: Toets op Inplanbaarheid van Rangeerbewegingen (in Dutch), M.Sc. thesis, Eindhoven University of Technology (2002)
2. Billionnet, A.: Using Integer Programming to Solve the Train-platforming Problem. Transportation Science 37, 213–222 (2003)
3. Blasum, U., Bussieck, M.R., Hochstättler, W., Moll, C., Scheel, H.-H., Winter, T.: Scheduling Trams in the Morning. Mathematical Methods of Operations Research 49, 137–148 (2000)
4. Bussieck, M.R., Winter, T., Zimmermann, U.T.: Discrete Optimization in Public Rail Transport. Mathematical Programming 79, 415–444 (1997)
5. Cordeau, J.F., Toth, P., Vigo, D.: A Survey of Optimization Models for Train Routing and Scheduling. Transportation Science 32(4), 380–404 (1998)
6. Cornelsen, S., Di Stefano, G.: Platform Assignment (to appear)
7. Di Miele, F., Gallo, G.: Dispatching Buses in Parking Depots. Transportation Science 35(3), 322–330 (2001)
8. Duinkerken, E.: Toets op Opstelcapaciteit, M.Sc. thesis (in Dutch), University of Amsterdam (2003)
9. Di Stefano, G., Koci, M.L.: A Graph Theoretical Approach to the Shunting Problem. In: Proceedings of the 3rd Workshop on Algorithmic Methods and Models for Optimization for Railways (ATMOS 2003) Electronic Notes in Theoretical Computer Science, vol. 92 (2004)
10. Freling, R., Lentink, R.M., Kroon, L.G., Huisman, D.: Shunting of Passenger Train Units in a Railway Station. Transportation Science 39, 261–272 (2005)
11. Hamdouni, M., Desaulniers, G., Marcotte, O., Soumis, F., Van Putten, M.: Dispatching Buses in a Depot Using Block Patterns. Technical report G-2004-051. GÉRAD, Montréal (2004)
12. Tomii, N., Zhou, L.J., Fukumara, N.: An Algorithm for Station Shunting Scheduling Problems Combining Probabilistic Local Search and PERT. In: Imam, I., et al. (eds.) Multiple approaches to intelligent systems: 12th international conference on industrial and engineering applications of artificial intelligence and expert systems, pp. 788–797. Springer, Heidelberg (1999)
13. Tomii, N., Zhou, L.J.: Depot Shunting Scheduling Using Combined Genetic Algorithm and PERT. In: Allen, J., et al. (eds.) Computers in Railways VII, pp. 437–446. WIT Press, Southampton (2000)
14. Winter, Th., Zimmermann, U.T.: Real-time Dispatch of Trams in Storage Yards, Annals of Operations Research 96, 287–315 (2000)
15. Zwaneveld, P.J.: Railway Planning, Routing of Trains and Allocation of Passenger Lines, Ph.D. thesis, Erasmus University Rotterdam (1997)
16. Zwaneveld, P.J., Dauzère-Pérès, S., van Hoesel, C.P.M., Kroon, L.G., Romeijn, H.E., Salomon, M., Ambergen, H.W.: Routing Trains Through Railway Stations: Model Formulation and Algorithms. Transportation Science 30, 181–194 (1996)

Railway Crew Pairing Optimization

Lennart Bengtsson, Rastislav Galia, Tomas Gustafsson,
Curt Hjorring, and Niklas Kohl*

Jeppesen AB, Odinsgatan 9, Göteborg, Sweden
curt.hjorring@jeppesen.com

Abstract. The use of automatic crew planning tools within the railway industry is now becoming wide-spread, thanks to new algorithm development and faster computers. An example is the large European railway Deutsche Bahn, which is using a commercial crew planning system developed by Jeppesen (formerly Carmen Systems). This paper focuses on the crew pairing problem that arises at major railways. Even though it is similar to the well-studied airline crew pairing problem, the size and complexity of the railway operation necessitates tailored optimization techniques. We show that a column generation approach to the pairing problem, which combines resource constraints, k-shortest path enumeration and label merging techniques, is able to heuristically solve a 7,000 leg pairing problem in less than a day.

1   Introduction

The process of crew planning at large transportation companies, such as railway and airline companies, is often very complex. Feasible work schedules have to comply with a large set of company rules and union agreements, and might also take into account preferences of individual crew members. The schedules should also take into consideration the available number of crew in each crew depot.

In order to manage the complexity and reduce the crew costs, some railways and most airlines have started to use automated crew planning tools. The Carmen crew scheduling system is successfully implemented and used by many of the world's largest transportation companies. Railway customers include the German state railways Deutsche Bahn, the Swedish State Railways and the freight operator Green Cargo. In the airline sector, the system is used by all major European airlines, as well as several operators in North America and Asia.

Compared to manual planning, automated planning tools have a number of advantages. As already mentioned, sophisticated optimization techniques can be used to reduce the crew costs. It is also possible for the planning department to produce crew schedules for several scenarios in parallel, and then pick the most suitable one for production. The ability to evaluate "what-if" scenarios can be used for strategic planning and timetable changes etc.

A challenge when going from manual planning to automatic, is to model the many rules which are soft in the sense that "they should not be broken unless it's

---

* Now at DSB Planning, Sølvgade 40, DK-1349 Copenhagen K, Denmark.

F. Geraets et al. (Eds.): Railway Optimization 2004, LNCS 4359, pp. 126–144, 2007.

necessary". Especially for railways, in manual planning there are frequently cases where the planner violates rules based on experience and "common sense". In an automated system, this means that we have to model and express the many "common sense" trade-offs with penalties, and end up with a highly complex cost-function for the optimizer.

A trend in the industry is a drive for incorporating timetable changes very late in the planning process. In addition, one would like to be able to better adjust the number of conductors on a train to the daily or weekly passenger demand. Altogether, this means that the turn-around time and flexibility of the automatic tool is very important and the performance of the optimizer is essential.

In the following we will focus on the railway crew pairing problem. The techniques described have been successfully applied to planning problems from several rail operators. We exemplify our approach with data from Deutsche Bahn (DB), which in our experience has been the most challenging with respect to both size and modeling complexity. DB, which is one of the world's largest transportation companies, consists of a number of partly independent companies which operate regional and commuter traffic, together with the long-distance operator DB Fernverkehr. The total number of crew (train drivers and conductors) is around 30,000, distributed across more than 100 crew bases. A train is operated by one train driver and from zero to seven conductors depending on the type of train, the expected number of passengers, and a number of other factors.

1.1   The Railway Planning Process

The operation of a passenger railway is defined by the timetable, the allocation of infrastructure such as tracks and platforms to the timetable and the allocation of rolling stock to the timetable. Timetable, infrastructure and rolling stock are typically planned well in advance before the actual operation and often a plan is valid for an entire timetable period of six months or more. However, modifications, often due to maintenance work or special events, must occasionally be introduced in to the plan.

The timetable and the rolling stock schedule together define the basic crew need to operate the trains. Basic crew need includes the locomotive driver and, for passenger trains, the minimum number of conductors required. In addition to the basic crew need, extra crew might be required, depending on factors such as the type and quantity of rolling stock, the type of service offered on the train and the expected number of passengers. These factors also determine the qualifications required by the crew members assigned to the train.

Because a train might depart early in the morning and arrive at its destination late in the evening, it is necessary to divide the trains into *legs* (pieces of work the cannot be divided). Legs start and end at stations where it is possible to change crew. The goal of railway scheduling is to assign the legs to qualified named individuals such that rules and regulations with respect to working time, rest time etc. are satisfied, and such that costs and other quality aspects of the solution are optimized. The problem naturally decomposes into the crew pairing problem, in which pairings, i.e. sequences of legs, starting and ending at a crew base are constructed, and the crew rostering problem, where the anonymous pairings are assigned to named individuals. Furthermore, the pairing problem can be decomposed into a daily problem, in which pairings that cover all legs that are operated more or less on a daily basis are constructed, and a roll-out process, in which the daily solution is "rolled out" over a larger planning period, followed by a second planning phase that takes care of any irregularities in the timetable.

On the level of detail introduced so far, the airline and the railway crew scheduling problems are very similar. This is especially true for the rostering problems, which mainly have to deal with rostering rules and crew qualifications, and are less exposed to the underlying transportation network. A more detailed comparison of the pairing problems reveals a lot of similarities, but also important differences.

1.2   Outline of This Paper

After briefly reviewing existing literature on crew scheduling problems in Sect. 2, we present a mathematical model of the general crew pairing problem in Sect. 3. This model applies to both airline and railway problems. In Sect. 4 we describe how the general pairing problem can be solved with a column generation approach. A more detailed description of a typical *railway* crew pairing problem can be found in Sect. 5. In order to successfully solve railway pairing problems, Jeppesen has introduced some modifications in the general column generation scheme; these are presented in Sect. 6 along with some illustrative results. Then we round off the paper with large-scale railway planning results in Sect. 7 and some conclusions in Sect. 8.

2   Previous Work — Literature Review

Among the crew scheduling problems within the transportation industry, the airline crew pairing problem has got a lot of attention in the operations research literature. Finding an optimal set of pairings can be formulated as an integer programming problem, in which each column represents a feasible pairing. Unfortunately, the number of columns is immense. Snowdon et al. ([12]) report that a daily problem for a large American carrier has roughly 10^{14} columns. Therefore, delayed column generation is frequently used. An early application of this is due to Minoux ([11]), in which it is shown that columns can be priced by solving a shortest path problem on a network with arcs representing flights and overnight connections. A limitation of the flight network approach is that the network only captures flight connection rules, so rules that affect working days as a whole, or the entire pairing, are ignored. Therefore only a small fraction of the paths in the flight network form legal pairings. In the airline business, the sequence of flights flown in a typical working day is known as a *duty period*, and for many rules the legality of a single duty period is independent of preceding or succeeding duties.

Lavoie et al. ([9]) take advantage of the structure and form a duty period network where nodes are duty periods with state information, and arcs represent legal overnights. A similar approach was later used by Desaulniers et al. ([5]) to solve crew pairing problems at Air France. Vance et al. ([13]) present results for both flight and duty based implementations. Resources are added to each node in the networks to track additional legality conditions, and *resource constrained shortest path* problems are solved. Comparisons between the two versions are difficult to make since they implement different variants of the rules, but in general the flight based version spends a larger portion of time in the pricing routine than does the duty based version. However, the duty version cannot solve as large problems as the flight version due to memory limitations. Storage of the duty periods and the duty period connections is prohibitive for large problems (there is one arc for each legal duty period connection).

The excessive memory usage of the duty network is addressed in Hjorring and Hansen ([7]) by creating an initially relaxed network where duty-duty connections are replaced by flight-flight connections. Portions of the network are dynamically refined when required, in order to capture additional cost and legality information. The method also uses a k-shortest path approach, instead of resource constraints. This allows rules and costs to be implemented in a separate module that presents a black box interface to the pricing subproblem. The rules module can then be implemented using a modeling language that allows the user to easily write their own rules and cost function.

Until recently, the existing literature on railway optimization problems was mainly focused on vehicle scheduling problems, see for example the review by Cordeau et al. ([4]). The large size and the complexity of the crew scheduling problems that arise in the railway business made it difficult to apply optimization methods for crew planning. This has now changed, thanks to new algorithm development and faster computers. An example is the project "Destination: Customer" at the Dutch railway operator NS Reizigers, which aims at increasing the quality and punctuality of passenger trains, see Kroon and Fischetti ([8]). The authors present two approaches for solving the crew pairing problem: a greedy approach based on constraint programming, and a column generation approach based on reduced cost generation of duties combined with a subgradient optimizer and a variable fixing scheme. Another example is the development of a combined crew pairing and rostering system for the Italian state railways, see Caprara et al. ([2,3]). The scheduling process is divided into three phases: enumeration of all feasible pairings, solution of a large set covering problem using Lagrangian heuristics and variable fixing, and the production of cyclic *rosters* (sequences of pairings and weekly rest periods) which are subsequently distributed among the crew members.

Wren et al. ([15]) focuses on the problem of scheduling drivers for short-distance trains and buses. They stay away from the column generation approach, because they believe that the cost function is not sufficiently decomposable. Instead they enumerate feasible duty days explicitly, and apply filtering heuristics to limit the number of duties that need to be sent to the optimizer. The planning system, called TRACS II, appears to be in use at a number of British bus and railway companies.

3   General Problem Definition and Terminology

In this section we will introduce some terminology and give a mathematical formulation of the general crew pairing problem.

Let $\mathcal{L}$ denote the set of (leg, function)-combinations. A leg may require several crew members, but each will be assigned in a unique function, so for each function-leg l, $l \in \mathcal{L}$, exactly one crew member should be assigned. In the literature function-legs are sometimes called "trips" or "increments of work". $\mathcal{P}$ denotes the set of legal pairings. A pairing p, $p \in \mathcal{P}$, is an ordered set of function-legs, which can legally be operated by a crew member. A pairing will also contain other activities, such as preparation and closing activities, and may contain so called deadheads. A deadhead is a passive transport, either on a leg on which other crew work, or by some other means of transportation. A pairing must start and end at the same crew base and must comply with all rules and regulations regarding work time, rest, qualifications, etc. The coefficient a_{lp} is 1 if pairing p contains function-leg l and 0 otherwise. The cost of a pairing is denoted by c_p and may be the sum of real costs associated with the operation of the pairing and penalties capturing robustness and social aspects.

The main constraint of the crew pairing problem requires all function-legs to be covered by exactly one pairing, but the solution space will also be limited by other constraints. For this purpose we introduce the set $\mathcal{H}$ of hard constraints and the set $\mathcal{S}$ of soft constraints. The contribution of pairing p towards a hard constraint h, $h \in \mathcal{H}$, and a soft constraint s, $s \in \mathcal{S}$, is denoted b_{hp} and b_{sp} respectively. The sum of constraint contributions over pairings in a solution is bounded by d_h and d_s for hard and soft constraints respectively. A soft constraint s can be violated at the cost of c_s. These types of constraints are often used to ensure a distribution of the pairings corresponding to the distribution of crew and to ensure an acceptable portion of pairings with particular unpopular properties such as night stops.

The decision variables of the model are x_p, which is 1 if pairing p is used in the solution and 0 otherwise, and y_s which determine the violation of soft constraint s. The problem can now be stated as

$$\min \sum_{p \in \mathcal{P}} c_p x_p + \sum_{s \in \mathcal{S}} c_s y_s, \quad \text{s.t.} \tag{1}$$

$$\sum_{p \in \mathcal{P}} a_{lp} x_p = 1, \; \forall l \in \mathcal{L} \tag{2}$$

$$\sum_{p \in \mathcal{P}} b_{hp} x_p \leq d_h, \; \forall h \in \mathcal{H} \tag{3}$$

$$\sum_{p \in \mathcal{P}} b_{sp} x_p - y_s \leq d_s, \; \forall s \in \mathcal{S} \tag{4}$$

$$x_p \in \{0,1\}, \; y_s \geq 0 \tag{5}$$

The objective function (1) specifies that we want to minimize the sum of the cost of selected pairings and the cost of violating soft constraints. Constraint set (2) requires each function-leg to be covered by exactly one selected pairing. Constraint set (3) requires all hard constraints to be satisfied and constraint set (4) ensures that y_s is at least equal to the violation, if any, of the soft constraint s. Constraint set (5) requires x variables to be binary and y variables to be non-negative. The model is a generalization of the well known set partitioning problem.

The model presented above is not exactly identical to the one solved by the Carmen Crew Pairing system. The modifications can be summarized as follows:

- In constraint set (2) equality is replaced by greater than equal, i.e. with covering constraints. This is a relaxation, but typically a covering solution, i.e. a solution where the left hand side is more than 1 for at least one l, can be transformed to a partitioning solution, by substituting a function-leg with a deadhead on the same leg. Generally, the less constrained covering formulation is easier to solve. In cases when this relaxation is not valid, for instance when a legal deadhead cannot be found, the penalty for overcovering the active leg can be gradually increased.
- All constraints are soft. For constraint sets (2) and (3) a very large penalty for violation is defined. This is because a solution that violates a hard constraint is much more useful from a practical point of view than no solution at all. Typically a solution with hard constraint violations will hint at problems with the input data.

4 Algorithms for the General Pairing Problem

The problem discussed in Sect. 3 is generic and can be applied to both airline and railway crew pairing problems. In this section we outline how Carmen Crew Pairing solves the general problem.

The set covering problem in Sect. 3 is difficult to solve in practice, for two reasons. To start with, the number of columns in the constraint matrix A tends to be huge, so it is not feasible to enumerate all of them. One either has to limit the enumeration to some cleverly chosen subset of the feasible pairings, or generate columns "on demand" using optimizer feedback. In addition, large set covering problems are difficult to solve with standard integer solvers, so one has to resort to heuristic techniques.

Carmen's approach to the pairing problem is to generate pairings dynamically using reduced cost feedback from the set covering optimizer, a technique sometimes referred to as Dantzig-Wolfe decomposition in the literature. The goal of the decomposition is to separate the "complicating" problem constraints (the pairing rules) from the "simple" constraints (2)– (5) above. The "simple" constraints form the *master* optimization problem, whereas the "complicating" constraints are hidden inside a column generation subproblem, usually called the *pricing* problem. In this context, the set covering problem in Sect. 3 and the master problem is the same.

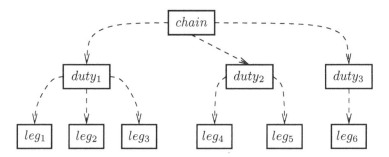

Fig. 1. Hierarchy of chain levels

4.1   The Pricing Problem

The goal of the pricing problem is to find legal pairings with negative *reduced cost*, for addition to the master problem. The reduced cost includes feedback from the optimizer (the pairing cost is reduced by the sum of the dual variables of the constraints that the pairing covers.) Because the pricing problem is intimately connected with the pairing rules and the cost function, we start with describing the general structure of the pairing rules and the pairing cost function.

The Carmen Crew Pairing system (CCP) contains a rule modelling language (Rave) that permits planners to easily add and modify the rules and the cost function. Rave is implemented as a black-box system, which is able to decide whether a pairing is legal with respect to the rules, and to evaluate the pairing cost. However, it does not expose the internal details of the evaluation process to the rest of the system.

For modelling purposes, a sequence of legs is usually grouped into a hierarchy of levels, as shown in Fig. 1. In the pairing problem there is usually at least one intermediate level of *duties* between the atomic level of the legs and the top–most pairing level. The levels are used when defining rules and costs: certain rules are leg dependent and need to be evaluated for each leg in the chain, whereas other rules may be duty- or chain-dependent.

A related concept is the range of objects that evaluated expressions depend on. For example, a rule that checks the min connection time between two legs has very short range: only two legs. In the other extreme, an expression counting the length of the pairing in days has chain range.

The column generator that solves the pricing problem relies on k-shortest-path enumeration within a network, in which the arcs are of two types: block arcs, which represent partial pairings, and connection arcs, which connect partial pairings together, see Fig. 2 for an example. The block is defined by an intermediate level in the Rave language. Preferably, the majority of the paths in the network should correspond to legal pairings; otherwise the efficiency of the column generator is reduced. Therefore the block level needs to be selected carefully. The most constraining pairing rules should have a range of at most one or two blocks; they will then be captured accurately by the network. Often, a suitable block level is equal to one day of work (a duty), but there are other possibilities.

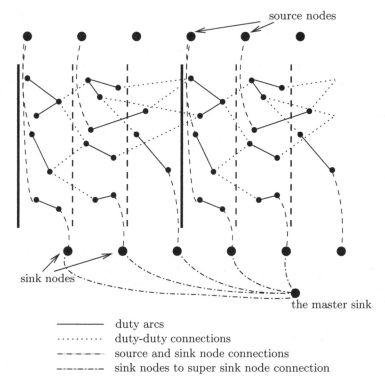

Fig. 2. Schematic view of the pricing network. There is a pair of source and sink nodes for each calendar day and homebase combination. For a given start day, the master sink node connects all sink nodes that represent legal ending days, assuming that there is a max calendar length rule in the ruleset.

In a similar fashion, we assume that the majority of the cost terms in the cost function can be decomposed into block and block-block contributions, which can be stored on the arcs in the network.

In one iteration of the column generator, paths with negative reduced costs are enumerated in increasing cost order, using a k-SP algorithm. For each path found, the Rave system evaluates the true pairing cost, which might differ from the network estimate, and verifies that the pairing is legal. If the pairing still looks good, it is added to the master problem. The process continues until no more attractive (negative cost) paths can be found, or if "sufficiently" many pairings have already been added. Because there are multiple source nodes in the network (c.f. Fig. 2), each base and starting day combination is treated separately.

4.2 The Master Problem

After each call to the pricing routine, the newly generated columns need to be incorporated into the master set covering problem defined in Sect. 3. The solution of the LP-relaxed master problem yields a set of dual variables that are sent to the pricing problem in the next iteration. We iterate over pricing calls and solutions of the master problem until no more attractive columns can be found. At that point, we have arrived at the optimal LP solution of the pairing problem. Typically, the LP solution is fractional, and only yields a lower bound of the pairing solution cost. Branching schemes or integer heuristics can be used to find good integer solutions, given the set of columns found so far.

Small set covering problems may be solved successfully by commercial IP solvers, such as CPLEX or XPRESS, but they tend to become unreasonably slow for problems of realistic size. The Carmen pairing optimizer relies entirely on specialized solvers. They are described in more detail in Sections 6.6 and 6.7.

5   The Railway Crew Pairing Problem

After the presentation of the general crew pairing problem in the previous sections, we now consider the *railway* pairing problem in more detail. This section focuses on the Deutsche Bahn planning problem, which is an example of the large and complex planning problems that arises at major railways.

5.1   Modelling of Rules, Regulations and Objectives

It is not possible to give a complete account of all rules, regulations and objectives of the DB crew pairing planning within the context of this paper. The purpose of this subsection is just to outline some of the more important rules, regulations and objectives. It is also the case that the labour agreements change from year to year, putting very high demands on the flexibility of the system.

In addition to function-legs and deadheads, a pairing will contain a number of derived preparation and closing activities. The derived activities can be calculated from the function-legs and deadheads and the rolling stock rotations.

The construction of individual pairings is mainly limited by a number of work, rest and connection time rules. There are three main different time concepts. There are activities associated with starting and ending a duty, starting and ending a train and walking times, for example between trains and the rest facilities. In addition to these times there is a required connection (buffer) time when crew change from one train to an other.

The *duty time* is the time from duty start to duty end. *Work time* is the time when the crew member carry out activities that according to German legislation are classified as work. *Paid time* is the time for which the crew member is paid. All work time is paid, but some non-work activities, primarily deadheading and paid rest, are paid without classifying as work time. Work time is generally limited to 10 hours per duty-day. A duty with at least 6 hours of work time requires at least 30 minutes of rest time, and a duty with at least 9 hours of work time requires at least 45 minutes of rest time. Rest must generally be allotted at a station with rest facilities, but can also be planned on-board a train en-route. The possibility of using on-board rests is constrained by several factors, but especially complicating is the fact that only a limited number of crew can rest at the same time. Required rest is usually unpaid, whereas any additional rest while connecting between trains is paid.

A pairing may contain up to two duty days. Nightstops must last for at least 5 hours, and are only permitted at certain stations. If the nightstop is shorter than 9 hours, the total work time of the pairing is limited to 10 hours.

Not all train products can be operated from all crew bases and certain trains must be operated from a particular base. This is in particular the case for some of the international traffic. There are also limitations on to which extent different function-legs can be mixed in the same pairing. It could otherwise happen that the rostering problem becomes infeasible, because no crew members at all are qualified to work on certain pairings.

The objective function contains monetary terms including paid time, hotel costs and deadheading costs as well as a large number of terms capturing operational stability and crew preferences. These include

- A train should not change crew too often, so a working period of less than e.g. two hours on a train is penalized.
- For the same reason, changing crew near the start or end of a train is penalized.
- Pairings with a long nightstop relative to the total paid time on the pairing are penalized.
- Pairings where the duty after the nightstop is longer than the duty before the nightstop are penalized.
- Changes between different functions and different train products are penalized, in order to reduce the number of different qualifications that are needed to work on individual pairings in the pairing solutions.
- Legs that only require a team of two conductors should preferably not be mixed with legs that require three conductors, because then the benefits of teaming will be reduced, c.f. Sect. 5.4.
- There is a fixed cost per duty day. This is to decrease the total number of duty days and consequently increase the paid time per duty day. Pairings with much paid time per duty day are easier to roster.

In total the DB ruleset contains about 100 separate rules. All rule and cost definitions are expressed in the Jeppesen Rave modelling language. Currently, the DB Rave code consists of $\approx 30,000$ lines distributed over 50 modules. Thanks to the black box system, there is a clear separation between the optimizer core and the Rave code, a fact that greatly simplifies maintenance and allows the Rave code to be developed independently of the optimizer. Integrating the rules and costs directly into the optimizer would have been a formidable task, with little hope of commercial success.

5.2 Base Constraints

Clearly, a solution must consist of legal pairings, but there are additional constraints on the total solution. These are modelled with constraints of type (3) or (4), c.f. the mathematical model discussed in Sect. 3. The most important of these are base constraints and constraints on the number and distribution of nightstops. The purpose of the base constraints is to ensure a distribution of pairings, and consequently of the workload, between the bases, which corresponds to the actual distribution of crew. In the long and mid-term run the distribution of crew can be influenced by operational efficiency, but in the short run the crew distribution is fixed. For each crew base there is a maximum and possibly a minimum amount of paid time, which can be assigned. There might also be separate base constraints for crew with special qualifications. Pairings with nightstops are permitted and economically quite attractive, but are unpopular with crew. Therefore it is necessary to introduce limits on the total number of pairings with nightstops as well as on the distribution of these between crew bases and between crew groups with different qualifications.

5.3 Variable Crew Need

In practice, the number of crew members that should work on a leg is not always given a priori. For service reasons, an additional crew member may be desirable, but not required. If the number of expected passengers vary substantially between legs on the same train, the required crew may also go up and down, but for practical reasons, the staffing is kept at a constant level, if this is not costly. This could be modelled by partitioning the set $\mathcal{L}$ in Sect. 3 into a required set of function legs, for which equality in constraint set (2) must be satisfied, and a set of optional function-legs, where relatively cheap slack variables are introduced.

5.4 Teaming Aspects

From an operational point of view, it is desirable that crew members work in teams, because then it becomes more likely that all required crew members arrive at a particular train departure station in time. Teaming is also preferred for social reasons. It is therefore necessary to take teaming aspects into consideration when solving the pairing problem. A typical approach is the following. Firstly team master pairings that cover all trains that require teams are created. For each pairing in the master solution, the minimum supplementary crew need is calculated, and then a number of copies of the master pairing, one for each extra crew member in the team, are added to the pairing solution. Finally, pairings for supplementary crew are created to cover any remaining crew need.

6   Algorithmic Contributions for the Railway Pairing Problem

6.1 The Pricing Network of Trains

As we have seen in Sect. 4.1, the pricing problem is transformed into a shortest path problem within a network of partial pairings ("blocks"). A question that arises when designing a pricing network for railway pairings is what a "block" in the network should represent. In the airline case the duty day is the natural candidate. However, it turns out that the number of legal duties in a typical railway problem is enormous, so it is not feasible to create the full duty network. Instead we choose to let the blocks in the network represent work periods on the same train. Whereas this approach leads to a manageable network size, duty dependent rules are not captured accurately in the network. Consequently, a large fraction of the paths found by the k-SP routine will violate duty time or work time constraints. This, in turn, could severely impact the performance of the column generator and the solution quality. In the following paragraphs we discuss how the performance degradation can be avoided.

6.2 Resource Dependent Cost Elements

The resource constrained shortest path problem has traditionally been used as a subproblem in applications of column generation to scheduling problems. The technique can be extended to model some classes of nonadditive costs as well.

Let us assume that the cost of a pairing can be written in the form $c(p) = c_a(p) + c_n(r(p))$, where $c_a(p)$ captures the additive parts of the cost function, and is additive with respect to block and block-block connections. Here c_n is a, possibly non-linear, function of the resource vector $r(p)$.

Not every cost function can be modelled using resources. The main limitation is that the resource vector $r(p)$ must itself be additive, because it needs to be modelled accurately by the network. Furthermore, the non-additive part of the cost c_n is required to be a non-decreasing function with respect the the individual resource components. The last *convexity* requirement enables certain short-cuts in the k-SP enumeration, as described below.

The current network topology also allows c_n to depend on some other characteristics of the pairings, such as the starting day of the pairing (determined by the source node of the path) and the length in calendar days. Some examples of costs modelled by resources:

- Penalize pairings for which the work time exceeds a limit by a constant value.
- If the number of short night stops in the path exceeds one, and the total work time of the pairing is above 10:00, add a very high penalty to the cost of the pairing.

The latter example illustrates how the rule concerning max work time in pairings with short night stops (c.f. Sect. 5.1) can be modelled using a pair of resources.

6.3 Extension of the k–SP Algorithm

The pricing network, as described in Sect. 4.1, needs to be modified in order to handle a resource-dependent non-additive cost function. Instead of storing the total reduced cost on the arcs, we store the additive part $\bar{c}_a$ of the reduced cost, together with the resource vector $r(p)$.

The pricing routine enumerates paths in *additive cost* order. For each path found, the resource vector $r(p)$ is summed up along the path and the non-additive part of the path cost is evaluated using a Rave definition. Only if the total path cost is attractive, will the cost and legality be evaluated by Rave.

Because the k-SP routine only "knows" about the additive part of the cost, a potentially large fraction of the enumerated paths might turn out to be non-attractive when the non-additive part has been added. We use the concept of *non-dominated* paths to significantly reduce the number of useless paths. Consider the graph in Fig. 3. Two sub-paths can be found from the source node n_s to the intermediate node n_i, with (cost, resource) sums $(-2, 50)$ and $(2, 60)$, respectively. Because the second sub-path has both higher cost and higher resource consumption than the first sub-path, we can conclude that the *total* cost of the second path, extended to the sink node n_d, is going to be higher than the total cost of the first path. Here we have to rely on the assumption that the non-additive function is non-decreasing. We say that the second path is *dominated* by the first path.

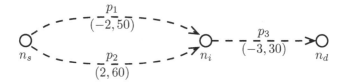

Fig. 3. Impact of the resources on the k–SP algorithm

More generally:

Definition 1. *The path p is said to be* dominated *by path p', if $\bar{c}_a(p) \geq \bar{c}_a(p')$ and $r_j(p) \geq r_j(p')$ for every resource r_j.*

Definition 2. *Let $P(n_s, n)$ be the set of all paths from the source node to node n. A path is called* Pareto-optimal, *if it is not dominated by any path in $P(n_s, n)$.*

We modify the k-SP routine so that it skips the dominated paths in the network. The dominance tests are done at all nodes in the network, and take advantage of the fact that paths are generated in increasing cost order. A variant of the algorithm has been described by Azevedo and Martins ([1]).

If the set $P(n_s, n_d)$ of paths from the source node n_s to the sink node n_d contains at least one attractive path, then it also contains at least one attractive Pareto-optimal path. This means that the pricing subproblem is guaranteed to find at least one attractive path, if any at all exist.

6.4 Label Merging

Because the number of arcs in the pricing network can exceed 1,000,000 for a typical daily DB pairing problem, the number of non-dominated path labels created by the k-SP routine might exceed memory limitations. A simple adaptive technique can be used to regulate the number of generated non-dominated paths in the network. The underlying idea is that we want to "merge" resource vectors in the Pareto-optimal set that are almost identical. To be more specific, path p is said to be dominated by path p' if $\bar{c}_a(p) + \kappa_a(n_s) \geq \bar{c}_a(p')$ and $r^{(j)}(p) + \kappa^{(j)}(n_s) \geq r^{(j)}(p')$ for each resource $r^{(j)}$, where $\kappa_a(n_s), \kappa^{(j)}(n_s)$ are positive parameters depending on the current source node n_s. The parameters $\kappa_a(n_s)$ and $\kappa^{(j)}(n_s)$ are adaptively changed after every pricing iteration, depending on the time spent in a pricing routine for source node n_s.

6.5 Impact of Improvements

The extensions to the network model and pricing subproblem solver that were mentioned in Sect. 6 have been tested on a number of problem instances, and we observe significant performance improvements. Fig. 4 depicts the effect of using resources to model rules that govern duty time and work time on a DB pairing problem. The input is a subset of the daily long-haul conductor problem, and contains 1,361 legs. We see that the k-shortest path approach without resources is unable to produce acceptable solutions for this problem. The reason is that the shortest path enumeration frequently times out after finding 10,000 paths, all of which are illegal. We also note that the resource merging technique not only reduces the runtime, but also improves the quality of the solution. This can be explained by the fact that the number of labels created by the resource-constrained shortest path solver may be limited by memory consumption when merging is not used.

6.6 Dual Strategy

As discussed in Sect. 4.2, one of the tasks of the master optimizer is to provide duals for the column generator. One approach to implementing the master optimizer is to relax the integrality conditions from the integer program, and solve the resulting LP, either using a simplex solver, or an interior point solver without crossover. A simplex solver returns values corresponding to extreme points of the optimal face, whereas interior point solutions are in the relative interior of the optimal face, which is usually a better representative of the possible dual solutions (Lübbecke and Desrosiers, [10]).

Another approach is to Lagrangian relax all the constraints and solve the relaxation using a subgradient approach. The details of this, along with a dual ascent approach to find integer solutions, are presented in Gustafsson ([6]). In brief, the integer heuristic and subgradient solver is highly integrated, and is run continuously with every call to the pricing module. A high focus is put on keeping the computational effort very low, and for example during the early stages of a complete column generation run, we accept sub-optimal subgradient duals as long as the integrality gap is reasonable.

Fig. 5 shows a comparison of these three approaches on a 530 leg problem. Fig. 5a tracks the progress of the lower bound with all integrality restrictions

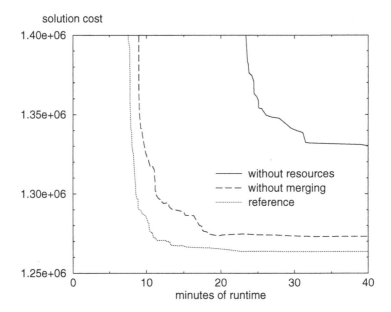

Fig. 4. Comparison of column generation runs. The reference run uses both resources and a label merging technique. None of the runs showed any further significant improvements after 40 minutes of runtime on a 2.8 GHz Pentium-4 server with 2 GB of memory.

removed. As expected, the LP solvers converge to the same lower bound, whereas the subgradient solver provides a significantly weaker lower bound (in this case about 0.5% lower). However, the simplex approach has a very long tail, requiring 1,800 iterations to reach convergence, whilst the interior point approach requires 800 iterations, and the subgradient only 500.

In Fig. 5b we also show the upper bounds coming from the dual ascent heuristic. To reduce the gap between the upper and lower bounds, we have implemented a simple integrality strategy. When no more negative reduced cost columns exist, a number of leg-leg connections are fixed to one, and the column generation process continues. The fix-and-regenerate steps continue until the lower bound to the integer restricted problem is fathomed by the best known solution. The upper bound converges fastest for the subgradient approach, and slowest for the simplex case. In fact the graph truncates the simplex run; convergence is reached after 5,000 major iterations, and then to a poorer quality solution.

Whilst Fig. 5 applies to one particular problem, the results are similar for other problems. As the size of the pairing problem grows, the difference between the approaches becomes even more significant. A further advantage of the subgradient approach is that the computations are much quicker. For the 530 leg problem, a subgradient major iteration is on average twice as fast as an interior point iteration, and four times faster than a simplex major iteration.

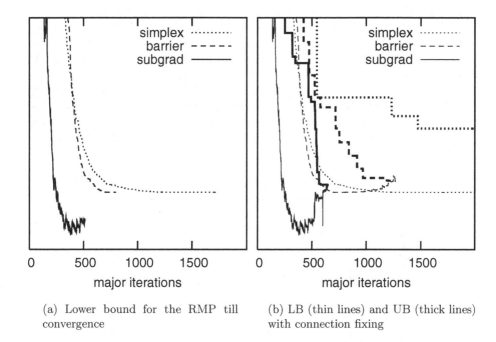

(a) Lower bound for the RMP till convergence

(b) LB (thin lines) and UB (thick lines) with connection fixing

Fig. 5. The effect of different master optimizers

6.7   Integer Strategy

The dual-ascent heuristic that provides integer solutions to the master problem was devised by Wedelin ([14]). This solver is fully integrated with the generation step, and typically finds new solutions every two or three pricing calls. Due to its probabilistic nature, not every new solution is the globally best so far; still, it is a very dynamic solver that produces close to optimal solutions throughout the run. This gives the end-user the ability to stop the run early (before full convergence), if the solution is deemed "good enough".

As mentioned in the previous section, a connection fixing strategy aids the search for high quality integer solutions. Starting with a fractional solution to the integrality relaxed master problem, we determine fractional leg-leg connections, and fix the connections that are closest to one. Columns and network arcs that violate fixed connections are removed, at least temporarily. When solving very large pairing problems (> 2,000 constraints), the time taken to calculate the fractional solution can be very large if standard LP solvers are used. Instead we have implemented an approximate solver, which is similar to the volume algorithm of Snowdon et al. ([12]).

We have further improved the connection fixing strategy by implementing *early branching* and by unlocking some or all connections when the lower and upper bounds meet, and then continuing. In early branching we no longer wait for LP convergence, but instead fix connections after a certain number of pricing iterations have passed. This has the effect of returning high quality integer solutions early in the search history, and can even reduce overall run time. If desired, a valid lower bound can be determined after the early branching by unlocking all connections, and generating columns until convergence is reached. Fig. 6 shows the effectiveness of this approach. In both cases an interior point solver was used as the LP optimizer. Each line shows the primal objective averaged over eight runs, each with a different starting solution produced by a very simple construction heuristic.

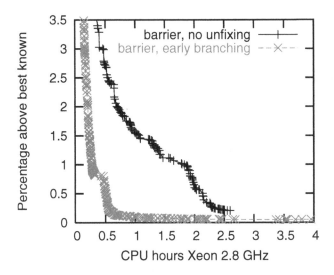

Fig. 6. Different integrality strategies

7 Large-Scale Results

We now integrate the new algorithmic contributions, and see how the resulting pairing optimizer performs on a range of railway problems. As a challenging test case, we have chosen the DB long distance planning problem, encompassing close to 7,000 daily legs and many more deadhead possibilities. There are 24 crew bases, 211 stations, and a number of different types of base constraints (Sect. 5.2), strongly affecting the nature of the solutions. Fig. 7 shows the results. We are able to solve the full problem in less than a day on standard PC hardware. The smaller problems presented in the graph were created by taking the solution from the full run, and selecting legs operated by a subset of the bases. The bases in the smaller problems are close geographically, making it non-obvious how the legs should be distributed across the bases.

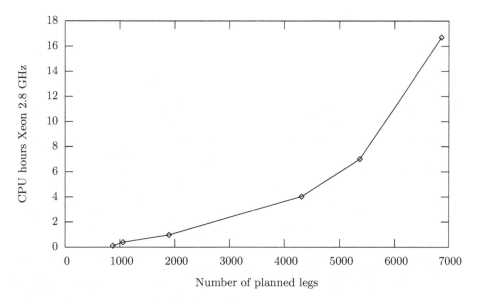

Fig. 7. Run times for various problem sizes

8   Conclusions and Future Work

Our results illustrate that it is possible to solve large and highly complex railway pairing problems in reasonable time using modern OR techniques. We believe that one of the key factors behind the success of our approach is the black box rule system, which separates the generic core optimization algorithms from the very user-specific problem definition, and makes it possible for planners to express rules and cost components in the tailor-made "Rave" language. An alternative approach, in which certain standard rules are "hard-coded" in the optimizer, would probably not have been flexible enough, given the complexity of the DB rule and cost structure. Another key factor is the column generation approach, which provides near-optimal solutions to the mathematical formulation of the pairing problem. Compared to manual planning, we have seen cost savings of 10–15% on large railway problems from several operators.

There are some aspects of the pairing optimizer that we are currently working on. Optimization speed is one of them, of course. Another is to improve the transportation leg search in the pricing problem. It is not feasible to insert all buses and trains in the whole region into the pricing network, especially not if the planning period stretches over a month. Finding the "right" set of transportation connections a priori is a non-trivial problem, however. Preferably the optimizer should find the right deadheads without manual intervention. Various preprocessing techniques can be used for this purpose.

Acknowledgements

We would like to thank our colleague Stefan Karisch, who helped in coordinating this work, and to the referees for their valuable comments.

References

1. Azevedo, J., Martins, E.: An algorithm for the multiobjective shortest path problem on acyclic networks. Investigacao Operacional 11(1), 52–69 (1991)
2. Caprara, A., Fischetti, M., Toth, P., Guida, P.L., Vigo, D.: Algorithms for railway crew management. Mathematical Programming B 79, 125–141 (1997)
3. Caprara, A., Fischetti, M., Toth, P., Vigo, D., Guida, P.L.: Solution of large-scale railway crew planning problems: the Italian experience. In: Wilson, N. (ed.) Computer-Aided Transit Scheduling. Lecture Notes in Economics and Mathematical Systems, pp. 1–18. Springer, Heidelberg (1999)
4. Cordeau, J.-F., Toth, P., Vigo, D.: A survey of optimization models for train routing and scheduling. Transportation Science 4, 380–404 (1998)
5. Desaulniers, G., Desrosiers, J., Dumas, Y., Marc, S., Rioux, B., Solomon, M.M., Soumis, F.: Crew pairing at Air France. European Journal of Operations Research 97, 245–259 (1997)
6. Gustafsson, T.: A Heuristic Approach to Column Generation for Airline Crew Scheduling. Lic. thesis, Chalmers University of Technology, Gothenburg, Sweden (1999)
7. Hjorring, C., Hansen, J.: Column generation with a rule modelling language for airline crew pairing. In: 34th Annual Conference of the Operational Research Society of New Zealand, pp. 133–142 (1999)
8. Kroon, L., Fischetti, M.: Crew scheduling for Netherlands Railways, Destination: Customer. In: Voss, S., Daduna, J. (eds.) Computer-Aided Scheduling of Public Transport. Lecture Notes in Economics and Mathematical Systems, pp. 181–201. Springer, Heidelberg (2001)
9. Lavoie, S., Minoux, M., Odier, E.: A new approach for crew pairing problems by column generation with an application to air transportation. European Journal of Operations Research 35, 45–58 (1988)
10. Lübbecke, M.E., Desrosiers, J.: Selected topics in column generation. Technical report, GERAD, Montreal, Canada (2002)
11. Minoux, M.: Column generation techniques in combinatorial optimization: A new application to crew pairing problems. In: XXIVth AGIFORS Symposium (1984)
12. Snowdon, J., Anbil, R., Pangborn, G.: The airline crew scheduling problem: Dual simplex, volume, and volume/sprint solutions. Technical report, IBM, T.J. Watson Research Center (2000)
13. Vance, P.H., Barnhart, C., Johnson, E.L., Nemhauser, G.L.: Airline crew scheduling: A new formulation and decomposition algorithm. Operations Research 45(2), 188–200 (1997)
14. Wedelin, D.: An algorithm for large scale 0-1 integer programming with application to airline crew scheduling. Annals of Operations Research 57, 283–301 (1995)
15. Wren, A., Fores, S., Kwan, A., Kwan, R., Parker, M., Proll, L.: A flexible system for scheduling drivers. Journal of Scheduling 6, 437–455 (2003)

Integer Programming Approaches for Solving the Delay Management Problem

Anita Schöbel

Institute for Numerical and Applied Mathematics, Georg-August University,
Göttingen
schoebel@math.uni-goettingen.de

Abstract. The *delay management problem* deals with reactions in case of delays in public transportation. More specifically, the aim is to decide if connecting vehicles should wait for delayed feeder vehicles or if it is better to depart on time. As objective we consider the convenience over all customers, expressed as the average delay of a customer when arriving at his or her destination.

We present path-based and activity-based integer programming models for the delay management problem and show the equivalence of these formulations. Based on these, we present a simplification of the (cubic) activity-based model which results in an integer *linear* program. We identify cases in which this linearization is correct, namely if the so-called *never-meet property* holds. We analyze this property using real-world railway data. Finally, we show how to find an optimal solution in linear time if the never-meet property holds.

1 Introduction

A major reason for complaints about public transportation is the missing punctuality, which—unfortunately—is a fact in many transportation systems. Since it seems to be impossible to avoid delays completely, it is a necessary issue in the operative work of a public transportation company to deal with delayed vehicles. In this paper we focus on the convenience of the customers and present a model for minimizing the average delay over all passengers.

Let us consider some vehicle (e.g., a train g) that arrives at a station with a delay. At the station, there are other vehicles (e.g., buses h and h') ready to depart, see Figure 1. What should each of these connecting vehicles do? There are two alternatives:

- A connecting vehicle h can **wait** to allow passengers to change from the delayed vehicle g to h.
- The connecting vehicle h can **depart** on time.

Unfortunately, both decisions have negative effects: In the first case, vehicle h causes a delay for passengers already within h, but also for customers who wish to board vehicle h later on, and possibly for subsequent other vehicles which

F. Geraets et al. (Eds.): Railway Optimization 2004, LNCS 4359, pp. 145–170, 2007.

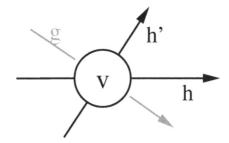

Fig. 1. The wait-depart decision at one single station

will have to wait for its delay. In the second case, however, all customers who planned to change from the delayed vehicle g into h will miss their connection.

In the first case the connecting vehicle h does not depart at its scheduled time, but with a delay. The new departure time of h is called its *perturbed* timetable. In the second case, the *perturbed* departure time of h at v equals the scheduled one.

In case of some known delays, the *delay management problem* is to find wait-depart decisions and a perturbed timetable for all vehicles in the network, such that the sum of all delays over all customers is minimized. The delay of a customer is defined as the delay he has when he reaches his destination. Recently the NP-completeness of this problem has been shown (see [9]).

Since in the delay management problem new departure times for each vehicle at each station have to be determined, it is related to finding timetables in public transportation. In this field, a lot of research has been done for periodic and non-periodic timetables. An excellent overview on periodic timetabling is given by [17]. We also refer to [15,4,13,26] and references therein. However, note the main difference between timetabling and delay management: In the timetabling problem the connections are given in advance, while in the delay management problem we have not only to find a (perturbed) timetable, but also to decide which connections should be maintained and which can be dropped.

How to *react* in case of delays has due to the size and complexity of the problem been mainly tackled by simulation and expert systems. We refer to [23,25] for providing a knowledge-based expert system including a simulation of wait-depart decisions with a *what-if* analysis. Simulation has also been used in [1,24].

In [11] the delay management problem has been formulated as a bicriteria problem, minimizing the number of missed connections and the delay of the vehicles simultaneously, and solved by methods of project planning. The weighted sum of these functions has been minimized in [18] by an enumeration procedure and a by greedy heuristic within a max-plus algebraic model, see also [22]. Dynamic programming has been used in [8] to identify polynomially solvable cases.

Integer programming formulations so far only exist as first attempts for the simple case without slack times, assuming that the customers on each edge are fixed (see the diploma theses of [14] and [21]). In Section 4 we are able to identify cases in which such models are correct. A first *exact* linear integer model for the delay management problem is presented in [19], and will be reviewed in this paper in a more convenient notation at the beginning of Section 3. A detailed description of the delay management problem will be published in [20]. Based on the formulation (TDM-B) presented in this text, [12] developed two new formulations reducing the number of variables in the models.

Related also work includes how to reduce delays by investing into new tracks ([7,6]), how to minimize the sum of waiting times of customers at their starting stations in a stochastic context ([2]), and a first on-line model of the problem along a line ([10]).

The aim of this paper is to present a new and more general integer programming formulation of the delay management problem, for which we are still able to develop solution approaches. Although our model can be applied to many different objective functions we specialize here on minimizing the sum of all delays over all customers. After introducing definitions and basic properties in Section 2 we develop a new integer programming formulation for the delay management problem in Section 3. In Section 4 we show that this formulation can be linearized if a special condition, called *the never-meet property* holds. We analyze this property in Section 5 using real-world data of the largest German railway company, *Deutsche Bahn*. In Section 6 we show how to solve the delay management problem in linear time in this case. The paper is concluded by some remarks on future research.

2   Notation, Concepts, and Basic Properties

We first introduce a new notation for the delay management problem, based on its representation as an *activity-on-arc project network* (see e.g. [15] for using this concept in timetabling).

We denote an arrival of a vehicle g at a station v as *arrival event* (g, v, arr), while a *departure event* (g, v, dep) describes the departure of some vehicle g at some station v. The *event activity network* is a graph $\mathcal{N} = (\mathcal{E}, \mathcal{A})$ where

- $\mathcal{E} = \mathcal{E}_{arr} \cup \mathcal{E}_{dep}$ is the set of all arrival and all departure events
- $\mathcal{A} = \mathcal{A}_{wait} \cup \mathcal{A}_{drive} \cup \mathcal{A}_{change}$ is a set of directed arcs, called *activities*, defined by

$$\mathcal{A}_{wait} = \{((g, v, \text{arr}), (g, v, \text{dep})) \in \mathcal{E}_{arr} \times \mathcal{E}_{dep}\}$$
$$\mathcal{A}_{drive} = \{((g, v, \text{dep}), (g, u, \text{arr})) \in \mathcal{E}_{dep} \times \mathcal{E}_{arr} : \text{vehicle } g \text{ goes}$$
$$\text{directly from station } v \text{ to } u\},$$
$$\mathcal{A}_{change} = \{((g, v, \text{arr}), (h, v, \text{dep})) \in \mathcal{E}_{arr} \times \mathcal{E}_{dep} : \text{a changing}$$
$$\text{possibility from vehicle } g \text{ into } h \text{ at station } v \text{ is required}\}.$$

The driving and waiting activities are performed by vehicles, while the changing activities are used by the customers. As an example, a small event-activity network is depicted in Figure 2.

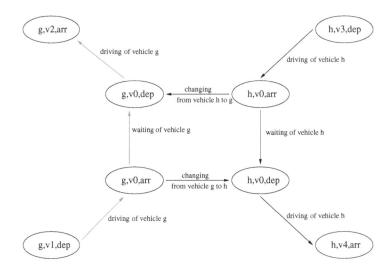

Fig. 2. An event-activity network

Note that $\mathcal{N}$ is a special case of a time-expanded network and hence contains no directed cycles. This means that a precedence relation $\prec$ between events (or activities) is canonically given, where $i \prec j$ hence indicates that there exists a path from i to j. We remark that for a given set of events (or of activities) a minimal element w.r.t. $\prec$ always exists, but it needs not be unique.

Using the notation of event-activity networks, a *timetable* $\Pi \in \mathbb{Z}^{|\mathcal{E}|}$ is given by assigning a time Π_i to each event $i \in \mathcal{E}$ (see [15]). Timetables are usually given in minutes and hence consist of integer values. The planned duration of activity $a = (i, j)$ is given by $\Pi_j - \Pi_i$. Furthermore, let $L_a \in \mathbb{N}$ be the minimal duration needed for performing activity a. We assume that the timetable is *feasible*, i.e.,

$$\Pi_j - \Pi_i \geq L_a \text{ for all } a = (i,j) \in \mathcal{A}.$$

We further assume that *source delays* are known at some of the events, where they might have occurred at the preceding activity or at the event itself. Let $SD \subseteq \mathcal{E}_{arr}$ denote the set of source-delayed events, and $d_i > 0$ indicate the delay they have. (For $i \notin SD$ the source delay $d_i = 0$.)

If source delays occur, some of the subsequent arrival and departure times Π_i can also not take place punctually, since the minimal durations L_a for subsequent activities have to be taken into account. The outcome $\Pi + y$ is called a *perturbed timetable*, and y_i is called the delay of event i. Such a perturbed timetable is feasible, if

- the source delays are taken into account, i.e., $\Pi_i + y_i \geq \Pi_i + d_i$, and
- the delay is carried over correctly from one event to the next, i.e.,
  $\Pi_j + y_j - (\Pi_i + y_i) \geq L_a$ holds for all driving and waiting activities $a = (i, j)$.

Defining the *slack time* s_a of an activity $a \in \mathcal{A}$ as the time which can be saved when performing activity a as fast as possible, i.e.,

$$s_a = \Pi_i - \Pi_j - L_a$$

we can equivalently restate the two above conditions in terms of the delay vector y as follows.

Definition 1. *A set of delays y_i for all $i \in \mathcal{E}$ is* **feasible,** *if*

$$y_i \geq d_i \text{ for all } i \in \mathcal{E} \quad and \tag{1}$$
$$y_i - y_j \leq s_a \quad for \text{ all } a = (i,j) \in \mathcal{A}_{wait} \cup \mathcal{A}_{drive}. \tag{2}$$

Condition (2) makes sure that the delay at the start of activity a is transferred to its end, where it can be reduced by the slack time of a.

In the following we only use the slack times s and the delays y instead of the minimal durations L and the timetable Π.

Definition 1 only takes the driving and waiting activities into account. However, in the delay management problem the goal is to identify which changing activities should be maintained and which ones can be dropped. For a changing activity we analogously require that

$$y_i - y_j \leq s_a \text{ if } a = (i,j) \text{ is maintained} \tag{3}$$

We are now in the position to specify feasible solutions of the delay management problem.

Definition 2. *A set of connections $\mathcal{A}^{fix} \subseteq \mathcal{A}_{change}$ together with a feasible set of delays y_i for all $i \in \mathcal{E}$ is a* **feasible solution of the delay management problem,** *if*
$$y_i - y_j \leq s_a \quad for \text{ all } a = (i,j) \in \mathcal{A}^{fix},$$

i.e., for all connections $a \in \mathcal{A}^{fix}$ which are maintained.

Note that a timetable would also be feasible if some vehicles depart or arrive late without any reason. Such solutions are clearly not optimal. The "most punctual" solutions are defined below.

Definition 3. *Let $(\mathcal{A}^{fix}, y)$ be a feasible solution of the delay management problem. The delay y is called* **time-minimal** *with respect to $\mathcal{A}^{fix}$ if all feasible solutions $(\mathcal{A}^{fix}, y')$ satisfy $y \leq y'$ (where as usual $\leq$ is meant component-wise). We write $y(\mathcal{A}^{fix})$.*

A time minimal solution with respect to each set $\mathcal{A}^{fix} \subseteq \mathcal{A}_{change}$ can be found efficiently by using the critical path method (CPM) of project planning.

To this end, we transform the event-activity network into a project network (as defined, e.g., in [5]) by introducing one super-source s and taking

$$\mathcal{A}(\mathcal{A}^{fix}) = \mathcal{A}_{wait} \cup \mathcal{A}_{drive} \cup \mathcal{A}^{fix}$$

and additional *timetable activities* $\{(s,i) : i \in \mathcal{E}\}$ as set of activities in the corresponding project network. The duration of an activity is set to L_a for $a \in \mathcal{A}$ and to the scheduled timetable Π_i if $a = (s,i)$. Then the earliest possible starting time of each activity is a time-minimal solution of the delay management problem. The following procedure uses the critical path method to determine the earliest starting times but is applied directly in the notation of slack times s and delays y according to Definition 1.

Algorithm 1: Calculating a time-minimal solution for a set $\mathcal{A}^{fix}$

```
Input: N, d_i, s_a, A^fix.
Output: Optimal (time-minimal) solution w.r.t. A^fix.
Step 1. Sort E = {i_1,...,i_{|E|}} according to ≺.
Step 2. For k = 1,...,|E|: y_{i_k} = max{d_{i_k}, max_{a=(i,i_k)∈A(A^fix)} y_i - s_a}
Step 3. Output: y_i, i ∈ E
```

By induction it is easy to show that the time-minimal solution $y(\mathcal{A}^{fix})$ with respect to each set $\mathcal{A}^{fix} \subseteq \mathcal{A}_{change}$ is unique, and that it has the following two properties:

1. $\mathcal{A}^1 \subseteq \mathcal{A}^2 \subseteq \mathcal{A}_{change} \implies y(\mathcal{A}^1) \leq y(\mathcal{A}^2)$. i.e. the delays get smaller if connections are dropped, and
2. $y = y(\mathcal{A}^{fix})$ satisfies $y_i \leq D = \max\{d_i : i \in \mathcal{E}\}$ for all $i \in \mathcal{E}$, i.e. the maximal delay of a single event in a time-minimal solution is bounded by the largest given source delay.

Other approaches for calculating time minimal solution sand the details of the proofs can be found in [20].

As mentioned before, our objective is to minimize the sum of all delays over all customers. To this end, we first specify the customers' data.

A customer's paths is given as a sequence of events, i.e.,

$$p = (i_1, i_2, \ldots, i_{p_L})$$

where $i_k \in \mathcal{E}$ are events, and $(i_k, i_{k+1}) \in \mathcal{A}$ are activities. We will write $a = (i_k, i_{k+1}) \in p$ in this case. Note that i_1 is a departure event, i_2 an arrival event, $i_3 \in \mathcal{E}_{dep}$ and so on. Furthermore, $i(p)$ denotes the last event on path p and w_p the number of passengers who want to use path p. We denote $\mathcal{P}$ as the set of all customers' paths.

To calculate the delay of a passenger on path p we need the following two basic assumptions:

1. There is one (common) time period T for all vehicles.
2. In the next time period all vehicles are on time.

In praxis, both assumptions are usually not satisfied. The first of them can be relaxed a bit, allowing different periods for each of the activities. Taking the largest of the periods of all lines overestimates the delay, but seems to be a reasonable approach. The second assumption is accepted by practitioners since the planning period in on-line disposition is usually less than the time period T (often one hour). It is an open problem to deal with future delays by using stochastic optimization.

To calculate the delay of a customer using some path $p \in \mathcal{P}$, we have to distinguish the following two cases.

Case 1: If all connections of path p are maintained (i.e., the path is *maintained*), the delay of a passenger on path p is the arrival delay $y_{i(p)}$ of his last event $i(p)$.

Case 2: If at least one connection of path p is missed, the delay of a passenger on path p is given by T.

We are finally in the position to define the **(total) delay management problem**.

(TDM): *Given $\mathcal{N} = (\mathcal{E}, \mathcal{A})$, slack times s_a for all $a \in \mathcal{A}$, source delays $d_i, i \in \mathcal{E}$ and a set of weighted paths $\mathcal{P}$, find a feasible pair $\mathcal{A}^{fix} \subseteq \mathcal{A}_{change}$ with delays y_i, $i \in \mathcal{E}$ such that the sum of all delays over all customers is minimal.*

3   Models for Delay Management

As first model we present a **path-oriented** description of (TDM) (based on the formulation in [19]) which uses the following variables

$$z_p = \begin{cases} 0 & \text{if all connections on path } p \text{ are maintained} \\ 1 & \text{otherwise} \end{cases}$$

(TDM-A)

$$\min f_{\text{TDM-A}} = \sum_{p \in \mathcal{P}} w_p (y_{i(p)}(1 - z_p) + T z_p)$$

such that

$$y_i \geq d_i \quad \text{for all } i \in \mathcal{SD} \tag{4}$$
$$y_i - y_j \leq s_a \quad \text{for all } a = (i, j) \in \mathcal{A}_{wait} \cup \mathcal{A}_{drive} \tag{5}$$
$$-M z_p + y_i - y_j \leq s_a \quad \text{for all } p \in \mathcal{P}, a = (i, j) \in p \cap \mathcal{A}_{change} \tag{6}$$
$$y_i \in \mathbb{N} \quad \text{for all } i \in \mathcal{E} \tag{7}$$
$$z_p \in \{0, 1\} \quad \text{for all } p \in \mathcal{P}, \tag{8}$$

where $M \geq D = \max\{d_i : i \in \mathcal{E}\}$.

The first two constraints (4) and (5) are the same as (1) and (2). Constraint (6) makes sure that all connections on a maintained path (i.e. a path with $z_a = 0$) satisfy (3). Finally, the objective function sums up the delay according to the two cases mentioned on page 151.

As already shown in [19], this formulation of model (TDM-A) can be linearized by substituting the quadratic terms $y_{i(p)}(1 - z_p)$ by additional variables q_p, leading to the following model.

(TDM-B)

$$\min f_{\text{TDM-B}} = \sum_{p \in \mathcal{P}} w_p (q_p + T z_p)$$

such that (4) – (8) hold, and such that

$$- M z_p + y_{i(p)} - q_p \leq 0 \quad \text{for all } p \in \mathcal{P} \tag{9}$$
$$q_p \geq 0 \quad \text{for all } p \in \mathcal{P} \tag{10}$$

The linear formulation is significantly weaker than the quadratic formulation (TDM-A), due to the fact that the feasible set of the linear programming relaxation increased. More intuitively, one would like to use variables $\bar{z}_a$ determining if a connection $a \in \mathcal{A}_{change}$ should be maintained or not. This yields a stronger activity-based formulation for (TDM) which is derived next.

In the **activity-based** model we use variables for each changing activity $\bar{z}_a$ describing if connection $a \in \mathcal{A}_{change}$ is missed ($\bar{z}_a = 1$) or maintained ($\bar{z}_a = 0$). The idea of the activity-based formulation is to calculate the total delay by summing up the *additional delays* over all activities $a \in \mathcal{A}$. To this end, let us first consider some activity $a \in \mathcal{A} \setminus \mathcal{A}_{change}$. We want to calculate the additional delay customers will get while using this activity. The delay customers already have at the start of $a = (i, j)$ is y_i, and at the end of a their delay is y_j. Hence, $y_j - y_i$ is the additional delay gained by the customers while performing activity a. Note that this additional delay can be negative, meaning that slack times are used to compensate an already existing delay. For changing activities we have to be more careful. Let $a = (i, j) \in \mathcal{A}_{change}$ and suppose first that a is maintained. Then the additional delay on a is again the tension $y_j - y_i$. On the other hand, if a is missed, the additional delay for the customers who planned to use activity a is given by $T - y_i = y_j - y_i + T - y_j$, since they now have to wait the remaining time period until the next (non-delayed) vehicle arrives for carrying on their journey.

We further need to extend the event-activity network by defining

$$\mathcal{E}^s = \mathcal{E} \cup \{s\}$$
$$\mathcal{A}^s = \mathcal{A} \cup \{(s, i) : i \in \mathcal{E}_{dep}\} \text{ and}$$
$$\mathcal{P}^s = \{(s, i_1^p, \ldots, i_L^p) : p \in \mathcal{P}\}.$$

The additional event s represents the arrival of the customers at their first station (by foot or by a means of transport which is not considered in the delay management problem). The extension makes sure that the delay of a customer waiting at some station for his first (delayed) vehicle to come, is taken into account. We always assume that customers reach their first station without any delay, i.e., $y_s = 0$.

Now we can present the new model. As before, we assume that $T, M \geq D$. The following additional variables are necessary for (TDM-C).

$$\tilde{z}_a^p = \begin{cases} 1 & \text{if activity } a \text{ is reached on path } p \text{ without any missed} \\ & \text{connection before} \\ 0 & \text{otherwise} \end{cases}$$

$w_a = $ number of customers who *really* use activity a

We stress that the number of customers w_a (really) using activity $a \in \mathcal{A}$ is a variable, since it depends on the wait-depart decisions whether customers using a path $p \in \mathcal{P}^s$ will reach all activities $a \in p$ or not.

(TDM-C)

$$\min f_{\text{TDM-C}} = \sum_{a=(i,j)\in\mathcal{A}^s} w_a(y_j - y_i) + \sum_{a=(i,j)\in\mathcal{A}_{change}} w_a \tilde{z}_a(T - y_j)$$

such that

$$y_i \geq d_i \quad \text{for all } i \in \mathcal{SD} \tag{11}$$

$$y_i - y_j \leq s_a \quad \text{for all } a = (i,j) \in \mathcal{A}_{wait} \cup \mathcal{A}_{drive} \tag{12}$$

$$-M\tilde{z}_a + y_i - y_j \leq s_a \quad \text{for all } a = (i,j) \in \mathcal{A}_{change} \tag{13}$$

$$\tilde{z}_a^p + \sum_{\substack{\tilde{a}\in p\cap\mathcal{A}_{change}: \\ \tilde{a}\prec a}} \tilde{z}_{\tilde{a}} \geq 1 \quad \text{for all } p \in \mathcal{P}^s \text{ and } a \in p \tag{14}$$

$$\tilde{z}_a^p + \tilde{z}_{\tilde{a}} \leq 1 \quad \text{for all } p \in \mathcal{P}^s \text{ and for all } a, \tilde{a} \in p$$
$$\text{with } \tilde{a} \in \mathcal{A}_{change} \text{ and } \tilde{a} \prec a \tag{15}$$

$$w_a = \sum_{p\in\mathcal{P}^s:a\in p} w_p \tilde{z}_a^p \quad \text{for all } a \in \mathcal{A}^s \tag{16}$$

$$y_i \in \mathbb{N} \quad \text{for all } i \in \mathcal{E} \tag{17}$$

$$\tilde{z}_a \in \{0,1\} \quad \text{for all } a \in \mathcal{A}^s \tag{18}$$

$$\tilde{z}_a^p \in \{0,1\} \quad \text{for all } p \in \mathcal{P}^s, a \in \mathcal{A}^s \tag{19}$$

$$w_a \in \mathbb{N} \quad \text{for all } a \in \mathcal{A}^s \tag{20}$$

In the objective function the additional amount of delay on each activity is multiplied by the number of customers *really* using it. Restrictions (11) and (12) again correspond to (1) and (2), while (13) models that (3) has to be satisfied exactly for maintained connections, i.e. connections a with $\tilde{z}_a = 0$. Restriction (14) defines the values of $\tilde{z}_a^p$ such that they are forced to be 1, if no connection on path p before a has been missed, and (15) makes sure that $\tilde{z}_a^p = 0$ for all activities a after a missed connection $\tilde{a}$ on path p. Finally, (16) determines the number of customers really using activity a.

Note that for technical reasons we need to be able to extend any feasible solution $y_i, i \in \mathcal{E}$ to a feasible solution $(y, C(y)) := (y, \tilde{z}(y), \tilde{z}(\tilde{z}), w(\tilde{z}))$ of

(TDM-C), where $(y, C(y))$ yields the same or a better objective function value for (TDM-C). This is done as follows.

$$\bar{z}_a(y) = \begin{cases} 0 & \text{if } y_i - y_j \leq s_a \\ 1 & \text{otherwise} \end{cases} \quad \text{for all } a = (i,j) \in \mathcal{A}_{change}, \tag{21}$$

$$\tilde{z}_a^p(\bar{z}) = \max \left\{ 1 - \sum_{\substack{a \in p \cap \mathcal{A}_{change}: \\ \tilde{a} \prec a}} \bar{z}_{\tilde{a}}, 0 \right\} \quad \text{for all } p \in \mathcal{P}^s, a \in p, \tag{22}$$

$$w_a(\bar{z}) = \sum_{p \in \mathcal{P}^s : a \in p} w_p \tilde{z}_a^p \quad \text{for all } a \in \mathcal{A}^s. \tag{23}$$

The main result of this section is the following.

Theorem 1. *(TDM-A) and (TDM-C) are equivalent. In particular, both models lead to the same set of optimal solutions $y \in \mathbb{R}^{|\mathcal{E}|}$.*

The proof can be found in the Appendix.

Using (TDM-C) we are able to derive the following reduction result. Assume that the slack times are so large that the delay disappears after a few activities. Then we need not consider events which can not gain any delay in the worst-case time-minimal solution.

Lemma 1. *Let $y = y(\mathcal{A}^{change})$ be a time-minimal solution w.r.t. $\mathcal{A}^{change}$. Then there exists an optimal solution $(y^*, \bar{z}^*, \tilde{z}^*, w^*)$ of (TDM-C) such that*

- *For all $i \in \mathcal{E}$: If $y_i = 0$ then $y_i^* = 0$.*
- *For all $a = (i,j) \in \mathcal{A}_{change}$: If $y_i = 0$ then $\bar{z}_a^* = 0$.*

The result shows that we need not consider events or activities which cannot gain a delay in the worst case. The *reduced set of events* is hence given as

$$\mathcal{E}_{relevant} = \{i \in \mathcal{E} : y_i(\mathcal{A}_{change}) > 0\}$$

and the subgraph induced by these events $\mathcal{E}_{relevant}$ is denoted by $\mathcal{N}_{relevant} = (\mathcal{E}_{relevant}, \mathcal{A}_{relevant})$. This kind of reduction leads to significantly smaller networks in real-world instances, see Table 1 in Section 5.

On a first glance, (TDM-C) does not seem to be useful for solving the delay management problem better than (TDM-B), since (TDM-B) is linear while (TDM-C) is cubic. Moreover, (TDM-C) is much larger in terms of variables, constraints, and non-zero entries of the coefficient matrix. However, it has some advantages. First, it is more general since it allows to replace the common time period T by time periods T_a for each changing activity $a \in \mathcal{A}_{change}$, which is a step to more realistic models and to relaxing our first assumption on page 150. Secondly, as the proof in the appendix shows, (TDM-C) is a stronger formulation than (TDM-A) and (TDM-B), since the decision variables $\bar{z}_a$ allow less freedom than the decision variables z_p. Hence, e.g., a classical branch-and-bound procedure using the variables $\bar{z}_a$ for branching can be easily implemented for (TDM-C) while for (TDM-A) other methods, e.g., constraint branching have to be investigated. Last, in the next section we utilize (TDM-C) to present a linear-time algorithm which solves the delay management problem exactly for a special class of problems.

4   Constant Weights and the Never-Meet Property

In order to solve (TDM-C) we fix the weights w_a as parameters instead of calculating them during the optimization. Doing so, we obtain the *total delay management problem with constant weights*. Its formulation is given by deleting constraints (14), (15), and (16) in (TDM-C), and fixing

$$w_a = \sum_{p \in \mathcal{P}^s : a \in p} w_p \quad \text{for all } a \in \mathcal{A}^s \tag{24}$$

as parameters, i.e., setting w_a as the planned "traffic load" on activity a. We obtain:

$$\min f_{\text{TDM-const}'} = \sum_{a=(i,j) \in \mathcal{A}^s} w_a (y_j - y_i) + \sum_{a=(i,j) \in \mathcal{A}_{change}} w_a \bar{z}_a (T - y_j)$$

such that (11),(12),(13),(17), and (18) hold.

We can further rewrite $f_{\text{TDM-const}'}$ as follows. For $i \in \mathcal{E}$ let

$$w_i = \sum_{p \in \mathcal{P} : i(p)=i} w_p \tag{25}$$

be the number of customers with final destination i. Since

$$\sum_{a=(i,j) \in \mathcal{A}^s} w_a (y_j - y_i) = \sum_{p \in \mathcal{P}^s} w_p \sum_{a=(i,j) \in p} y_j - y_i$$

$$= \sum_{p \in \mathcal{P}} w_p (y_{i(p)} - y_s)$$

$$= \sum_{i \in \mathcal{E}} \sum_{\substack{p \in \mathcal{P}: \\ i(p)=i}} w_p y_i = \sum_{i \in \mathcal{E}} w_i y_i \tag{26}$$

we rewrite

$$f_{\text{TDM-const}'} = \sum_{i \in \mathcal{E}} w_i y_i + \sum_{a=(i,j) \in \mathcal{A}_{change}} w_a \bar{z}_a (T - y_j).$$

First, we show that in general, we make a mistake by fixing the weights as above, which has not been realized in several previous attempts or simulation approaches for the delay management problem.

We assume there are three vehicles 1, 2, and 3, where vehicle 1 and vehicle 3 reach the stations v_2 and v_3 with a delay, see Figure 3. We consider a customers'

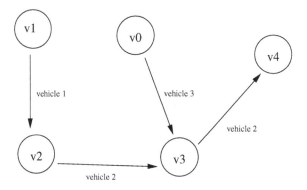

Fig. 3. An example in which fixing the weights is not correct

path $p = (v_1, v_2, v_3, v_4)$ using vehicle 1 until station v_2, changing to vehicle 2 and passing via v_3 to its destination v_4. Suppose that vehicle 2 is not waiting for vehicle 1 at station v_2, such that the path p is not maintained. Assume further that vehicle 2 waits for the delayed vehicle 3 at station v_3. If we have not adapted the weights, the customers on path p are counted twice in the objective function: First, since they missed their connection at station v_2, and secondly, since they reach their final destination v_4 with a delay. This double counting can in general lead to wrong decisions. Another example, depicted in Figure 5 will be further analyzed in Section 5.

Fortunately, there are problem instances for which the model with constant weights is correct, apart from the trivial case in which no customer changes at all. For example, it can be shown that the model with constant weights is correct, if we only allow paths of the form $p = (i_1, i_2, \ldots, i_{L-2}, i_{L-1}, i_L)$ where p contains at most one changing activity (i_{L-2}, i_{L-1}) followed by not more than one driving activity, see [20]. A more interesting case, in which we make no mistake by using the constant weights will be described next.

Since $f_{\text{TDM−const}'}$ still is no linear function we further simplify the model. In the following we simply forget about subtracting y_j in the second part of the objective, to obtain the **linear** program (**TDM-const**).

$$\min f_{\text{TDM−const}} = \sum_{i \in \mathcal{E}} w_i y_i + \sum_{a \in \mathcal{A}_{change}} w_a T \bar{z}_a$$

such that

$$y_i \geq d_i \quad \text{for all } i \in \mathcal{SD} \tag{27}$$

$$y_i - y_j \leq s_a \quad \text{for all } a = (i, j) \in \mathcal{A}_{wait} \cup \mathcal{A}_{drive} \tag{28}$$

$$-M\bar{z}_a + y_i - y_j \leq s_a \quad \text{for all } a = (i, j) \in \mathcal{A}_{change} \tag{29}$$

$$y_i \in \mathbb{N} \quad \forall i \in \mathcal{E}$$

$$\bar{z}_a \in \{0, 1\} \quad \text{for all } a \in \mathcal{A}_{change}$$

Each feasible solution of (TDM-const) yields an upper bound on (TDM). But the main advantage of (TDM-const) is due to the surprising fact that (TDM-const) is equivalent to (TDM) in a large class of practical instances. We denote

$$\mathcal{E}(i) = \{j \in \mathcal{E} : \text{ there exists a (directed) path from } i \text{ to } j\}$$

as the set of all events that can be reached from i, and $\mathcal{N}(i)$ as the subgraph induced by the events in $\mathcal{E}(i)$. Note that for all $j \in \mathcal{E}(i)$ we have $i \preceq j$. Furthermore, let $\mathcal{E}(\mathcal{SD}) = \bigcup_{i \in \mathcal{SD}} \mathcal{E}(i)$ and $\mathcal{N}(\mathcal{SD})$ be the subgraph containing the subgraphs $\mathcal{N}(i)$ for all $i \in \mathcal{SD}$, i.e. the graph consisting of all events and activities that can be reached by a path starting at a source-delayed event. Obviously, $\mathcal{E}_{relevant} \subseteq \mathcal{E}(\mathcal{SD})$.

Definition 4. *The delay management problem has the* **never-meet property** *if the following two conditions hold.*

1. $\mathcal{N}(i) \cap \mathcal{N}_{relevant}$ *is a forest for all* $i \in \mathcal{SD}$, *and*
2. $\mathcal{E}(i) \cap \mathcal{E}(j) \cap \mathcal{E}_{relevant} = \emptyset$ *for all* $i, j \in \mathcal{SD}$ *with* $i \neq j$.

Note that $\mathcal{N}(\mathcal{SD}) \cap \mathcal{N}_{relevant}$ is a forest, whenever the never-meet property holds, i.e. it is not allowed to contain cycles (neither directed nor undirected cycles).

The interpretation of the never-meet property is the following: By calculating the time-minimal solution (w.r.t. $\mathcal{A}^{fix} = \mathcal{A}_{change}$), but without using slack-times, we can find out how far the effects of the source delays can spread out in the worst case. The never-meet property requires that in **no** feasible solution of (TDM) the paths of two delayed customers will meet. Note that the formulation includes that source delays can only occur after non-delayed events.

If the never-meet property holds, however, we will show the following: In every time-minimal solution all events following a non-maintained connection are punctual, and all changing activities following a non-maintained connection are maintained. This property will be important for proving Theorem 2.

Lemma 2. *Let (TDM) have the never-meet property and let* $(y, C(y))$ *be a feasible (time-minimal) solution of (TDM-C). Let* $\tilde{a} = (\tilde{i}, \tilde{j}) \in \mathcal{A}_{change}$. *If* $\tilde{z}_{\tilde{a}} = 1$ *(i.e. $\tilde{a}$ is not maintained) we have the following.*

1. $y_i = 0$ *for all* $i \in \mathcal{E}(\tilde{j})$, *i.e. all events following* $\tilde{j}$ *are on time, and*
2. $\tilde{z}_a = 0$ *for all* $a = (i, j)$ *with* $i \in \mathcal{E}(\tilde{j})$, *i.e., all connections following* $\tilde{j}$ *are maintained.*

Proof. From $\tilde{z}_{\tilde{a}} = 1$ we know from (21) that $y_{\tilde{i}} > 0$. Hence there exists a source-delayed event $i_1 \in \mathcal{SD}$ such that $\tilde{i} \in \mathcal{E}(i_1)$. Now suppose there exists $i \in \mathcal{E}(\tilde{j}) \subseteq \mathcal{E}(i_1)$ with $y_i > 0$. Since $\tilde{z}_{\tilde{a}} = 1$ the delay of i is not transferred from i_1 to i via $\tilde{a}$. Hence

- either there is another path from i_1 to i, meaning that $\mathcal{N}(i_1) \cap \mathcal{N}_{relevant}$ is not a tree, or
- the delay of y_i is caused by another source-delayed event $i_2 \in \mathcal{E}$, meaning that $i \in \mathcal{E}(i_1) \cap \mathcal{E}(i_2) \cap \mathcal{E}_{relevant}$.

In both cases we have a contradiction to the never-meet property. Finally, consider $a = (i,j)$ with $i,j \in \mathcal{E}(\tilde{j})$. From part 1 we know that $y_i = y_j = 0$, hence (21) yields $\bar{z}_a = 0$.   □

We can now present our main result.

Theorem 2. *Model (TDM-const) is correct if the never-meet property holds.*

Proof. We show that (TDM-C) and (TDM-const) are equivalent if the never-meet property holds. Clearly, a feasible solution $(y, \bar{z})$ of (TDM-const) can be extended to a feasible solution $(y, C(y))$ of (TDM-C) with equal or better objective value, see (21), (22), and (23).

The other direction is the interesting one: We show that each feasible solution of (TDM-C) corresponds to a feasible solution of (TDM-const) with the same or better objective value. More precisely, given some feasible solution of (TDM-C) with delay y, let $(y, C(y)) = (y, \bar{z}, \tilde{z}, w^{real})$ denote a (maybe better) feasible solution of (TDM-C). We show that $(y, \bar{z})$ is a feasible solution of (TDM-const) with the same objective value as $(y, C(y))$. Feasibility of $y, \bar{z}$ for (TDM-const) is trivially satisfied. It remains to show that

$$f_{\text{TDM-C}}(y, \bar{z}, \tilde{z}, w) = f_{\text{TDM-const}}(y, \bar{z}).$$

To this end, suppose that for some $\bar{a} = (\bar{i}, \bar{j}) \in \mathcal{A}$ we made a mistake by fixing the weights, i.e., the number of customers $w_{\bar{a}}$ who planned to use $\bar{a}$ does not equal the number of customers $w_{\bar{a}}^{real}$, really using $\bar{a}$. To compare the objective functions of (TDM-const) and (TDM-C) we replace the first term of (TDM-const) by equation (26) and see that in this case it suffices to show that

$$y_{\bar{j}} - y_{\bar{i}} = 0,$$

and that, if $\bar{a} \in \mathcal{A}_{change}$

$$z_{\bar{a}} = 0,$$

This means that the error we make by using the wrong weights does not influence the value of the objective function. From $w_{\bar{a}} \neq w_{\bar{a}}^{real}$ we get (by comparing (23) and (24)), that

$$\sum_{p \in \mathcal{P}^s : \bar{a} \in p} w_p = w_{\bar{a}} \neq w_{\bar{a}}^{real} = \sum_{p \in \mathcal{P}^s : \bar{a} \in p} w_p \tilde{z}_{\bar{a}}^p.$$

Hence there exists some path $p \in \mathcal{P}$ containing $\bar{a}$ such that $\tilde{z}_{\bar{a}}^p = 0$. Due to (22) there exists $\tilde{a} \in p$ with $\tilde{a} \prec \bar{a}$ and $\bar{z}_{\tilde{a}} = 1$. Without loss of generality let us take $\tilde{a} = (\tilde{i}, \tilde{j})$ minimal with this property, i.e., we choose the first changing activity on path p that is marked as missed. For an illustration, see Figure 4.

Since $\bar{i}, \bar{j} \in \mathcal{E}(\tilde{j})$ we derive from Lemma 2 that

- $y_{\bar{i}} = y_{\bar{j}} = 0$, and
- if $\bar{a} \in \mathcal{A}_{change}$ then $\bar{z}_a = 0$.

Hence, $y_{\bar{j}} - y_{\bar{i}} = 0$, and if $\bar{a} \in \mathcal{A}_{change}$ we have that $z_{\bar{a}} = 0$, which completes the proof.   □

Fig. 4. The path p in the proof of Theorem 2. The grey events belong to $\mathcal{E}(\tilde{j})$.

5   The Never-Meet Property in Practice

To investigate the never-meet property in practice, we used real-world data of a part of the German railway network, namely around the region of Harz in Germany. The data we used consists of 158 railway stations, and 1101 different trains running at one particular day.

Since the never-meet property does not depend on the set of paths $\mathcal{P}$, we used two different sets $\mathcal{U}_{30}$, and $\mathcal{U}_{60}$ to generate potential connections. The set $\mathcal{U}_{30}$ contains reasonable connections within a scheduled waiting time between 3 and 30 minutes, while we allow a waiting time between 3 and 60 minutes in $\mathcal{U}_{60}$. By "reasonable" we mean that we do not consider connections where a transfer results in going directly back to the previous station. The size of U_{30} is 5567, while U_{60} contains 11229 connections. The resulting event-activity network of the public transportation network on our particular day has a size of 10492 events. The number of activities on this day depends on the allowed transfer time and varies between 13359 and 17616. In our numerical study we generated 500 example sets of delays with up to 5 source delays.

As shown in Table 1 the event-activity network can be drastically reduced if we delete all events that can never gain a delay. The table demonstrates the results of the reduction for different numbers of source delays. Each row contains the average number of events for 100 different delay scenarios. In column $|\mathcal{E}(\mathcal{SD})|$ the number of events that can be reached by a path from one of the source delays is given, while column $|\mathcal{E}(\mathcal{SD}) \cap \mathcal{E}_{relevant}|$ shows how many of these events can gain a delay and hence have to be considered in the optimization. The percentage of reduction is given in the last column.

The never-meet property cannot be sharpened by further reducing the sets $\mathcal{E}(i) \cap \mathcal{E}_{relevant}$ and only looking at the smaller sets $\mathcal{E}'(i) \subseteq \mathcal{E}(i) \cap \mathcal{E}_{relevant}$

Table 1. Reduction of the original set of 10492 events for different delay scenarios in the case of a transfer time up to 30 minutes

no. of source delays	$|\mathcal{E}(\mathcal{SD})|$	$|\mathcal{E}(\mathcal{SD}) \cap \mathcal{E}_{relevant}|$	reduction to
1	2668	228	8.5%
2	4172	460	11.0%
3	5029	599	11.9%
4	5344	801	15.0%
5	5495	948	17.3%

containing only those events that can gain a delay originating at $i \in SD$. This is demonstrated in the following example with two source delays at events v_1 and v_2. Source delay 1 spreads out to the dashed nodes $\mathcal{E}'(v_1)$, while the grey nodes (denoted by $\mathcal{E}'(v_2)$) show which events can gain a delay from source delay 2. Suppose that – due to sufficiently large slack times – no other events can be affected. Although the dashed and the grey nodes form trees, and their intersection is empty, the example does not have the never-meet property (since $v_4 \in \mathcal{E}(v_1) \cap \mathcal{E}(v_2) \cap \mathcal{E}_{relevant}$)! But this is what we want, since Theorem 2 does also not hold in this example: Consider path P from v_3 to v_4. Assume that a is a changing activity and it is missed. Then customers traveling along P never reach node v_4 and would be (wrongly) counted there.

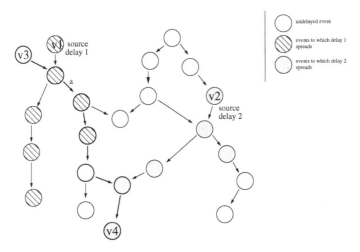

Fig. 5. Although source delay 1 disappeared before reaching an event which has a delay coming from source delay 2, this example does not have the never-meet property. In particular, Theorem 2 is not true in this example.

We tested the never-meet property in practice, which can be done efficiently by the forward phase of the critical path method (with zero slack times and $\mathcal{A}^{fix} = \mathcal{A}_{change}$). To analyze the results, let us call an event i *in conflict with the never-meet property*, if it can be reached by more than one path originating in a source-delayed event. The number of all events which are in conflict with the never-meet property is called the number of *node conflicts* of the problem. The events which are in conflict with the never-meet property can be determined by looking at their in-degrees within the graph $\mathcal{N}(SD)$. More precisely,

- an event $i \in \mathcal{E}_{relevant} \setminus SD$ is in conflict with the never-meet property, if its in-degree in the graph $\mathcal{N}(SD)$ is at least 2. The in-degree minus 1 is called its *degree of conflicts*.
- Event $i \in SD$ is in conflict with the never-meet property, if its in-degree in the graph $\mathcal{N}(SD)$ is at least 1. In this case, its in-degree equals its *degree of conflicts*.

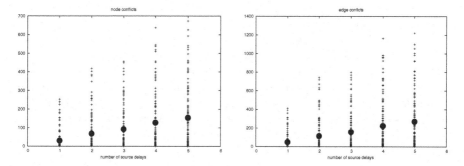

Fig. 6. The number of node conflicts (left) and edge conflicts (right) as a function of the number of source delays

The sum of all degrees of conflict will be called the number of *edge conflicts* with the never-meet property. It equals the number of edges which have to be deleted to ensure that the never-meet property holds.

In Figure 6 the number of conflicts with the never-meet property (node and edge conflicts) is depicted as a function of the number of delayed vehicles. Note that we considered scenarios with 1,2,3,4, and 5 source delays and generated 100 examples for each of these scenarios, with different amounts of source delay. Each example is given by a "+" in the figure. The average values for each amount of source delay are given by circles. As expected, the average number of conflicts with the never-meet property is relatively small for only one source delay, while it increases when more than one source delay is considered. Furthermore, the variance increases with the number of source delays: There are still many examples with 5 source delays which lead to very few conflicts, but there are also examples with many conflicts.

Figure 7 shows the same data, but here we graph the number of edge and node conflicts with the never-meet property as a function of the *amount* of the source delays. We observe that the number of conflicts increases if the source

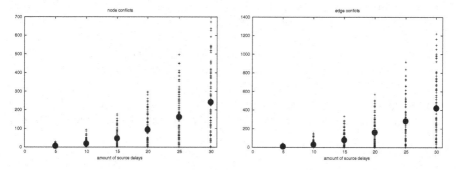

Fig. 7. The number of node conflicts (left) and edge conflicts (right) as a function of the amount of delay

delays increase, and that small source delays are very likely to generate nearly no conflicts with the never-meet property.

The reason for the relatively small number of conflicts in practice is in particular due to the fact that we only consider events in $\mathcal{E}_{relevant}$, i.e. only events that can gain a delay. As expected, most conflicts with the never-meet property arise at the larger stations, while the never-meet property is more likely to hold for smaller stations in a rural environment. But all this is only helpful if we can draw advantage of the simplified model with constant weights in terms of efficiently solving it. This will be investigated in the next section.

6   Solving (TDM-const)

The main goal of this section is to solve (TDM) in case of the never-meet property. Using Theorem 2 it is enough in this case to develop an algorithm for (TDM-const).

The first approach is to just use an integer programming solver. We solved our example problems using GLPK (GNU linear programming kit). The results for the examples with one source delay are illustrated in Figure 8. As before, each "+" refers to one example. The first coordinate shows the the number of conflicts with the never-meet property (node and edge conflicts, respectively), while the second coordinate represents the number of pivot operations needed for solving the corresponding program (TDM-const) to optimality. We observe that the number of pivot operations for solving (TDM-const) increases with the number of conflicts with the never-meet property, but not as badly as one could have expected.

To understand the reason for this behavior we first look at the following special case of (TDM) with the never-meet property, in which

- all source delays have the same amount, i.e., $d_i \in \{0, D\}$ for all $i \in \mathcal{E}$, and
- all slack times are equal to zero, i.e., $s_a = 0$ for all $a \in \mathcal{A}$.

Let y be a time-minimal solution of this problem. Then $y_i \in \{0, D\}$ for all $i \in \mathcal{E}$. This means that we can use binary variables y_i instead of integer ones, with

$$y_i = \begin{cases} 1 & \text{if event } i \text{ is delayed by } D \\ 0 & \text{if event } i \text{ is not delayed.} \end{cases}$$

Consequently, $M = 1$ is large enough and (TDM-const), even with the first objective $f_{\text{TDM-const}'}$ introduced on page 155, simplifies to the following **linear** program.

(TDM-const-zero)

$$\min \sum_{a=(i,j)\in\mathcal{A}^s} w_a D(y_j - y_i) + \sum_{a=(i,j)\in\mathcal{A}_{change}} w_a \bar{z}_a (T - D)$$

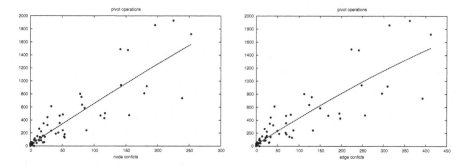

Fig. 8. The number of pivot operations needed for solving (TDM-const) versus the number of node conflicts (left) and edge conflicts (right)

such that

$$- y_i \leq -1 \quad \text{for all } i \in \mathcal{SD} \tag{30}$$
$$y_i - y_j \leq 0 \quad \text{for all } a = (i, j) \in \mathcal{A}_{wait} \cup \mathcal{A}_{drive} \tag{31}$$
$$\bar{z}_a + y_i - y_j \leq 0 \quad \text{for all } a = (i, j) \in \mathcal{A}_{change} \tag{32}$$
$$y_i \in \{0, 1\} \quad \forall i \in \mathcal{E}$$
$$\bar{z}_a \in \{0, 1\} \quad \forall a \in \mathcal{A}_{change},$$

where $w_a = \sum_{p \in \mathcal{P}^s : a \in p} w_p$ for all $a \in \mathcal{A}^s$ are given parameters as before (see, e.g., (24)). The following result explains the good behavior of mixed integer programming for (TDM-const).

Theorem 3. *The coefficient matrix of (TDM-const-zero) is totally unimodular.*

Proof. Let $C = |\mathcal{A}_{change}|$, $\bar{C} = |\mathcal{A}_{drive} \cup \mathcal{A}_{wait}|$ and $\bar{D} = |\mathcal{SD}|$. Moreover, let I_K denote the unit matrix of size $K \times K$ and $O_{K,L}$ the zero matrix of size $K \times L$. Then the coefficient matrix of (TDM-const-zero) is

$$\Phi = \left(\begin{array}{c|c} -I_{\bar{D}} & 0_{\bar{D},C} \\ \hline \Theta^T & \begin{array}{c} 0_{\bar{C},C} \\ \hline I_C \end{array} \end{array} \right),$$

where the $|\mathcal{A}| \times |\mathcal{E}|$-matrix Θ^T is the transposed of the node-arc-incidence matrix Θ of $\mathcal{N}$, and hence totally unimodular. Consequently, Φ is also totally unimodular. □

We remark that (TDM-const-zero) is equivalent to the models developed independently in diploma theses by Kliewer [14] and Scholl [21], where the latter author also recognized the total unimodularity of the model.

The lemma gives the explanation for the graphics of Figure 8: In case of zero slack times the LP-relaxation of (TDM-const-zero) yields an integer solution (see e.g., [16]), which makes the problem efficiently solvable in this case. Since the structure of the problem does not change by introducing slack times, one can hope for an efficient algorithm also in the general case. We therefore turn our attention back to the case of non-zero slack times and will develop an efficient algorithm (running in $O(|\mathcal{A}|)$ time) for solving (TDM-const). In contrast to the case with zero slack times, this approach relies on the never-meet property. In particular we will utilize the two facts listed below.

- First, if we fix $\bar{z}_a = 1$ for some $a = (i, j)$, we can set $y_{i'} = 0$ for all $i' \in \mathcal{E}(j)$ and know that all subsequent connections are maintained (Lemma 2).
- Secondly, the problem can be decomposed into at most $|\mathcal{A}_{change}|$ independent subproblems due to the following lemma, which also follows directly from the never-meet property.

Lemma 3. *Let $i, j \in \mathcal{E}$, $i \neq j$, and let $(y, \bar{z})$ be a feasible solution of (TDM-const) with $y_i > 0, y_j > 0$. If the never-meet property holds, exactly one of the following three cases occurs.*

$$\mathcal{E}(i) \subseteq \mathcal{E}(j) \quad or \quad \mathcal{E}(j) \subseteq \mathcal{E}(i) \quad or \quad \mathcal{E}(i) \cap \mathcal{E}(j) \cap \mathcal{E}_{relevant} = \emptyset.$$

The idea of the algorithm is to decompose the problem iteratively into subproblems, and solve them bottom-up. A subproblem P_a is identified by a changing activity $a = (i, j)$ and represents the delay management problem on the subgraph $\mathcal{N}(i)$ (recall the notation on page 157) with a single source delay at event i. P_a might be decomposable into subproblems itself. Formally, we define

$$\mathrm{SP}(a) = \{a' \in \mathcal{A}_{change} : \text{ there exists a directed path from } a \text{ to } a' \text{ not containing}$$
$$\text{any other changing activity}\}$$

The subproblems of the problem itself are collected in $\mathrm{SP}(a_0)$, and can be derived by taking all changing activities reachable directly from one of the source-delayed events.

We remark that all subproblems within the same set $\mathrm{SP}(a)$ are independent of each other due to the never-meet property.

In Algorithm 2, subproblems that might further be decomposed are stored in "Decompose", and if a subproblem cannot be decomposed any more it is collected in "Compose". Moreover, at the end of Step 2 of the algorithm, for each subproblem identified by some changing activity a,

- maintain(a) contains the value of the objective function of the subproblem if a is maintained, and
- miss(a) contains the objective value if a is missed.
- $f(a)$ contains the minimum of maintain(a) and miss(a).

To compute maintain(a) we need to calculate the minimum delay which occurs if a is maintained. In contrast to $\mathcal{E}(i)$ which is the set of events that can be reached from i, if *all* $a \in \mathcal{A}_{change}$ can be used we now define

$$\mathcal{G}(i) = \{j \in \mathcal{E} : \text{ there exists a path from } i \text{ to } j \text{ with activities in } \mathcal{A}_{wait} \cup \mathcal{A}_{drive}\}$$

as the set of events that can be reached from i without passing any changing activity. Furthermore, assume that $i \in \mathcal{E}$ has a delay $d_i > 0$, and let y be a time-minimal solution. The minimum delay that will be caused by d_i independent of any wait-depart decision is then given by

$$G(i, d_i) = \sum_{j \in \mathcal{G}(i)} w_j y_j.$$

The algorithm can now be stated.

Algorithm 2: Enumeration for (TDM-const)

Input: $\mathcal{N}, \mathcal{P}, w_p, d_i, s_a, T$.
Output: Optimal solution of (TDM), if the never-meet property holds.
Step 0.
   1. Calculate the time-minimal solution $y(\mathcal{A}_{change})$ if all connections are maintained by Algorithm 1.
   2. (Initializations) Let a_0 denote (TDM-const), set $\mathrm{SP}(a_0) = \emptyset$, $f(a_0) = 0$, Decompose $= \emptyset$, Compose $= \emptyset$, $\bar{z}_a = 0$ for all $a \in \mathcal{A}_{change}$.
   3. (Calculate $\mathrm{SP}(a_0)$) For all $i \in \mathcal{SD}$:
     (a) $f(a_0) = f(a_0) + G(i, d_i)$
     (b) For all $a = (j_1, j_2) \in \mathcal{A}_{change}$ with $j_1 \in \mathcal{G}(i)$: If $y_{j_1} > 0$ then $\mathrm{SP}(a_0) = \mathrm{SP}(a_0) \cup \{a\}$, and Decompose $=$ Decompose $\cup \{a\}$
   4. (Optimality test) If $\mathrm{SP}(a_0) = \emptyset$ stop: f is the optimal objective value, $\bar{z}_a = 0$ for all $a \in \mathcal{A}_{change}$
Step 1. While Decompose $\neq \emptyset$
   1. Choose $a = (i_1, i_2) \in$ Decompose
   2. $\mathrm{SP}(a) = \emptyset$, $\mathrm{miss}(a) = w_a T$, $\mathrm{maintain}(a) = G(i_2, y_{i_2})$
   3. (Calculate $\mathrm{SP}(a)$) For all $a' = (j_1, j_2) \in \mathcal{A}_{change}$ with $j_1 \in \mathcal{G}(i_2)$: If $y_{j_1} > 0$ then $\mathrm{SP}(a) = \mathrm{SP}(a) \cup \{a'\}$, Decompose $=$ Decompose $\cup \{a'\}$
   4. (Update Compose) If $\mathrm{SP}(a) = \emptyset$ then Compose $=$ Compose $\cup \{a\}$.
   5. (Update Decompose) Decompose $=$ Decompose $\setminus \{a\}$.
Step 2. While Compose $\neq \emptyset$.
   1. Choose $a \in$ Compose. Let $\tilde{a}$ be parent of a, i.e. $a \in \mathrm{SP}(\tilde{a})$
   2. (Solve subproblem P_a) $f(a) = \min\{\mathrm{maintain}(a), \mathrm{miss}(a)\}$,

$$\bar{z}_a = \begin{cases} 0 & \text{if } \mathrm{maintain}(a) \leq \mathrm{miss}(a) \\ 1 & \text{if } \mathrm{maintain}(a) > \mathrm{miss}(a) \end{cases}$$

   3. (Update values for parent problem $\tilde{a}$)
     $\mathrm{SP}(\tilde{a}) = \mathrm{SP}(\tilde{a}) \setminus \{a\}$, $\mathrm{maintain}(\tilde{a}) = \mathrm{maintain}(\tilde{a}) + f(a)$
   4. (Update Compose)
     Compose $=$ Compose $\setminus \{a\}$
     If $\mathrm{SP}(\tilde{a}) = \emptyset$ and $\tilde{a} \neq a_0$ then Compose $=$ Compose $\cup \{\tilde{a}\}$
Step 3.
   1. (Correct values for $\bar{z}_a$) For all $a \in \mathcal{A}_{change}$: If $\bar{z}_a = 1$ then set $\bar{z}_{a'} = 0$ for all $a' \neq a$ with $a \prec a'$.
   2. Output: $f(a_0) := \mathrm{maintain}(a_0)$, $\bar{z}$

Theorem 4. *Algorithm 2 is correct and runs in time $O(|\mathcal{A}|)$.*

Proof. We show by induction over all $a \in \mathcal{A}_{change} \cup \{a_0\}$ that $f(a)$ contains the objective value for the subproblem P_a at the end of Algorithm 2.

Start: Let $a = (i, j)$ be a maximal element of $\mathcal{A}_{change}$ (with respect to $\prec$). The subproblem with respect to a is (TDM-const) in the small network $\mathcal{N}(i)$, which does not contain any changing activity except a itself, hence $\text{SP}(a) = \emptyset$ in step 2 of the algorithm. Furthermore,

$$\text{maintain}(a) = \sum_{i' \in \mathcal{G}(j)} y_{i'} w_{i'}, \text{ and}$$
$$\text{miss}(a) = T w_a$$

give the objective values of this small network when maintaining or not maintaining activity a. To see the correctness of $\text{miss}(a)$ we note that due to Lemma 2 $y_{i'} = 0$ for all $i' \in \mathcal{E}(j)$ (which equals $\mathcal{G}(j)$ in this case).

Since $a \in$ Compose we compare both values $\text{maintain}(a)$ and $\text{miss}(a)$ in step 2, and choose the better as (correct) objective value, which is then stored in $f(a)$.

Conclusion: Now take any $a = (i, j)$ and let the induction hypothesis be true for all a' with $a \prec a'$.

- If a is not maintained, we know from Lemma 2 that all connections $a' \in \mathcal{N}(j)$ are maintained and all $i' \in \mathcal{E}(j)$ satisfy $y_{i'} = 0$, i.e., the objective value is given by $\text{miss}(a)$ as calculated in step 2.
- If a is maintained, the algorithm calculates in step 2 the delay which will be gained in any case, i.e., the delay of all events $i' \in \mathcal{G}(i)$ that can be reached without passing any changing activity, and store it in $\text{maintain}(a)$. All changing activities a' that can be reached from j without passing any other changing activity are stored in $\text{SP}(a)$. Due to Lemma 3 the corresponding subproblems $P_{a'}$ for $a' \in \text{SP}(a)$ are independent and have objective value $f(a')$ due to the induction hypothesis, such that $\text{maintain}(a) + \sum_{a' \in \text{SP}(a)} f(a')$ calculated in step 2.3 finally is the correct value of maintain.

Comparing $\text{maintain}(a)$ with $\text{miss}(a)$ and choosing the smaller of both gives the best possible choice for activity a assuming the delay y_i as given.

Finally, in step 0, the problem with the given source delays is decomposed into a set of subproblems $\text{SP}(a_0)$. All these subproblems are independent due to Lemma 3, and they are all solved optimally due to the claim above. Adding up these optimal values and adding the delay of all events which are reached before entering one of the subproblems gives the optimal objective function value $f(a_0)$.

For the time complexity we note that the number of subproblems equals the number of changing activities, which in a tree is the same as the number of events. For the decomposition steps we have to process each activity and each event exactly once, and in the composition step we need one comparison and one summation for each subproblem, and again a visit of all events. The overall time complexity is hence linear in $|\mathcal{A}|$. □

7   Future Research

Algorithm 2 relies on the fact that each activity $a \in \mathcal{A}_{change}$ appears in exactly one list, i.e., for each $a \in \mathcal{A}_{change}$ there exists a unique $\tilde{a}$ such that $a \in SP(\tilde{a})$, or $a \in SP(a_0)$. If the never-meet property is not satisfied, this needs not be the case, and hence Algorithm 2 cannot be applied to (TDM) for general problems. To resolve this problem (and to obtain a heuristic by applying Algorithm 2) one can either allow that the same element is added more than once to *Compose* in step 2 (this would mean to duplicate activities until the never-meet property is satisfied), or to update the values of *maintain* to the larger one, if an element which is already contained is added.

(TDM-const) and (TDM) can both be solved by branch and bound, taking $\bar{z}_a$ as branching variables and reducing the number of conflicts with the never-meet property in each node. Lower bounds are derived in [20]. Details and implementations are under research.

Two other directions of future research in delay management should be mentioned. First, it is a challenging task to apply delay management approaches in railway transportation. The drawback here is that capacity constraints have to be taken into account on the tracks. Different possibilities how such constraints can be included in the models are under research, see [3]. Second, it is an open field to deal with the stochastic nature of the delays instead of assuming that the source delays are fixed.

References

1. Ackermann, T.: Die Bewertung der Pünktlichkeit als Qualitätsparameter im Schienenpersonennahverkehr auf Basis der direkten Nutzenmessung. PhD thesis, Universität Stuttgart (1999)
2. Anderegg, L., Penna, P., Widmayer, P.: Online train disposition: to wait or not to wait? Electronic Notes in Theoretical Computer Science, 66(6) (2002)
3. Bissantz, N., Güttler, S., Jacobs, J., Kurby, S., Schaer, T., Schöbel, A., Scholl, S.: DisKon - Disposition und Konfliktlösungs-management für die beste Bahn. Eisenbahntechnische Rundschau (ETR) (in German) 45(12), 809–821 (2005)
4. Carey, M.: Ex ante heuristic measures of schedule reliability. Transportation Research 53B(3), 473–494 (1999)
5. Elmaghraby, S.E.: Activity Networks. Wiley Interscience Publication, Chichester (1977)
6. Engelhardt-Funke, O.: Stochastische Modellierung und Simulation von Verspätungen in Verkehrsnetzen für die Anwendung der Fahrplanoptimierung. PhD thesis, Univerität Clausthal (2002)
7. Engelhardt-Funke, O., Kolonko, M.: Simulating delays for realistic timetable-optimization. In: Operations Research Proceedings 2001, pp. 9–15. Springer, Heidelberg (2001)
8. Gatto, M., Glaus, B., Jacob, R., Peeters, L., Widmayer, P.: Railway delay management: Exploring its algorithmic complexity. In: Hagerup, T., Katajainen, J. (eds.) SWAT 2004. LNCS, vol. 3111, pp. 199–211. Springer, Heidelberg (2004)

9. Gatto, M., Jacob, R., Peeters, L., Schöbel, A.: The computational complexity of delay management. In: Kratsch, D. (ed.) WG 2005. LNCS, vol. 3787, Springer, Heidelberg (2005)

10. Gatto, M., Jacob, R., Peeters, L., Widmayer, P.: On-line delay management on a single train line. In: Algorithmic Methods for Railway Optimization. LNCS, Springer, Heidelberg (2006) (presented at ATMOS 2004, to appear)

11. Ginkel, A., Schöbel, A.: The bicriterial delay management problem. Technical report, Universität Kaiserslautern (2002)

12. Giovanni, L., Heilporn, G., Labbé, M.: Optimization models for the delay management problem in public transportation. European Journal of Operational Research (to appear, 2006)

13. Goverde, R.M.P.: The max-plus algebra approach to railway timetable design. In: Computers in Railways VI: Proceedings of the 6th international conference on computer aided design, manufacture and operations in the railway and other advanced mass transit systems, Lisbon, 1998, pp. 339–350 (1998)

14. Kliewer, N.: Mathematische Optimierung zur Unterstützung kundenorientierter Disposition im Schienenverkehr. Master's thesis, Universität Paderborn (2000)

15. Nachtigall, K.: Periodic Network Optimization and Fixed Interval Timetables. Deutsches Zentrum für Luft– und Raumfahrt, Institut für Flugführung, Braunschweig, Habilitationsschrift (1998)

16. Nemhauser, G.L., Wolsey, L.A.: Integer and Combinatorial Optimization. Wiley, Chichester (1988)

17. Peeters, L.: Cyclic Railway Timetabling Optimization. PhD thesis, ERIM, Rotterdam School of Management (2002)

18. De Schutter, B., de Vries, R., De Moor, B.: On max-algebraic models for transportation networks. In: Proceedings of the International Workshop on Discrete Event Systems, Cagliari, Italy, pp. 457–462 (1998)

19. Schöbel, A.: A model for the delay management problem based on mixed-integer programming. Electronic Notes in Theoretical Computer Science 50(1) (2001)

20. Schöbel, A.: Customer-oriented optimization in public transportation. In: Applied Optimization, Springer, New York (to appear, 2006)

21. Scholl, S.: Anschlusssicherung bei Verspätungen im öPNV. Master's thesis, Universität Kaiserslautern (2001)

22. De Schutter, B., van den Boom, T.: Model predictive control for railway networks. In: Proceedings of the 2001 IEEE/ASME International Conference on Advanced Intelligent Mechatronics, Como, Italy, pp. 105–110 (2001)

23. Suhl, L., Biederbick, C., Kliewer, N.: Design of customer-oriented dispatching support for railways. In: Voß, S., Daduna, J. (eds.) Computer-Aided Transit Scheduling. Lecture Notes in Economics and Mathematical systems, vol. 505, pp. 365–386. Springer, Heidelberg (2001)

24. Suhl, L., Mellouli, T.: Managing and preventing delays in railway traffic by simulation and optimization. In: Mathematical Methods on Optimization in Transportation Systems, pp. 3–16. Kluwer, Dordrecht (2001)

25. Suhl, L., Mellouli, T., Biederbick, C., Goecke, J.: Managing and preventing delays in railway traffic by simulation and optimization. In: Pursula, M., Niittymäki (eds.) Mathematical methods on Optimization in Transportation Systems, pp. 3–16. Kluwer, Dordrecht (2001)

26. van Egmond, R.J.: An algebraic approach for scheduling train movements. In: Proceedings of the 8th international conference on Computer-Aided Transit Scheduling, Berlin, 2000 (2001)

Appendix

Proof of Theorem 1: (TDM-A) and (TDM-C) lead to the same set of optimal solutions $y \in \mathbb{R}^{|\mathcal{E}|}$.

Proof. First, using (16) the objective function of (TDM-C) can be reformulated to

$$
\begin{aligned}
f_{\text{TDM-C}} &= \sum_{a=(i,j)\in\mathcal{A}^s} w_a(y_j - y_i) + \sum_{a=(i,j)\in\mathcal{A}_{change}} w_a \bar{z}_a(T - y_j) \\
&= \sum_{a=(i,j)\in\mathcal{A}^s} \sum_{p\in\mathcal{P}^s:a\in p} w_p \tilde{z}_a^p(y_j - y_i) + \sum_{a=(i,j)\in\mathcal{A}_{change}} \sum_{p\in\mathcal{P}^s:a\in p} w_p \tilde{z}_a^p \bar{z}_a(T - y_j) \\
&= \sum_{p\in\mathcal{P}^s} w_p \left(\sum_{a=(i,j)\in\mathcal{A}^s:a\in p} \tilde{z}_a^p(y_j - y_i) + \sum_{\substack{a=(i,j)\in\mathcal{A}_{change} \\ a\in p}} \tilde{z}_a^p \bar{z}_a(T - y_j) \right) \\
&=: \sum_{p\in\mathcal{P}^s} w_p C_p.
\end{aligned}
$$

For the objective of (TDM-A), we define

$$
A_p = y_{i(p)}(1 - z_p) + T z_p.
$$

(TDM-C) $\implies$ (TDM-A): Let $(y, \bar{z}, \tilde{z}, w)$ be feasible for (TDM-C). Define $z_p = z_p(\bar{z})$ as follows:

$$
z_p(\bar{z}) = \begin{cases} 0 & \text{if } \bar{z}_a = 0 \text{ for all } a \in p \cap \mathcal{A}_{change} \\ 1 & \text{otherwise} \end{cases} \tag{33}
$$

Then (4) holds due to (11), (5) holds due to (12), and (6) is trivially satisfied, if $z_p = 1$, and for $z_p = 0$ we know that $\bar{z}_a = 0$ for all $a \in p$ and hence (6) holds because of (13). This means (y, z) is feasible for (TDM-A). It remains to show that $A_p \leq C_p$. To this end, let $p = (s, i_1, \ldots, i_L) \in \mathcal{P}^s$ be a path with $i(p) = i_L$.

Case 1: $\bar{z}_a = 0$ for all $a \in p \cap \mathcal{A}_{change}$. Then, we define $z_p = 0$. From (14) we get that $\tilde{z}_a^p = 1$ for all $a \in p$. Since $y_s = 0$ we conclude that

$$
C_p = \sum_{a=(i,j)\in\mathcal{A}^s:a\in p} y_j - y_i = y_{i_L} - y_s = A_p.
$$

Case 2: There exists $a \in p \cap \mathcal{A}_{change}$ with $\bar{z}_a = 1$. Choose a minimal with respect to $\prec$ with this property, say $\bar{a} = (i_{\bar{k}-1}, i_{\bar{k}})$. Then, since $\bar{z}_a, \tilde{z}_a^p$ satisfy (14) and (15) we obtain

$$
\tilde{z}_a^p = 0 \text{ for all } a \in p \text{ with } \bar{a} \prec a
$$
$$
\tilde{z}_a^p = 1 \text{ for all } a \in p \text{ with } a \preceq \bar{a}.
$$

Hence, for all $a \in \mathcal{A}_{change} \cap p$ we get

$$z_a^p \bar{z}_a = \begin{cases} 1 & \text{if } a = \bar{a} \\ 0 & \text{otherwise} \end{cases}$$

This yields

$$C_p = \sum_{\substack{a=(i,j) \in \mathcal{A}^s : a \in p \\ \text{and } a \preceq \bar{a}}} y_j - y_i + (T - y_{i_{\bar{k}}})$$

$$= y_{i_{\bar{k}}} - y_s + T - y_{i_{\bar{k}}} = T = A_p,$$

and consequently, $f_{\text{TDM-C}}(y, \bar{z}, \tilde{z}, w) = f_{\text{TDM-A}}(y, z(\bar{z}))$.

(TDM-A) $\implies$ **(TDM-C):** Now let a feasible solution $(\tilde{y}, z)$ of (TDM-A) be given. Using Algorithm 1 we may replace $\tilde{y}$ by a time-minimal solution y which satisfies $y_i \leq T$ for all $i \in \mathcal{E}$, and has equal or better objective value. Since y satisfies (4) and (5) we can construct a feasible solution for (TDM-C) according to (21),(22), and (23).

For the objective value of this solution we again compare C_p and A_p for a path $p = (s, i_1, \ldots, i_L) \in \mathcal{P}^s$ and get:

Case 1: If $z_p = 0$, we get from (6) that $y_i - y_j \leq s_a$ for all $a = (i, j) \in p$. Hence, due to the definition of $\bar{z}_a$ we conclude that $\bar{z}_a = 0$ for all $a \in p \cap \mathcal{A}_{change}$, yielding $C_p = y_{i(p)} = A_p$ analogously to Case 1 of the first part of the proof.

Case 2: Now consider the case $z_p = 1$.

Case 2a: $y_i - y_j \leq s_a$ for all $a = (i, j) \in p$, yielding that $\bar{z}_a = 0$ for all $a \in p$ and hence $C_p = y_{i(p)} \leq T = A_p$.

Case 2b: There exists $a = (i, j) \in p$ such that $y_i - y_j > s_a$. This gives us $\bar{z}_a = 1$. Choose $\bar{a} = (i_{\bar{k}-1}, i_{\bar{k}})$ minimal with respect to $\prec$ with this property. Then, from the definition of $\tilde{z}_a^p$ we get

$$\tilde{z}_a^p = 0 \text{ for all } a \in p \text{ with } \bar{a} \prec a$$
$$\tilde{z}_a^p = 1 \text{ for all } a \in p \text{ with } a \preceq \bar{a}$$

and analogously to Case 2 of the first part of the proof $C_p = T = A_p$.

Together, $f_{\text{TDM-A}}(\tilde{y}, z) \geq f_{\text{TDM-A}}(y, z) \geq f_{\text{TDM-C}}(y, C(y))$.

Combining both directions yields that there exists an optimal solution for (TDM-A) with delays y if and only if there exists an an optimal solution for (TDM-C) with the same delays y.   □

Decision Support Tools for Customer-Oriented Dispatching

Claus Biederbick and Leena Suhl

University of Paderborn, Decision Support & OR Lab and
International Graduate School for Dynamic Intelligent Systems
{biederbick,suhl}@dsor.de

Abstract. Unavoidable disturbances induce the necessity of operations control and dispatching tasks in the timetable-driven rail traffic. Therefore, dispatching becomes one of the most relevant challenges for the economic success, since it directly affects the timeliness of passengers, which is a core indicator of customer satisfaction. Thus, the integration of passenger preferences into dispatching strategies is a reasonable extension of conventional dispatching algorithms. In this paper we discuss decision support tools to be used by dispatchers in order to achieve customer orientation. The tools are based on an agent-based simulation system modelling the complete German railway network with about 30000 trains and millions of passengers per day. We report tests of various dispatching strategies considering passenger information and aiming to reduce passenger waiting times. We show that even simple heuristics produce better results than the rule based dispatching strategies currently in use.

1 Ideas and Basics of Customer-Oriented Dispatching

Unavoidable disturbances induce the necessity of operations control and dispatching tasks in timetable-driven rail traffic. Dispatchers of a railway have to make decisions about changes to the original train schedule often in a matter of minutes.

Traditionally, railways often emphasise timeliness of trains and cost minimisation within the dispatching process. However, this tends to be suboptimal from the customer point of view. We argue that timeliness of customers is more important for the long-term economic success of a railway than timeliness of trains.

With customer-oriented dispatching we mean dispatching strategies that give customer timeliness a higher priority than train timeliness. Obviously, we first have to determine measures how to judge customer timeliness, which certainly means different things to different people, thus being an extremely multi-faceted and complex aspect.

Mainly, there are two different aspects of customer oriented dispatching: a) regarding wishes of passengers in the dispatching process through usage of passenger information, and b) proactive and individualised information of passengers using modern communication networks and mobile technology like cell

F. Geraets et al. (Eds.): Railway Optimization 2004, LNCS 4359, pp. 171–183, 2007.
© Springer-Verlag Berlin Heidelberg 2007

phones, personalised digital assistants (PDA) or so called "smartphones" which are widely spread within industrialised countries.

Both aspects can be implemented and established independently, but obviously they are closely related: The more information there is available, the better the quality of dispatching that can be achieved. In this paper, we focus on aspect a), because it seems to us that there is a lot of room to improve dispatching quality even with little information about passengers.

Although customer orientation is getting more and more important in the competitive passenger traffic market, little is known about how to reliably measure customer satisfaction in railways and how to design dispatching strategies that maximise customer satisfaction. There are several established simulation tools available for railway traffic (cf. [12], as an example), however, they are usually technology-oriented aiming at optimal dispatching of trains and other equipment. In this paper, we show how simulating railway traffic in a complex network using intelligent software agents results in real-time decision support tools significantly enhancing the on-trip support for passengers thus reducing unscheduled waiting times during their trips. Furthermore, we show how the simulation system can be used to figure out good online-dispatching strategies in order to maximise customer satisfaction.

With "dispatching strategy" we mean every algorithm capable of calculating dispatching decisions consisting of waiting instructions for single *connecting trains* waiting for late *feeder trains* in case of a *connection conflict*. In other words, a train – the feeder – is delayed by a disturbance and could not reach its scheduled connectors in time. If a connector does not wait for the corresponding feeder, all transit passengers of this specific connection will be delayed, if it waits, all passengers in the connector get into the risk of missing their connections.

The implementation and test described in this paper were carried out with data from Deutsche Bahn AG. The complete German railway network with about 30000 trains and about five million passengers daily was implemented within our simulator. With this system, we were able to test various dispatching strategies considering passenger information and aiming in reduction of passenger waiting times.

Topics related to railway dispatching and reliability have been discussed in literature during recent years (see [2, 5, 6, 7], for example), however, in this paper we take a more explicit view considering customer satisfaction as the main goal of disposition activities.

The paper is outlined as follows: In the second section we describe our agent-based simulation test bed for simulating large railway networks, which leads to a simple architecture of co-operating decision support components for dispatchers. Section 3 describes the enhanced dispatching process and the components necessary for this; Section 4 presents some numerical results. In Section 5 the new information process is described, and in Section 6 we close with some conclusions drawn from extensive experimentation with this system.

2 Intelligent Software Agents for Simulation of Large Railway Networks

A railway as a system consists of autonomous mobile actors, such as passengers, trains, and personnel. The system components act on a stable infrastructure, such as tracks, stations, and so on. A simulator of railway traffic needs to model all relevant groups of actors.

A natural way to describe mobile autonomous components is to use *intelligent software agents*. Software agents are (broadly speaking) computer programs which are able to behave autonomously in a certain sense, interact with other software agents, thus building communities, and move within a digital network [12]. Because there are usually several interacting agents, we often speak about multi-agent systems. It is conceptually appropriate to use software agents to implement a microscopic view of the railway system, because an agent is usually configured to represent a microscopic item like one certain train or station, even one certain customer.

From the computer science point of view, multi-agent technology can be understood as an advancement of object-oriented programming, thus presenting a new programming paradigm. Especially, the concepts of autonomous behaviour and intentionality provide new dimensions to model objects with control functionality; this is not possible with traditional object-oriented programming techniques.

We generally distinguish between two basic variants of architecture models for multi-agent systems: *deliberative* and *reactive* architecture. Agents in the deliberative model behave analogously to expert systems as they are known in artificial intelligence. They possess an explicit symbolic model of their environment, together with logical inference mechanisms in order to be able to derive conclusions and evaluate possible actions. A well-known deliberative agent architecture model is the "Belief, Desire, Intention" model of [1]. Reactive agents react on certain stimuli from their environment with executing specified actions. The easiest ways of implementation are if-then rules, e.g., *if* obstacle ahead *then* go left. Agent goals are not given explicitly, intelligence arises incrementally a non-centralised way, leading to a new modelling paradigm for distributed systems.

In our opinion, the agent concept is very well suited for modelling parallel and distributed simulation systems running on low-cost hardware. Furthermore, the agent concept enables us to integrate (autonomous) real world players, such as customers and dispatchers, trains and stations, and more specialised agents[1], into the simulation, thus supporting customer information and distributed real-time dispatching as well.

Using the agent paradigm, the simulation model was divided into several regions by simply building disjoint subsets of network vertices (e.g. stations) and arcs (tracks) as described in [4,8,11]. Every region contains one dedicated central dispatching agent, each of them assisted by several task agents, e.g.,

---

[1] The "special agents" will be discussed in forthcoming publications.

for carrying out more complex strategy calculations or to re-route passengers missing their connections (cf. Section 3). Then, the size of a (virtual) region is only limited by the dispatching productivity, i.e., the number of decisions computed per time unit. Of course, the regions should be chosen in such a way that communication costs are low within the cluster, and that there are enough communication resources available for peak demands (many delays in the network causing much dispatching activity).

To ensure consistency, a central data storage for infrastructure data (especially the network topology) as well as timetable data (static and dynamic, i.e., with delays during simulation) were chosen. The system is based on the general system architecture introduced in [10].

Between regions, only border crossing trains and passengers are exchanged. Furthermore, dispatchers can receive expected delays and passenger data from trains in neighbouring regions by asking their "colleagues". This becomes necessary when a connector waits for a feeder train still located in a neighbouring region.

In our simulation model, every complex dispatching strategy is implemented as an agent as well, even when it has no typical agent properties (such as proactiveness or mobility), thus enabling us to use the load balancing mechanisms of the agent-based simulation environment.

The dispatching process is modelled as follows:

- Trains send disturbances occurring on their route to a central server. If a dispatcher has to decide, he will ask this server to receive the latest information.
- A departing train asks the responsible dispatcher for permission.
- The dispatcher computes a decision using different strategies (cf. Section 3) and assisting agents, e.g., the passenger router (cf. Section 4). He also determines whether the train has to wait for some other (technical or security) reason, e.g., congested tracks.
- Resulting decisions will be sent to affected entities in the network, both trains and passengers.
- Additionally, every train and station performs "passenger administration". All passengers are continuously transferred from one administrative unit to the next one during the course of their trip.

Of course, the dispatcher is the core component of this system. From an agent technology point of view he could be regarded as a deliberative information agent. This agent has to watch the state of the network and to act autonomously in case of present or predictable conflicts. Even if no feasible plan (with no missed connections) could be found, the dispatcher can use the passenger router to compute new routes for all affected passengers.

The passenger agents then communicate with their real-world counterparts enabling on-trip-interaction with passengers. In other words, they are the distributed user-interface of the system, giving specific information to "their" passengers on the one hand, and collecting useful data from them to support dispatching decisions on the other.

3 Components of Customer-Oriented Dispatching for Enhancing the Dispatching Process

Each dispatcher in the system is responsible for a disjunctive region of the railway network, calculating dispatching decisions for each train waiting for departure, based on the states of all entities within the system. At first, the dispatcher determines the set of actual and potential, i.e., probably forthcoming, conflicts in the controlled area. Then, the time window in which a solution has to be found is determined implying a set of possible decision strategies. The best strategy builds the recommendation for a human dispatcher if the system is used within real world. If it runs in simulation mode, the best calculated plan will be executed without intervention of a dispatcher.

The internal architecture of the dispatching agent is outlined in Figure 1 It is based on the typical architecture of knowledge based systems. The dispatcher uses the passenger router in two ways:

1) Before a decision is made, alternative routes for affected passengers are determined. If all those passengers can be re-routed, no further delay is necessary.
2) After a decision, the router can be used by the agents representing passengers missing their connection to re-route their owners.

Dispatching strategies are of crucial importance, because they have direct impact on customer satisfaction as measured by delays and missed connections. In order to rate passenger strategies we chose a simple measure which is directly influenced by dispatching decisions: We measured the individual and total passenger waiting time. Although there are many criteria defining customer satisfaction (cf. [9]), we think (passenger) timeliness to be the most important factor.

To properly weight waiting times of customers we have to consider the circumstances under which the waiting takes place. Therefore, we define a parameter called *waiting time quality*: Does the customer wait in the train or at the station? If the latter, what type of station is it, i.e., is there any pastime, or is he or she forced to wait at the perhaps cold and windy platform? A second criterion influencing the quality of waiting time is the type of a trip. Waiting a few minutes longer may not be arduous if the customer is on a holiday trip, while it could be crucial for reaching a business appointment, while in both cases longer waiting times lead to disproportionate annoyances. Thus, passengers should not wait "too long" if dispatching can avoid it. This leads to waiting time costs for each passenger, which could be represented by utility functions (cf. [9]). Many other criteria are possible as well. In reality, the final decision has to be made by the (human) dispatcher: the role of the computer-based system is to provide optimal decision support.

In general, we may categorise dispatching strategies according to the amount and quality of input information they require, as well as according to the art and amount of their running time which may be deterministic or stochastic. Usually, the amount or information required is closely related to the expected running

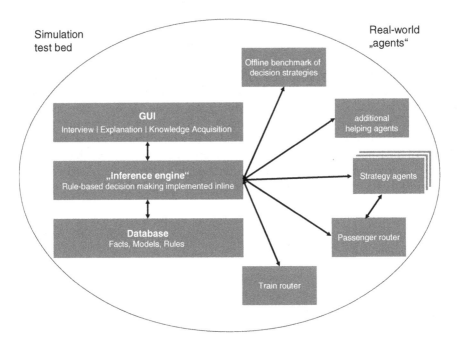

Fig. 1. Internal architecture of a dispatching agent

time. Within the study described in the next section we compared the following basic dispatching strategies:

- General regulations (Do not regard any individualities)
  Strategy #1. Do not wait at all (except for security reasons).
  Strategy #2. Wait until every feeder train has arrived.
- Decisions regarding categories of feeder trains
  Strategy #3. Wait t minutes if the feeder train is of higher category than you. The existing waiting time rules of Deutsche Bahn AG are of this type, too: Only if t is exceeded, a conflict arises and a dispatcher has to decide what to do.
- Time table dependent decisions
  Strategy #4. Wait maximally as long as no connection on the route is endangered considering the scheduled arrivals of your own connectors.
- Decisions including passenger information
  Strategy #5. Wait, if there are "many" changers in specific feeders. This may be compared with the total number of waiting passengers in the connector, e.g., a given rate q is exceeded, where q is (#changers/#waiting passengers). In real world, this information is usually not available. With simulation, however, we are able to show how useful it could be.
  Strategy #6. Variation of #5: Regard the *sum of all changers in all* delayed feeders. If q is exceeded, wait for all feeder trains.

Strategy #7. Variation of #5, but without individual routes. A certain rate of the passengers in a feeder is assumed to change, e.g., based on experienced numbers. This strategy is certainly more easily practicable than the previous ones, because the total number of passengers in a train is fairly easy to estimate.

Strategy #8. Variation of #7 similar to #6.

Strategy #9. Extension of #5: All waiting passengers *at the following* stations are taken into account additionally.

Strategy #10. Extension of #6 in the same way #16 extended #12.

Strategy #11. Extension of #7, again not regarding individual information.

Strategy #12. Analogous to extension of #8.

- Combined strategies

Strategy #13. Use of passenger router: Wait, if at least one passenger in the feeder could not reach the target station in a reasonable amount of time. For every changing passenger alternative routes are computed. If the estimated new arrival time is close enough to the planned arrival time for all passengers, the connector leaves.

Strategy #14. Compute the total consequences of both the decision that the connector waits for the delayed feeder and that it does not wait, using online passenger re-routing. Choose the alternative with less accumulated total waiting time along all routes and all passengers.

In order to be able to construct an exact model we would need the specific route of each individual passenger over one day. Since this information was not available, we decided to generate "artificial passengers" based on known information about average train contacts per day. We believe that the approximation is good enough to judge effects of various dispatching strategies.

4 Comparison of Dispatching Strategies

The strategies described above were tested within the simulation environment. The more promising strategies may be suitable for decision support systems for the human dispatcher in real-life situations as well.

Test Environment

The time horizon of the simulation is six hours, i.e., about one million passengers and nearly 9000 trains. Within a simulation run, exponentially distributed delays with expected length of 5 minutes are generated for trains which were chosen arbitrarily. Inter-arrival time of delays is exponentially distributed as well. The delays were chosen in such a way that about 10 % of the simulated trains will be delayed. Note that these are only primary delays; secondary delays could easily be induced by dispatching strategies: a train waiting for its feeders is of course delayed, too.

Our experimentation shows that most simulation runs could be executed in less than two hours on a small cluster of three personal computers (ca. 1.4 GHz and 512 MB RAM). When the passenger router is in use, it runs on a dedicated computer. One single dispatching decision can be computed in a few seconds, even when more complex strategies are in use. (Of course, run-time is mainly influenced by the calculations necessary for the used strategy.)

Some strategies demand the specification of additional parameters, especially those strategies that integrate passenger information comparing numbers of passengers in trains etc. All simulations were replicated 5 times.

Computational Results

First of all we compared strategies without re-scheduling passengers with missed connections, but penalising a missed connection with a high value in our passengers' utility functions, arguing that a missed connection implies not only delays for affected passengers, but also disproportionately reduces the quality of the product the customer paid for.

Table 1. Percentage of passengers missing a connection in strategies without the passenger router

RWZ	1	2	3	4	5	6	7	8	9	10	11	12
0,59	0,79	6,10	1,58	3,43	0,37	0,38	1,35	0,90	0,36	0,57	0,58	0,70

The percentage of passengers missing a connection in our model is comparatively small; due to the settings of our delay parameters (see Table 1). This is intended, because no valid estimation of this number in reality was available. In this way, we tend to underestimate the real number which was approved by experts of Deutsche Bahn. Therefore, it is unlikely that we overestimate the impact of online passenger re-scheduling.

A comparison of the strategies described above clearly proves our hypothesis, that the strategies including passenger information in many cases achieve much better results than waiting time rules of Deutsche Bahn AG do (see Figure 2). Although the regular waiting time (rwt) of Deutsche Bahn AG performs well, there are strategies integrating passenger information which reduce the weighted passenger waiting time down to about 60 %.

Remarkably, our model showed that rwt in general achieves only about 70 % of passengers reaching their destination on time. Ceteris paribus, passenger-oriented strategies achieved an average rate of about 90 % (with low variance) in our simulation runs.

Impact of Passenger Re-routing

Figure 3 demonstrates the effect of using the passenger router (strategies 13 and 14 obviously cannot be simulated without passenger router). It is obvious that strategies utilising passenger information perform much better than others.

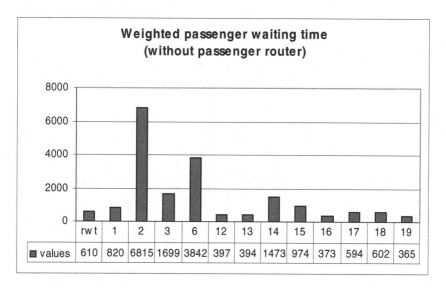

Fig. 2. Weighted passenger waiting time of strategies without using the passenger router

The passenger router clearly is an important part of the system as it is, in a way, the bridge between the two areas of customer-oriented dispatching – a) on-trip information and re-direction of passengers and b) calculating certain strategies. The passenger routing agents were developed jointly with T. Mellouli, J. Goecke, and DB Systems, (system vendor of Deutsche Bahn AG).

The passenger routers carry out the task of computing alternative routes through the dynamic connection network in case of missed connections. Therefore, a router has to update its internal network model whenever a delay occurs.

There are two ways to use passenger routers in customer-oriented dispatching:

- To determine an optimal passenger re-routing strategy in case of a disturbance taking into account several decision alternatives. Thus, we determine in real-time whether it is better to let some passengers in feeder trains miss their connector or to let some passengers in the connector train miss their connections later on. A dispatching strategy might for example minimise the induced total waiting time in both cases.
- To compute re-routing proposals for passengers already affected by missed connections – this might be helpful for passengers standing on a railway station.

If a disturbance is known before a trip starts, such as the case of a major construction site within the route, the router can be used already before a trip starts.

The internal network representation of a passenger router uses topological sorting and a very efficient update routine. This enables us to include individual

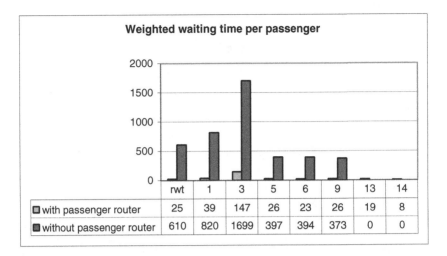

Fig. 3. Comparison of weighted passenger waiting time with and without passenger router

restrictions during the search for alternative routes, e.g., personalised minimum transit times for passengers, preferred train categories, or minimum number of train changes.

5   Information of and Communication with Passengers

An elegant way to communicate and interact with passengers is to make use of their very own mobile communication devices such as cell-phones (via SMS), PDAs (via Wireless LAN and internet) or all kinds of mobile computers (dito).

For this work we assume these technologies to be wide-spread in industrialised countries. Only this enables us to distribute individualised information and interact with every single passenger on trip and in real time, which can not be achieved using conventional information devices in stations or trains.

We assume furthermore that we are able to track passengers on their routes since we initially know the scheduled route and the position of all trains the passengers use. Even if a passenger is re-routed, it can be argued that the systems advice will be taken thus providing the new position, because he or she wants to reach his/her destination.

The complex task of collecting data of single passengers and their routes is not only helpful for customer-oriented dispatching. It is also very important as a source of information from a strategic point of view, e.g., for facilitating product planning and timetable optimization. The absence of confirmed data made these steps very difficult in the past.

One of the most important applications is *revenue management* (formerly known as *yield management*). In [3] yield management is defined as follows: "yield management is [...] to optimize total revenues at the site where a service is provided. This optimization is achieved by adopting techniques to manage the capacities marketed by a service company". Deutsche Bahn AG recently tried to introduce some revenue management instruments, e.g., giving special discounts on dedicated trains, so demonstrating their huge interest in such applications.

In our system, every agent with real-world counterpart has a communication interface with its owner, at least on the conceptual layer. This makes it easy to implement this system in real-world contexts: An earlier version of this part of our system was already installed in 2001 in co-operation with DB Systems. Passengers booked the service via World Wide Web and then got information about train states via SMS.

From a technical point of view, our system simply uses existing protocols for communication with passengers as it is shown in Figure 4. Passenger agents just monitor the network status in order to determine whether their principals are affected. If necessary, they just use a conventional web server (farm) as gateway.

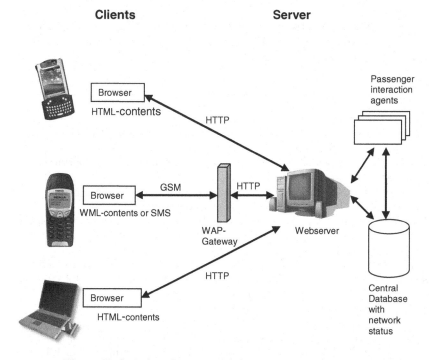

Fig. 4. Technical architecture of the passenger information system

We believe that the huge amount of short messages sent every day proof the ability of such architectures to handle all communication activities in our system, even, if millions of passengers use it.

6   Conclusions

With this paper we show how agent-based simulation under usage of estimated or exact passenger information could be used to improve the quality of the railway dispatching process both for the provider and the passengers.

We have chosen a customer oriented approach, emphasising the timeliness of customers in the connection network: some simple dispatching strategies were presented, proving that even the use of estimated passenger data could enhance the process from the customers' point of view.

However, although the simulation model was validated carefully, it is not totally sure that the presented "best" strategies would perform in reality as well as they do in simulation, because all passenger routes are not taken from real-world data, but generated by an algorithm under certain (plausible) assumptions. Therefore, tests with more reliable passenger data would be necessary to evaluate strategies correctly, i.e., we explicitly do not claim to have determined the best possible strategy, but we provide a system which enhances testing of strategy candidates.

Therefore, modern proactive information systems pushing individualised information are an important field of further research in order to maintain competitiveness of rail traffic providers.

References

1. Bratman, M.: Intention, plans and practical reason. Harvard University Press, Cambridge, MA (1987)
2. Carey, M.: Ex ante heuristic measures of schedule reliability. Transportation Research, Part B 33, 473–494 (1999)
3. Daudel, S., Vialle, G.: Yield Management: Applications to air transport and other service industries. Institut du Aérien, Bayeux (1994)
4. Goecke, J., Biederbick, C.: Simulation in Konfliktmanagement in großen Transport-Netzwerken. In: Inderfurth, et al. (Hrsg.) (eds.) Operations Research Proceedings 1999, Springer, Berlin (2000)
5. Goverde, R.M.P.: Synchronization control of scheduled train services to minimize passenger waiting times. In: Conference Proceedings, Part 2, 4th TRAIL Year Congress 1998, TRAIL Research School, Delft (1998)
6. Martin, U.: Die dispositive Lösung von Fahrweg- und Belegungskonflikten. Leipzig Annual Civil Engineering Report No. 3, 311–333 (1998)
7. Shen, S., Wilson, N.: An optimal integrated real-time disruption control model for rail transit systems. In: Voß, S., Daduna, J. (eds.) Computer-Aided Transit Scheduling. LNEMS, vol. 505, pp. 335–364. Springer, Berlin (2001)
8. Suhl, L., Biederbick, C.: Verteilte Simulation mit Softwareagenten zur Unterstützung von Dispositionsentscheidungen im schienengebundenen Personenverkehr. In: Panreck, T.K., Dörrscheidt, F. (Hrsg.), ASIM Symposium Simulationstechnik 2001, Fortschrittsberichte Simulation, SCS-Europe, Gent, Belgium (2001)
9. Suhl, L., Biederbick, C., Kliewer, N.: Design of customer-oriented dispatching support for railways. In: Voß, S., Daduna, J. (eds.) Computer-Aided Transit Scheduling. LNEMS, vol. 505, pp. 365–386. Springer, Berlin (2001)

10. Suhl, L., Mellouli, T.: Requirements for, and Design of, an Operations Control System for Railways. In: Wilson, N. (ed.) Computer-Aided Transit Scheduling. LNEMS, vol. 471, pp. 371–390. Springer, Berlin (1999)
11. Suhl, L., Mellouli, T., Biederbick, C., Goecke, J.: Managing and preventing delays in public transportation by simulation and optimization. In: Pursula, M., Niittymäki, J. (eds.) Mathematical Methods on Optimization in Transportation Systems, Kluwer Academic Publishers, Dordrecht (2001)
12. Woolridge, M., Jenkins, N.: Intelligent Agents: Theory and practice. The Knowledge Engineering Review 12(10), 115–152 (1995)
13. Zhu, P., Schnieder, E.: Determining Traffic Delays through Simulation. In: Voß, S., Daduna, J. (eds.) Computer-Aided Transit Scheduling. LNEMS, vol. 505, pp. 387–399. Springer, Berlin (2001)

Part II

Proceedings of ATMOS 2004

An Integrated Methodology for the Rapid Transit Network Design Problem

Gilbert Laporte[1], Ángel Marín[2], Juan A. Mesa[3], and Francisco A. Ortega[4]

[1] Canada Research Chair in Distribution Management, HEC Montréal
gilbert@crt.umontreal.ca
[2] Departamento de Matemática Aplicada y Estadística,
Universidad Politécnica de Madrid
amarin@dmae.upm.es
[3] Departamento de Matemática Aplicada II, Universidad de Sevilla
jmesa@us.es
[4] Departamento de Matemática Aplicada I, Universidad de Sevilla
riejos@us.es

Abstract. The Rapid Transit System Network Design Problem consists of two intertwined location problems: the determination of alignments and that of the stations. The underlying space, a network or a region of the plane, mainly depends on the place in which the system is being constructed, at grade or elevated, or underground, respectively. For solving the problem some relevant criteria, among them cost and future utilisation, are applied. Urban planners and engineering consulting usually select a small number of corridors to be combined and then analysed. The way of selecting and comparing these alternatives is performed by the application of the four-stage transit planning model. Due to the complexity of the overall problem, during last ten years some efforts have been dedicated to modelling some aspects as optimisation problems and to provide Operations Research methods for solving them. This approach leads to the consideration of a higher number of candidates than that of the classic corridor analysis. The main aim of this paper is to integrate the steps of the transit planning model (trip attraction and generation, trip distribution, mode choice and traffic equilibrium) into an optimisation process.

Keywords: Network Design, Rapid Transit Systems.

1 Introduction

Increasing mobility, longer trips caused by the enlargement of the urbanised areas and the reduction of average ground traffic speed are some of the reasons why during last 30 years new lines of rapid rail transit systems (metro, light rail, people mover, monorail, etc.) have opened in some agglomerations, while in others, systems are being constructed or planned.

F. Geraets et al. (Eds.): Railway Optimization 2004, LNCS 4359, pp. 187–199, 2007.
© Springer-Verlag Berlin Heidelberg 2007

The minimum threshold of population of a city or of a metropolitan area for needing a rapid rail transit system depends on density, private vehicle availability, traffic congestion, environmental aspects and other characteristics. This figure has diminished from about two millions inhabitants during the sixties to half a million at the end of last century. As a consequence the number of cities interested in such systems is increasing.

Because of the very large cost of constructing and operating rapid transit systems, it is important to pay close attention to their efficiency and effectiveness (Karlaftis, 2004, [7]). A crucial part of the planning process is the underlying network design, which consists of two intertwined problems: the determination of alignments and the location of stations.

The methodological support for forecasting the travel demand is a multi-stage process where different techniques can be used at each stage. Once the area under consideration has been adequately broken into study zones, the classical four-stage model is applied in order to finally obtain a prediction of the travel demand for the proposed transportation system.

1. Trip Generation Analysis: computation of the number of trips starting in each zone for each particular trip purpose.
2. Trip Distribution Analysis: production of a table containing the number of trips starting in each zone and ending in each other zone.
3. Modal Choice Analysis: allocation of trips among the currently available transportation systems (bus, train, pedestrian and private vehicles).
4. Trip Assignment Analysis: assignment of trip flows for the specific routes on each transportation system that will be selected by the users.

The four-stage process for selecting a network of lines of a mass transit system leads to the identification of a list of potential rapid transit corridors, which are assessed on the basis of several factors, among which the expected future ridership, computed by taking into account the modal split, is the most important. Corridors are then ordered and those that are selected are combined into several networks giving rise to different scenarios. Note that, since the number of alternatives is reduced to a very short list of corridors, this approach does not need to apply optimisation methods. However, it is probable that good candidates would be eliminated at an early stage or not considered at all, especially when the network can be underground.

During the last ten years some aspects of the planning problem have been modelled by optimisation models and solved by Operations Research techniques, thus allowing the consideration of a higher number of candidates than for the classical four-stage corridor analysis. The main efforts in this line of research have been oriented toward the determination of a single alignment and the location of stations given an alignment. Examples of the first type of application are the papers by Dufourd, Gendreau and Laporte (1996, [3]) and Bruno, Gendreau and Laporte (2002, [2]) when it is desired to maximise the coverage, and those by Bruno, Ghiani and Improta (1999, [1]) and Laporte, Mesa and Ortega (2005, [11]) which incorporate origin-destination matrix data, thus integrating the second stage of the classic model. García and Marín (2001, [4]; 2002, [5]) studied the transit network design problem using bilevel programming. They considered the multimodal traffic assignment problem with combined mode at the lower level.

Since an objective of the paper consists of encouraging the introduction of optimisation methods in the process of designing a rapid transit system, three stages are going to be identified below:

S1. Selecting key nodes
The network design will be based on the knowledge of the location of the main sites where trips are generated as origin or as destination.

S2. Designing the core network
A short list of lines, which are supported by the key nodes previously selected, will be determined so that the system effectiveness is maximised.

S3. Locating secondary stations
Once lines are broadly decided new secondary stations will be located, along the edges determined by pairs of key stations, in order to optimally increase the total coverage of the line.

This paper primarily addresses the second stage, although some comments about the first stage are also included in order to present their associated optimisation problems. The papers by Laporte, Mesa and Ortega (2002, [10]) and Hamacher, Liebers, Schöbel, Wagner and Wagner (2001, [6]) deal with the problem of locating stations on a given alignment; these are useful references for the study of the third stage.

The paper is organised as follows. In the next section the problem of selecting the main stations is discussed. Section 3 is dedicated to the problem of deciding how to connect the selected stations. Section 4 describes some computational experience based on an example.

2   Selection of Key Station Sites

As above mentioned above, once the area under consideration has been partitioned into zones, the number of trips that each zone will produce or attract must be quantified taking into account land use activities and the socio-economic characteristics of potential users.

As a general rule, zones are designed to include city blocks relatively homogeneous with respect to their urban activities (residential, commercial or industrial) and size. For instance, Figure 1 represents the city of Sevilla (Spain) split into 47 transportation zones. The two grey bands correspond to the branches of Guadalquivir river.

Some zones produce a high number of trips since they are densely populated (e.g., darker zones in Figure 1) and situated far away from the central area of the city (trip-generator zones). On the other hand, those that provide a large number of jobs (such as office zones or industrial areas) or contain important facilities (such as commercial zones, universities and hospitals) must be considered as trip-attractor zones.

Fig. 1. Transportation zones in Sevilla

Figure 2 illustrates those sites in Sevilla which produce a high number of trips. Moreover, the most used roads for getting into the city from the towns of the metropolitan area have been drawn with a triangle indicating an access flow. Figure 2 also shows one line of the Metro of Sevilla (currently under construction; more details in http://www.urbanrail.net/eu/sev/sevilla.htm) based on the key node set which was selected.

Therefore, this simplified scenario with nineteen sites which produce a high number of trips contains:

- three centroids of zones densely populated (e.g., node WEST);
- four filled triangles point out those roads where the density of trips by public bus is higher. The number associated to each triangle is measured in thousands of users;
- twelve key nodes where the main facilities of the city are located: universities (nodes labelled by US1, US2, US3, US4 and UPO), hospitals (nodes HOS1 and HOS2), train stations (ST1,ST2), etc;
- an approach to the real drawing of Line 1 based on the earlier nodes.

Often, the trip-producing zones cannot be represented as a point in the model (as, for instance, a large university campus or a distant and dense area);

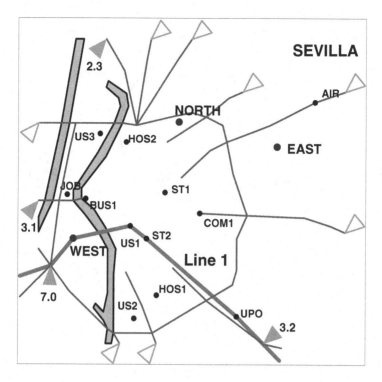

Fig. 2. Key node set and Line 1 of the Metro of Sevilla

therefore, some of these connecting centroids will be considered in the model as candidate sites for locate future stations of the new rapid transit network, channelling the zone demand total or partially. In order to obtain the most effective placement of the stations, a location problem on the zone should be solved.

Hence, let each demand node be denoted by $\mathbf{a}_k$, $k \in \mathcal{K} = \{1, \ldots, M\}$ in area A. We assume that each node $\mathbf{a}_k$, $k \in \mathcal{K} = \{1, \ldots, M\}$ is weighted with an average number w_k of inhabitants or visitors attracted by the service. Let $d_A(\mathbf{x}, \mathbf{y})$ be a planar distance measure, which is the best one fitted to the existing urban structure in zone A with $\| \cdot \|_A$ as associated norm: $d_A(\mathbf{x}, \mathbf{y}) = \| \mathbf{x} - \mathbf{y} \|_A$. Let $B_A(\mathbf{x}, r)$ denote the set of points in the plane whose distance to the station $\mathbf{x}$ is not greater than r (usually, called *ball* of radius r).

1. If the attraction model is assumed to be all-or-nothing and the solution set is continuous (all points in A are candidate sites), then a maximal covering location problem with fixed capture radius $r > 0$ must be solved to determine the best location for concentrating the area demand. For this purpose, let $\mathcal{I}(\mathbf{x}, r) = \{k \in \mathcal{K} : \mathbf{a}_k \in B_A(\mathbf{x}, r)\}$ be the index set necessary to formulate the maximal covering problem with fixed radius:

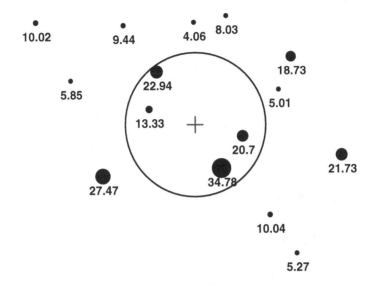

Fig. 3. A solution of the maximal covering problem

$$\max_{\mathbf{x}\in A} \sum_{k\in\mathcal{I}(\mathbf{x},r)} w_k.$$

Figure 3 illustrates a solution for the maximal covering problem where building blocks have been replaced by centroids; the radii of nodes are proportional to their weights which appear below.

The corresponding method which leads to the solution requires a number of steps is directly related to set cardinality M. A discussion on the continuous covering location problems can be found in Plastria (2002, [12]). The distance between the potential users and the station must be mathematically formulated by accurately fitting a representative sample of real travel times in the area under study. The distance type used in the approach can determine the solution to be adopted; see Laporte, Mesa and Ortega (2002, [10]) for an application arising from the planning of the Sevilla metro.

A more realistic approach to assess the catchment level around each station of an alignment, consists of establishing concentric geometrical shapes with decreasing attraction factors (Figure 4) which can vary following continuous or discrete models. A discrete version of this approach was developed by Dufourd, Gendreau and Laporte (1996, [3]) and by Bruno, Gendreau and Laporte (2002, [2]). On the other hand, a continuous model for the catchment problem, using a gravity model in order to maximize coverage, was applied by Laporte, Mesa and Ortega (2002, [10]) to determine the most effective pair of sites to locate two stations in a section of the metro of Sevilla.

2. If the set $\{s_1, s_2, \ldots, s_L\}$ of candidate sites for the station is considered as discrete, then a discrete version of the covering problem can be used. Namely,

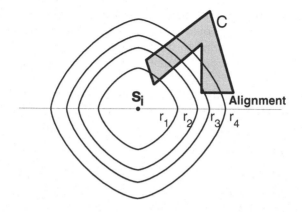

Fig. 4. Catchment area in census tract C from station S_i

$$\max_{l=1,\ldots,L} \sum_{k\in\mathcal{I}(S_l,r)} w_k.$$

Since the location of candidate sites is previously known, the attractiveness for the users can be better estimated by taking other additional considerations into account (like the transportation cost along the street network, the cost of establishing a facility, or the penalty cost for providing poor service to the users). This requires sophisticated integer linear programming models. Useful references on this topic are Kolen and Tamir (1990, [8]) and Schilling, Jayaraman and Barkhy (1993, [13]).

3   Core Network Design

After having located the stations that must belong to some of the lines of the network, the problem of connecting them with a small number of alignments $\mathcal{A} = \{A_l : l = 1,\ldots,L\}$, with origins o_l and destinations d_l given, in competition with the private mode PRIV, is tackled. For this purpose and as a first approach, some geometric models like the Minimum Spanning Tree, the Minimum-Diameter Spanning Tree and the Steiner Tree problems could be applied. However, these approaches do not take into account the mode and route user decisions. For this reason an integer network design formulation considering the user and location decisions will be developed in this section.

3.1   Data

First of all, we will assume that the data required for the model are known, namely:

- the set $N = \{n_i, i = 1, 2, \ldots, I\}$ of potential locations for key stations. Typically for a medium size agglomeration, $5 \leq I \leq 20$;

- the matrix $d = (d_{ij})$ of distances between pairs of points of N. Note that the entries of matrix d could correspond to (almost) Euclidean distances if the system were designed to be underground; otherwise, i.e. for a grade or elevated system, the data of matrix d should reflect the street network distances;
- the travel patterns given by the origin-destination matrix: $F = (f_p)$; $p = (q, r) \in P$, where P is the set of ordered pairs of demand points.

3.2   Formulating the Model

Let then E be the set of feasible edges linking the key stations. Therefore, we have a network $\mathcal{N} = (N, E)$ from which the core network is to be selected. For each node $n_i \in N$ denote by $N(i)$ the set of nodes adjacent to n_i. Let c_{ij} and c_i denote the costs of constructing a section of an alignment on edge ij and that of constructing a station at node n_i. The generalised routing cost (under demand point of view) of satisfying the demand of pair p through the private and the public network are c_p^{PRIV} and c_p^{PUB}, respectively. The first one is a given value, but the latter cost depends on the final topology of the public network and therefore on the edges that are selected; for this reason a generalised cost c_{ij}^{PUB} is given for each edge. This value is taken equal to the distance between i and j. Depending on the available budget for the total construction cost of each alignment, bounds $c_{\min}^l$, $c_{\max}^l$, $l = 1, \dots, L$, on the construction cost of each alignment and bounds $c_{\min}$ and $c_{\max}$ on the total construction network cost are known.

The problem we are dealing with consists of choosing a low number L of lines (typically $1 \le L \le 5$) covering as much as possible the travel demand between the points of N, subject to constraints on the construction cost. The model has four decision variables:

1. the first variable is the binary variable which represents the selection of stations for each alignment: $y_i^l = 1$ if $n_i \in A_l$; $y_i^l = 0$ otherwise;
2. the second one is the binary variable associated to the selection of the specific edge used to build the alignment: $x_{ij}^l = 1$, if edge $ij \in E$ is selected for the alignment l; $x_{ij}^l = 0$ otherwise;
3. since edge ij will be included in the design of the public network depending on if the demand between pairs of origin-destination nodes is satisfied, a third variable is used for modelling that decision: $u_{ij}^p = 1$, if the demand of pair p of origin-destination nodes would use edge ij in the public network; $u_{ij}^p = 0$ otherwise;
4. the objective consists of deciding what particular pair of origin-destination nodes will be included in the final design. For assessing that decision $z_p = 1$ if the generalised cost for the demand of node pair p, through the public network $\mathcal{A}$, is less than that of the private mode; $z_p = 0$ otherwise.

Thus the problem can be stated in the following terms:

- Objective function: Trip covering

$$\max \sum_{p \in P} f_p\, z_p$$

- Cost constraints

$$\sum_{ij \in E} c_{ij}\, x_{ij}^l + \sum_{i \in N} c_i\, y_i^l \in [c_{\min}^l, c_{\max}^l], \quad l = 1, 2, \ldots, L \tag{1}$$

$$\sum_{l=1}^{L} \sum_{ij \in E} c_{ij}\, x_{ij}^l + \sum_{i \in N} c_i \sum_{l=1}^{L} y_i^l \in [c_{\min}, c_{\max}] \tag{2}$$

- Alignment location constraints

$$\sum_{j \in N(o_l)} x_{o_l j}^l = 1, \quad l = 1, 2, \ldots, L \tag{3}$$

$$\sum_{i \in N(d_l)} x_{id_l}^l = 1, \quad l = 1, 2, \ldots, L \tag{4}$$

$$x_{ij}^l = x_{ji}^l \quad ij \in E, \quad l = 1, 2, \ldots, L \tag{5}$$

$$y_{o_l}^l = y_{d_l}^l = 1, \quad l = 1, \ldots, L \tag{6}$$

$$\sum_{j \in N(i)} x_{ij}^l = 2 y_i^l, \quad i \in N \setminus \{o_l, d_l\}, \; l = 1, \ldots, L \tag{7}$$

- Routing demand constraints

$$\sum_{j \in N(q)} u_{qj}^p = 1, \quad p = (q, r) \in P \tag{8}$$

$$\sum_{i \in N(q)} u_{iq}^p = 0, \quad p = (q, r) \in P \tag{9}$$

$$\sum_{i \in N(r)} u_{ir}^p = 1, \quad p = (q, r) \in P \tag{10}$$

$$\sum_{j \in N(r)} u_{rj}^p = 0, \quad p = (q, r) \in P \tag{11}$$

$$\sum_{i \in N(j)} u_{ij}^p - \sum_{k \in N(j)} u_{jk}^p = 0, \quad j \in N, \; p = (q, r) \in P \tag{12}$$

- Splitting demand constraints

$$\sum_{ij \in E} c_{ij}^{PUB}\, u_{ij}^p - c_p^{PRIV} - M\,(1 - z_p) \leq 0, \quad p \in P \tag{13}$$

- Location-Allocation constraints

$$u_{ij}^p + z_p - 1 \leq \sum_{l=1}^{L} x_{ij}^l, \quad p \in P, \; ij \in E \tag{14}$$

$$x_{ij}^l, y_i^l, u_{ij}^p, z_p \in \{0, 1\}$$

3.3   Description of the Constraints

Constraints [1] and [2] impose lower and upper bounds on the cost of each line and on the overall network, respectively. They could be simplified by considering only the individual and the total line lengths instead of costs.

- Length constraints

$$\sum_{ij \in E} d_{ij}\, x_{ij}^l \in [\text{length}_{\min}^l, \text{length}_{\max}^l], \quad l = 1, 2, \ldots, L \tag{1'}$$

$$\sum_{l=1}^{L} \sum_{ij \in E} d_{ij}\, x_{ij}^l \in [\text{Tlength}_{\min}, \text{Tlength}_{\max}] \tag{2'}$$

Constraints [3] and [4] guarantee that each line starts and ends at its specified origin and destination. Note that, although the trip flow is directed, the network in undirected; therefore, the decision of connecting two nodes must involve flows in both directions, as constraints [5] express. Constraints [6] ensure that all origins and destinations belong to $\mathcal{A}$. Constraints [7] impose that each line must be a path between the corresponding origin and destination.

Constraints [8], [9], [10], [11] and [12] guarantee demand conservation. Constraints [13] were introduced to ensure that $z_p = 1$ when the demand of pair p goes through the public network and $z_p = 0$ if it uses the private network. Finally, constraints [14] guarantee that a demand is routed on an edge only if this edge belongs to the public system.

Note that this formulation does not include the common subtour elimination constraints, but when a solution contains a cycle, then such a constraint is imposed. However well developed networks (e.g. Paris, London, Moscow, Tokyo and Madrid) often contain circular lines. It has also been proved by Laporte, Mesa and Ortega (1997, [9]) that the inclusion of a circle line often increases the effectiveness of the network and thus the inclusion of cycles may be interesting.

4   Computational Experiments and Conclusions

The integer model of the latter section was implemented in GAMS 2.0.27.7, which calls CPLEX 9.0, and tested on the 6-node network $\mathcal{N} = (N, E)$ of Figure 5.

Each node n_i has an associated cost c_i and each edge ij has a pair (c_{ij}, d_{ij}) of associated weights: the first coordinate is the cost c_{ij} of constructing the edge and the second is the distance d_{ij} (which in the experiments is also the generalised public cost of using the edge) between nodes n_i and n_j.

The origin-destination demands f_p; $p = (q, r) \in P$ and the private cost c_p^{PRIV} for each demand pair $p \in P$ are given in the following matrices.

$$
F = \begin{pmatrix}
- & 9 & 26 & 19 & 13 & 12 \\
11 & - & 14 & 26 & 7 & 18 \\
30 & 19 & - & 30 & 24 & 8 \\
21 & 9 & 11 & - & 22 & 16 \\
14 & 14 & 8 & 9 & - & 20 \\
26 & 1 & 22 & 24 & 13 & -
\end{pmatrix}
; \quad
C^{PRIV} = \begin{pmatrix}
- & 1.6 & 0.8 & 2 & 2.6 & 2.5 \\
2 & - & 0.9 & 1.2 & 1.5 & 2.5 \\
1.5 & 1.4 & - & 1.3 & 0.9 & 2 \\
1.9 & 2 & 1.9 & - & 1.8 & 2 \\
3 & 1.5 & 2 & 2 & - & 1.5 \\
2.1 & 2.7 & 2.2 & 1 & 1.5 & -
\end{pmatrix}
$$

Although the levels of congestion are not uniform in a real context (in fact, they depend on the time of the day, the day of the week, the month, the weather condition as well as the area of the city), different congestion coefficients (which are contributing factors to the private costs) have been considered. The following tables show, for each case, the lines that compose the optimal network and the corresponding values of the objective function. For Table 1, the total length of the network is an active constraint, whereas a maximum location cost is the additional constraint considered for Table 2.

Times spent in the transfers was not taken into account in the model. Location cost for multiple stations (stations with line transfers) was considered proportional to the location cost of a single station, with a proportionality factor equal to the number of lines crossing the corresponding station.

The results are quite similar for the two different constraints considered: the optimal network is composed of a single line when the restriction level is higher,

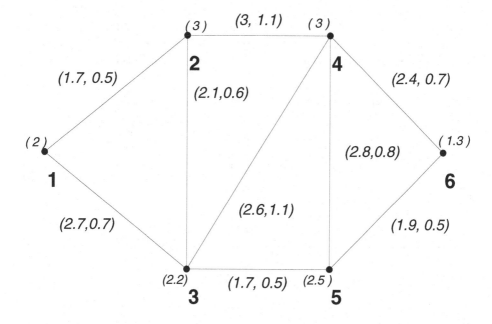

Fig. 5. Network

or composed of two lines if the constraint is less restrictive. In some cases more than one configuration reaches the maximum covering for the demand given. The accuracy of the model can be confirmed by observing that when congestion increases, the value of the objective function also increases, giving rise to a more important use of the public transit system.

The experiment was carried out on a Pentium IV laptop computer at 2.56 Mghz. Since all the results were obtained within seconds we hope the model would be useful for larger instances. In particular, for a 9-node network the computation time, once the origin and destination nodes have been previously fixed, is 27.19 seconds, whereas the exploration for all possible solutions for takes 195905.91 seconds (more than five hours).

Further computational experiments should be done for real-world contexts. In order to obtain valuable results, good data are required and some relaxation on the 0/1 allocation of the demand should be done. However, there exist several factors such as security, convenience and availability of private cars, all having some influence on the potential ridership, which can hardly be taken into account in models.

On the other hand, an interesting extension of the previous maximal covering problems with fixed radius, dealt with in Section 2, arises when an attraction gravity model is combined in the objective:

$$\max_{\mathbf{x} \in A} \sum_{k \in \mathcal{I}(\mathbf{x}, r)} \frac{\rho w_k}{d^*(\mathbf{x}, \mathbf{a}_k)^2}$$

where parameter $\rho > 0$ represents some proportional constant and $d^*(\cdot, \cdot)$ is an hyperbolic approach to the Euclidean distance, in order to avoid the singularities of the function. As far as the authors are aware, no work on this gravitational version of the maximal covering problem with fixed radius has been carried out.

Table 1. Maximum total length constraint [2'] active

TABLE I	$\text{Tlength}_{max} = 2$		$\text{Tlength}_{max} = 4$	
Congestion	Line(s)	Value	Line(s)	Value
0.75	$n_2 - n_3 - n_5 - n_6$	168	$n_1 - n_3 - n_5 - n_6 - n_4$ $n_2 - n_3 - n_5$	319
1	$n_1 - n_3 - n_5 - n_6$	216	$n_1 - n_2 - n_3 - n_5 - n_6$ $n_2 - n_4 - n_5$	446
1.5	$n_1 - n_3 - n_5 - n_4$	227	$n_1 - n_2 - n_4 - n_5 - n_6$ $n_2 - n_3 - n_5$	496

Table 2. Maximum location cost [2] active

TABLE II	$c_{max} = 15$		$c_{max} = 30$	
Congestion	Line(s)	Value	Line(s)	Value
0.75	$n_2 - n_3 - n_5 - n_6$	168	$n_1 - n_2 - n_3 - n_4 - n_6$ $n_3 - n_5 - n_6$	327
1	$n_1 - n_3 - n_5 - n_6$	216	$n_1 - n_3 - n_2$ $n_4 - n_3 - n_5 - n_6$	430
1.5	$n_1 - n_3 - n_5 - n_6$	216	$n_3 - n_4 - n_6$ $n_1 - n_2 - n_3 - n_5 - n_6$	496

Acknowledgements

This work has been supported in part by a grant from the Spanish research Projects BFM2003-04062/MATE (Ministerio de Ciencia y Tecnología) and 2003/1360 (Ministerio de Fomento). Thanks are due to Armando Garzón-Astolfi for his valuable contributions to the implementation of the computational experiments.

References

1. Bruno, G., Ghiani, G., Improta, G.: A Multi-modal Approach to the Location of a Rapid Transit Line. European Journal of Operational Research 104, 321–332 (1998)
2. Bruno, G., Gendreau, M., Laporte, G.: A Heuristic for the Location of a Rapid Transit Line. Computers & Operations Research 29, 1–12 (2002)
3. Dufourd, H., Gendreau, M., Laporte, G.: Locating a Transit Line Using Tabu Search. Location Science 4, 1–19 (1996)

4. García, R., Marín, A.: Urban Multimodal Interchange Design Methodology. In: Pursula, M., Niittymäki, J. (eds.) Mathematical Methods on Optimization in Transportation Systems, pp. 49–79. Kluwer Academic Pub. Dordrecht (2001)

5. García, R., Marín, A.: Parking Capacity and Pricing in Park'n Ride Trips: A Continuous Equilibrium Network Design Problem. Annals of Operations Research 116, 153–178 (2002)

6. Hamacher, H., Liebers, A., Schöbel, A., Wagner, D., Wagner, F.: Locating New Stops in a Railway Network. Electronic Notes in Theoretical Computer Science, vol. 50, pages 11 (2001)

7. Karlaftis, M.G.: A DEA Approach for Evaluating the Efficiency and Effectiveness of Urban Transit Systems. European Journal of Operational Research 152, 354–364 (2004)

8. Kolen, A., Tamir, A.: Covering Problems. In: Michandani, P.B., Francis, R.L. (eds.) Discrete Location Theory, pp. 263–304. Wiley, Chichester (1990)

9. Laporte, G., Mesa, J.A., Ortega, F.A.: Assessing the Efficiency of Rapid Transit Configurations. Top 5, 95–104 (1997)

10. Laporte, G., Mesa, J.A., Ortega, F.A.: Locating Stations on Rapid Transit Lines. Computers & Operations Research 29, 741–759 (2002)

11. Laporte, G., Mesa, J.A., Ortega, F.A., Sevillano, I.: Maximizing Trip Coverage in the Location of a Single Rapid Transit Alignment. Annals of Operations Research 136, 49–63 (2005)

12. Plastria, F.: Continuous Covering Location Problems. In: Drezner, Z., Hamacher, H. (eds.) Facility Location, pp. 37–79. Springer, Heidelberg (2002)

13. Schilling, D.A., Jayaraman, V., Barkhi, R.: Review of Covering Problems in Facility Location. Location Science 1, 25–55 (1993)

A Simulation Approach of Fare Integration in Regional Transit Services

Domenico Gattuso and Giuseppe Musolino

Mediterranea University of Reggio Calabria
Department of Computer Science, Mathematics, Electronics and Transportation
Feo di Vito, 89100 Reggio Calabria, Italy
{giuseppe.musolino,domenico.gattuso}@unirc.it

Abstract. The paper presents a general procedure, supported by a system of simulation models, to estimate the effects due to the implementation of an integrated fare system in regional transit services. The effects on users (demand level variations) and on transit companies (management revenues) are considered. In particular, the demand is assumed to be elastic with respect to the fare at the modal and path choice dimensions. The general procedure is applied to the transit system of an Italian regional area.

1 Introduction

In the last years, many regional and metropolitan areas successfully introduced integrated fare systems in transit services, due to the modifications of European and national legislations. The implementation of an integrated fare system involves decisional implications concerning the necessity to activate comprehensible and easy usable fares for the user, to identify the management service revenues and their allocation among transit companies. The actual tendency is to move from non-integrated fare systems, where each transit company has his own fare structure and levels, towards integrated zone fare systems, where a common zone fare structure is defined, accepted and adopted by all transit companies operating in the area. A systematic description of fare integration experiences in transit services of some European metropolitan and regional areas is reported in [1], [2], [3].

Methods for designing integrated fare systems for transit services can be classified according to a "what if" (simulation) approach and a "what to" (optimization) approach (Figure 1).

In the first case, many alternative fare system scenarios (zone partition and fare structure and levels) are exogenously defined on the basis of experiences or specific knowledge of the examined area. Such configurations are simulated through a system of simulation models and compared with the support of some evaluation indicators. In the second case, fare system scenarios are generated automatically by means of an optimization model, which is usually composed by a zone partition model and a fare level model. The solutions of the optimization

F. Geraets et al. (Eds.): Railway Optimization 2004, LNCS 4359, pp. 200–218, 2007.

model are analyzed and evaluated through an iterative procedure. Both approaches have, in a general case, the minimum fare increment for the users and the minimum revenue reduction for the transit companies as objectives. The decision variables are the zone partition of the area and the fare structure and levels, and the constraints are given by the territorial integrity of the zones, some thresholds for maximum variations of fare levels and revenues, or fare monotonicity (increasing fares with trip length).

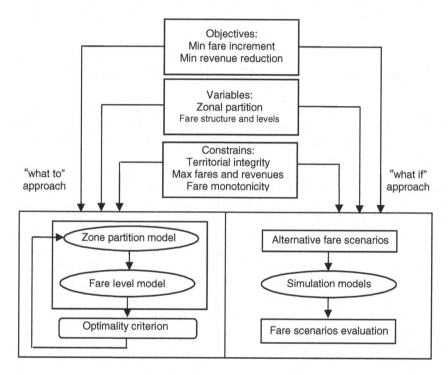

Fig. 1. "What to" and "what if" approaches for integrated zone fare system design

Some work based on the "what if" approach are proposed in [4] and more recently in [5], [6] and [7]. In the latter publications a procedure for management revenues estimation deriving from the definition of alternative fare systems in a regional area is proposed. Some work based on the "what to" approach are focused to the optimal definition of a zone fare system, with the objectives to minimize fare increments for the users and revenue reductions for the operating companies ([8], [9], [10], [11], [12] and [13]).

In all the above work the effects of the implementation of an integrated zone fare system are simulated and evaluated, assuming rigid demand.

The paper presents a general procedure, which uses a system of simulation models to estimate the effects on users (demand level variations) and on transit companies (management revenues), due to the implementation of an integrated fare system in regional transit services. In particular, the demand is assumed to be elastic with respect to the fare at the modal and path choice dimensions.

The paper is organized in three sections. In the first section, the general procedure for the simulation of fare systems and the system of simulation models for the estimation of demand levels and management revenues in a regional area are presented. The second section describes an application of the procedure to the Province of Reggio Calabria (Italy); the effects deriving from the definition of different fare systems have been simulated and compared in terms of demand levels and management revenues. Finally, the third section reports the conclusions and the research perspectives.

2   General Procedure for Fare Systems Simulation

The proposed procedure, based on the "what if" approach, allows to simulate the effects of the implementation of an integrated fare system in regional transit services on users (in terms of demand levels) and on transit companies (in terms of management revenues).

The procedure is presented according to the scheme shown in Figure 2 and involves specifications of a supply model, a fare model (it is separated from other components of the supply model), a demand model, a transit assignment model and a revenues model.

The starting step concerns the definition of the integrated fare system scenario, simulated through a fare model. Road facilities and transit services are simulated through a supply model. The demand can be simulated through a system of models that, from user specific, level-of-service and cost attributes (fares), provides modal origin-destination (O/D) matrices. At this point, the modal demand levels are evaluated with reference to prefixed constrains (i.e., minimum thresholds given by the demand levels in the current transit system modes), described in detail below. If such constrains are not satisfied the fare system scenario can be modified, otherwise modal O/D matrices are assigned to the network through an assignment model. Cost attributes are taken into account in the definition of systematic utility (or cost) connected to each path. So, it is possible to simulate the effects of a fare system scenario on the demand in the dimensions of mode and path choice. The assignment model (demand-supply interaction) gives back the flows (or loads) on each element of the network (link), that represent the input, together with fare matrices, for the revenues model. The estimated revenues are evaluated with reference to prefixed constrains (i.e., minimum thresholds of revenues/cost rates imposed from legislation, as in Italy). If such constrains are satisfied the procedure ends, otherwise, there is a new feedback to the starting step.

The decision variables are represented by the fare system (zone partition, fare structure and levels), assuming the network topology and performances to be constant. The objectives can be the minimum fare increment for the users and/or the minimum revenue reduction for the operating companies. The constrains are given by thresholds for demand levels and fare revenues, for example:

- estimated demand levels for every existing transit mode for the fare system scenario $(d^{T,S})$ must be not less $(\lambda \geq 1)$ than demand levels for the current fare system $(d^{T,C})$:

$$d^{T,S} \geq \lambda d^{T,C} \qquad (1)$$

- estimated fare revenues for the fare system scenario $(R)^S$ must be not less than prefixed thresholds of the revenues/costs rate (i.e., $\eta \geq 0.35$, as in Italy):

$$R^S \geq \eta MC \qquad (2)$$

where: T, transit system; S, fare system scenario; C, current fare system; R, fare revenues; MC, operating management costs.

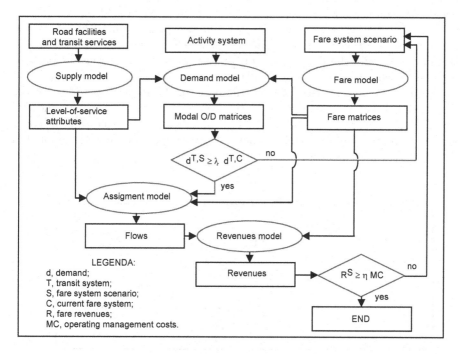

Fig. 2. Graphical scheme of the general procedure

2.1 Supply Model (Road Facilities, Transit Services)

Road facilities are modeled through a network model, composed by a topological graph. Transit low-frequency services (as in regional areas) are represented by a run-based supply model, where the graph is made up by a service sub-graph (timetable) and an access/egress sub-graph. Cost functions for links on the two sub-networks are assumed to be separable and not flow-dependent (not congested networks).

Different run-based supply models for low-frequency transit services are presented in [14], [15] and [16]. In this work, the mixed line-based/database supply model from [15] is considered. It uses a line-based network model together with a timetable database; in particular, it uses a line-based approach to describe the spatial service network topology in terms of routes, lines and stops and data associated to nodes and links to define runs on the network.

The run-based supply model is able to simulate each mode/service at a run level. It is also able to simulate a "mixed" mode, which is composed by two (or more) transit modes, through the explicit representation of intermodal transfer nodes. Two approaches exist in literature [7] to simulate intermodal transfers. According to the first one, a supply model is implemented, which is able to integrate all transport modes in a single network and to represent, through specific nodes and links, intermodal transfers. The network is quite complex and needs the definition of some criteria in order to prevent the choice of infeasible paths. According to the second one, a network is implemented for each transport mode. The connection between these networks is ensured by a fictitious origin and/or destination node in each of them, which represents the intermodal transfer nodes. So, the path on the mixed mode is composed from the path on the first network from the origin node to the intermodal transfer node (fictitious destination on the first network) and the path on the second network from the intermodal transfer node (fictitious origin on the second network) to the real destination (Figure 3). The network is less complex, if intermodal transfer nodes are not numerous and are explicitly defined. This second approach is adopted in this work for the representation of the mixed mode.

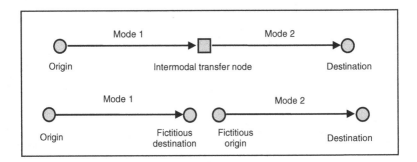

Fig. 3. Intermodal transfer node representation

2.2 Fare Model

The fare model is a component of the supply model, but it is separately treated here because cost attributes (fare) are the ones that change due to the implementation of an integrated fare system.

Different fare models can be specified to represent different fare systems. Generally, fare (p) is composed of an access fare to the service (p_0) and a variable on-board fare (p_v), which depends on the use of service. The last one has different specifications which depend on the fare structure. Fare models can be linear or non-linear; in the following some models are presented which are able to simulate different fare systems:

- constant fare model ($p_v = 0$):

$$p = p_0 \tag{3}$$

- distance (or time) fare model, when p_v depends on traveled distance (d) between origin and destination (or travel time t);

$$p = p_0 + p_v(d) \tag{4.a}$$

$$p = p_0 + p_v(t) \tag{4.b}$$

- zone fare model, when p_v depends on the number of crossed zone boundaries (n):

$$p = p_0 + p_v(n) \tag{5}$$

- origin-destination fare model, when fare depends on each origin-destination (O/D) couple; according to the service provided, distance, specific commercial strategies:

$$p = p_{O/D} \in P_{O/D} \tag{6}$$

where $P_{O/D}$ is a fare matrix, having as elements fare associated to O/D couple.

Fare models (4) and (5) have a linear specification and simulate additive fares independent from O/D couple and path, while fare model (6) simulate a non-additive fare which depends on each O/D couple. In this work, only linear fare models are considered.

2.3 Demand Model

Transport demand can be obtained from a direct estimation, from model estimation, from traffic counts. Model estimation provides parameters estimation of the sub-models related to different choice dimensions and allows to forecast demand variations in short, medium and long term.

In this work, in order to simulate the effects deriving from the implementation of an integrated fare system, mode and path choices are explicitly simulated. To such purpose, modal and path choice sets and choice models are specified.

Modal choice set includes individual (car) and transit modes available for users for extra urban trips. Among transit modes, a mixed mode is dealt as a specific mode and is considered as a specific alternative in the choice set. Attributes (especially level-of-service attributes) related to mixed mode derive from the composition of attributes of the first mode from origin to intermodal transfer node and of attributes of the second mode from intermodal transfer node to destination. Such attributes are determined through network loading models with deterministic path choice for both transport modes.

Mode choice model is based on random utility theory ([17] and [18]); it provides O/D matrices for each transport mode.

Modal O/D matrices must be segmented into more detailed time-varying matrices, in order to be consistent to run-based supply model. Demand segmentation is performed according to users desired departure time (DDT) from origin and desired arrival time (DAT) at destination. Usually, DAT is related to home-living trips and DDT to returning trips.

In low-frequency transit systems, path choice set can be composed of selected paths, according to some criteria ([19] and [16]), which lead to the generation of a reduces set of feasible paths.

Path choice model (determinist or stochastic) is completely pre-trip and depends on the systematic (or perceived) utilities of each path equal to the opposite of the path costs. Such costs are equal to the sum of two components: additive costs and non-additive costs. The first ones derive from the sum of costs (generally time) of links of the path, while the second ones are specific of the path and/or the origin-destination couple.

Access/egress time, earliness/lateness arrive at stop, waiting time, boarding time, on-board time, alighting time are additive costs (in not-congested networks), while fare can be an additive or not additive cost, depending on the fare systems (respectively linear or not linear). For linear fare structures (Eq. 4 and 5), it is possible to convert fare monetary cost in temporal cost, assigning the access fare (p_v) to the boarding link and the variable on-board cost (p_v) to the (or some) on-board links. For non-linear fare structures (Eq. 6), fare cost represents a non-additive path cost that can be considered only after the explicit enumeration of all paths.

2.4   Transit Assignment Model

In low-frequency transit systems, where all runs must be explicitly simulated, transit assignment of demand flow to the network must be performed through a dynamic schedule-based approach, which allows to obtain disaggregate results in terms of on-board loads on each vehicle. Different transit assignment models based on schedule-based approach are proposed in literature ([15], [19], [20], [21] and [16]). For not-congested networks, transit assignment is simulated with a whit-in day dynamic network loading model which allows to simulate at each time load on links representing services.

2.5   Revenues Model

Management revenues are estimated through revenues models which differ according to the fare system, that can be linear or not and can operate in an integrated context or not ([5] and [6]).

Table 1 presents revenues models for different fare systems. Models (8) and (12) are applicable having as input demand flows ($d_{O/D}$), while models (7), (9), (10) and (11) are applicable after performing the transit assignment and, generally, need as input demand flows ($d_{O/D}$), flows (f) on some links, the two components of the fare model (p_0 and p_v).

Model (7) is applicable for revenues calculation for a not integrated constant fare system, simulated by equation (3). Model (9) is related to a not integrated distance linear fare system (Equation 4); it needs as input, for every run (r), flows on boarding links (f_{tr}), flows on on-board links (f_{br}), additive link fare associated to each on-board link depending on link length (p_{vr}). Model (10) is referred to an integrated distance linear structure and it does not require flows on boarding links. Model (11) is applicable for integrated zone linear fare systems and needs as input flows on on-board link crossing zone boundaries (f_w) and additive variable fare associated to each on-board link crossing zone boundaries (p_{vw}).

Table 1. Revenues models and required inputs

Fare system	Integr.	Input	Model	
Constant	No	p_0, f_{tr}	$\sum_r \sum_t p_0 f_{tr}$	(7)
	Yes	$p_0, d_{O/D}$	$p_0 \sum_{O/D} d_{O/D}$	(8)
Distance	No	$p_0, p_{vr}, f_{tr}, f_{br}$	$\sum_r \sum_t p_0 f_{tr} + \sum_r \sum_b f_{br} p_{vr}$	(9)
	Yes	$p_0, p_{vr}, d_{O/D}, f_{br}$	$p_0 \sum_{O/D} d_{O/D} + \sum_r \sum_b f_{br} p_{vr}$	(10)
Zone	Yes	$p_0, p_{vw}, d_{O/D}, f_w$	$p_0 \sum_{O/D} d_{O/D} + \sum_w f_w p_{vw}$	(11)
O-D	Yes	$p_{O/D}, d_{O/D}$	$\sum_{O/D} p_{O/D} d_{O/D}$	(12)

t, boarding link; w, on-board link crossing zone boundaries; b, on-board link; r, run; $d_{O/D}$, demand flow on O/D couple; f, link flow.

3   Application to the Province of Reggio Calabria

The general procedure is applied to the transit system of the Province of Reggio Calabria, located in the south of Italy. The general objective, in this case, is the simulation of effects on users and on fare revenues deriving from the modification of a season-ticket (monthly ticket) for users who travel systematically inside the Province for work and study purposes. In particular, two fare systems are simulated:

- the current not integrated distance fare system, where each company sells its own monthly ticket (the price depends on traveled distance between origin and destination) to users not valid for services provided for other companies (users which make modal/service transfers need to buy multiple tickets);
- an integrated zone fare system, where companies sell a monthly ticket (the price depends on the number of crossed zone boundaries between origin and destination), valid for all services/modes connecting origin and destination.

The steps for the application of the procedure are: supply representation, through the network graph (nodes for spatial location of bus stops and rail stations, links for spatial connection of nodes, intermodal transfer nodes) and the service timetable (temporal representation of runs: leaving time from the terminus and arriving/leaving time at stops/stations); fare system definition, through the specification and calibration of fare models simulating the two fare systems; demand estimation for each mode during an average working day and time-varying O/D demand matrices estimation, transit assignment of time-varying O/D demand matrices to the run-based network; revenues estimation.

3.1 Supply Representation

In the examined area, the road system is composed by local and regional roads, while a highway runs along the Thyrrenian coast. The total length of road network is 1097 kilometers. Many transit companies (27) operate in the area with 505 lines/day and 1344 runs/day, providing connections inside the province. Two transit modes are present: bus and rail, providing mono-class services. No intra/intermodal integration (fare and timetable) exists, however a small number of intermodal transfers are present in the major coastal towns. Rail lines run along the coast connecting all costal towns, while bus lines connect each other all coastal towns and hilly villages (Figure 4). Some characteristics of the two transit modes are described in Table 2.

A network graph for each mode provides the spatial representation of transit services, while timetable provides their temporal representation. Mixed mode representation is relatively easy, due to the small number of intermodal transfer nodes and the transit system structure (Figure 4). Therefore, mixed mode is derived from the composition of the rail mode (running along the coast) and of the bus mode (running from sea to mountains and vice versa). Four intermodal transfer nodes are defined (Reggio Calabria, Gioia Tauro, Melito Porto Salvo and Locri), where it is possible the transfer from bus system to rail one and vice versa. Intermodal transfer nodes are selected after an empirical evaluation of the number of bus and train runs stopping at each bus stop and rail station of the network.

Table 2. Bus and rail characteristics

	Bus	Rail
Network length [km]	1097	203
Number of stops	269	36
Number of lines/day	428	77
Number of runs/day	1132	212
Commercial speed [km/h]	56.5	36.2
Number of operators	26	1

3.2 Definition of Fare Systems

Two different fare systems are simulated. The first is a non-integrated distance linear fare system. A fare model is specified and calibrated in a previous study ([5] and [6]) and parameters (Table 3) are calibrated for the two existing modes (bus and train) for a monthly season-ticket:

$$p = p_0 + \chi d \tag{13}$$

Table 3. Parameters of the non-integrated linear distance fare model

Bus		Rail	
p_0 (€)	χ (€/km)	p_0 (€)	χ(€/km)
0.147	0.019	0.352	0.028

The second is an integrated zone linear fare system, in which zone fares are defined according to a previous zone partition. The fare model is specified as follows:

$$p = p_0 + \beta_n \qquad (14)$$

Three fare levels scenarios are defined assigning different values to parameters p_0 and β, as shown in Table 4.

Table 4. Parameters for the integrated zone linear fare model

Scenario	p_0(€)	β(€/n)
1	0.35	0.40
2	0.45	0.50
3	0.55	0.60

The Province is divided in 33 fare zones with equal medium diameter of 10 km (Figure 4). Fare zones are designed in order capture the sea-mountain and coastal trips. Each zone has a convex shape and encloses one or more municipalities.

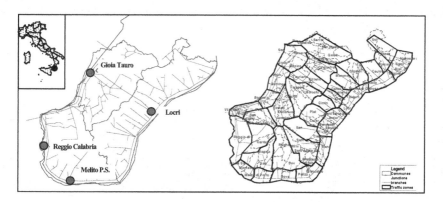

Fig. 4. Transit network and zone partition of the Province of Reggio Calabria

The specification and calibration of the fare models (13) and (14) allow to determine a fare matrix for each fare system, which is necessary in next steps concerning modal demand O/D matrices estimation and transit assignment of time-varying demand flows to the run-based network.

3.3   Demand Estimation

The area, with a population of 580000 inhabitants, has been divided in 97 traffic zones, corresponding to the municipalities. The city of Reggio Calabria is the main attraction/emission center and minor towns are located along the coast. The total daily O/D matrix of systematic extra urban trips in the area for all modes and two purposes (work and study) is provided by national statistics institute ISTAT in 1991.

A disaggregate mode choice model is specified and calibrated in order to split the total daily O/D among transport modes. Modal choice set is composed by four transport modes: car, bus, rail, mixed. The specified mode choice model is a multinomial logit:

$$p(\text{m/od}) = \exp(V_{\text{m/od}})/\sum_{\text{m'}} \exp(V_{\text{m'/od}}) \tag{15}$$

with: $V_{\text{m/od}} = \sum_j \beta_j X_{\text{m}j}$, systematic utility function associated to mode m; β_j, parameters to be calibrated; $X_{\text{m}j}$, attributes of mode m.

The systematic utility function is a linear combination of level-of-service attributes, defined in Table 5.

The values of attributes are average daily values for each O/D couple. Car on-board travel times are determined through a stochastic equilibrium assignment model. Cost for car mode is fuel cost, assuming a unit fuel consumption of 10 Km/liter and a unit fuel cost of 1 €/liter. Bus and rail on-board travel times are determined through a network loading model with a determinist choice of hyper path. Cost for transit modes (fare) is obtained from models (13) and (14) for each O/D couple (fare matrices). For mixed mode, three more average daily attributes are estimated: average headway between two runs (Inter), average daily intermodal transfer time (t_t) in the four transfer nodes, percentage use of rail mode (%rail).

Table 5. Attributes in the systematic utility function

Symbol		Definition
$t_{a/e}$	[hour]	Access/egress time to/from the bus stop or rail station
t_b	[hour]	On-board travel time
C	[€]	Cost for the monthly season-ticket (bus, rail) or fuel cost (car)
Inter	[hour]	Average daily headway between two runs
t_t	[hour]	Average daily intermodal transfer time
%rail	[%]	Percentage use of rail mode in the mixed mode
ASA		Alternative specific attributes

Observed data are obtained from a survey executed during a working day at bus stops, rail stations and on-board. A random sample of more than 500 interviews with workers and students performed in a revealed preference way is available.

Model calibration is performed through the maximum-likelihood method. The results of two different specifications related respectively to work and study purposes are presented in Table 6.

Concerning the model related to work purpose, all parameters are negative (unless the %rail one), as expected, and statistically significant, as t-student statistics shows. The positive sign of %rail parameter expresses the users appreciation for the higher reliability and speed of rail services.

On-board and access/egress parameters respect the ratios that it is possible to find in literature. The value of on-board time is satisfactory, while the value of access/egress time seems to be slightly high, due to the relative high average distances of the access terminal to the service from the origin (and of the final destination from the egress terminal). Finally, the goodness of fit statistic is acceptable ($\rho^2 = 0.482$).

Table 6. Parameters calibration of modal choice models for work and study purposes

Mode	Attributes	Model Work		Study	
		β	t-statistics	β	t-statistics
Bus	$t_{a/e}$	−1.437	−12.8	−3.687	−13.5
	t_b	−0.477	−8.0	−1.775	−13.1
	C	−0.054	−4.9	−0.408	−12.7
	Inter	−0.030	−8.0	−0.026	−6.6
	ASA	−0.749	−12.0	- - -	- - -
Rail	$t_{a/e}$	−1.437	−12.8	−3.687	−13.5
	t_b	−0.477	−8.0	−1.775	−13.1
	C	−0.054	−4.9	−0.408	−12.7
	Inter	−0.030	−8.0	−0.026	−6.6
	ASA	−1.147	−16.4	−0.883	−16.7
Mixed	$t_{a/e}$	−1.437	−12.8	−3.687	−13.5
	t_b	−0.477	−8.0	−1.775	−13.1
	C	−0.054	−4.9	−0.408	−12.7
	Inter	−0.030	−8.0	−0.026	−6.6
	t_t	−0.922	−3.4	−0.458	−2.5
	%rail	1.737	2.8	- - -	- - -
	ASA	−3.977	−7.4	−3.533	−14.2
Car	t_b	−0.477	−8.0	- - -	- - -
	C	−0.054	−4.9	- - -	- - -
ρ^2		0.482		0.388	
V.O.T.(a/e)		25.75		9.04	
V.O.T.(b)		7.50		4.35	

As far as concern the model for study purpose, the access/egress and cost parameters are greater than the previous case. The values of the time are lower, as expected. Finally, the goodness of fit statistic is less than the previous one ($\rho^2 = 0.338$).

The two specified and calibrated modal demand models are applied to estimate the daily O/D demand matrices for the three transit modes with the current fare system (not integrated linear distance) and for the three fare scenarios of the integrated zone linear fare system. The results are presented in Table 7. The first observation is that mixed mode in all cases attracts an extremely small number of users; this is due to the lack in the area of any modal/service integration, that make transfers very burdensome. Transit daily demand for current distance fare system is equal to 23128 pax/day. Transit daily demand for scenario 1 ($p_0 = 0.35€; \beta = 0.40€/n$) of zone fare system is 23438 pax/day; while, for scenarios 2 ($p_0 = 0.45€; \beta = 0.50€/n$) and 3 ($p_0 = 0.55€; \beta = 0.60€/n$), demand is respectively 23136 pax/day and 22805 pax/day. Table 7 reports also the demand for each mode and percentage demand variation related to the current fare distance system.

Table 7. Modal demand for each simulated fare systems

Fare system			Demand [pax/day]			
			Bus	Rail	Mixed	Total
Distance		Abs	15067	7681	380	23128
Zone	1	Abs	15262	7780	396	23438
		$\Delta\%$	+1.29	+1.29	+4.21	+1.34
	2	Abs	15026	7724	386	23136
		$\Delta\%$	−0.27	+0.56	+1.58	+0.03
	3	Abs	14815	7621	369	22805
		$\Delta\%$	−1.70	−0.80	−2.90	−1.40

Abs = absolute demand values,
$\Delta\%$ = percentage demand variation related to current distance fare system.

In order to obtain time-varying O/D matrices to be assigned to the run-based network, the simulation period of an average working day (from 4:30 to 20:00) is discretized into 64 time-slices of 15 minutes. A desired arrival time (DAT) at destination and a desired departure time (DDT) from the origin are associated to each time-slice. Then, DAT and DDT distributions for each travel purpose (work and study), assumed to be rigid to the simulated fare systems, are obtained on the base of previous researches on the same area [22] and on a similar Italian regional area [23]. DAT and DDT distributions for work and study purposes are presented respectively in Figures 5 and 6.

In Figure 5, DAT distribution for work purpose has a morning peak value between 7:00 and 7:30 and has low values after 14:00, while DDT distribution has two peaks at 14:00 and 18:00. Figure 6 shows that DAT and DDT distributions for study purpose are very concentrated with peak values respectively at 8:00 and 13:00.

At this point, as home-living and returning trips are executed inside the average working day, each daily modal O/D matrix is equally divided into two sub-matrices for home-living trips and for returning trips. At the end, each sub-matrix is splitted into 64 time-slice O/D matrices for trip purpose, according to

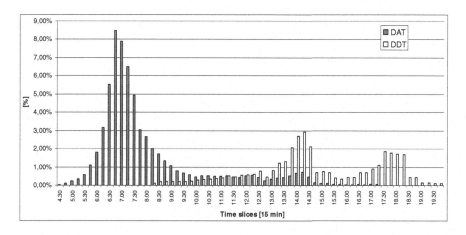

Fig. 5. DAT and DDT distributions for work purpose

distributions of Figures 5 and 6. For each simulated fare scenario, it is necessary to obtain totally 256 time-slice O/D matrices.

Concerning path choice simulation, path choice set is composed of paths selected according to some criteria related to DDT from origin, DAT at destination, maximum earliness and lateness accepted from users [19]. Systematic utility for each path is a generalized cost given from a weighted sum of time and monetary additive costs: access and egress times $(t_{a/e})$, on-board time (t_b), boarding/transfer time (t_t), schedule penalty, fare (p). Generalized cost for path k (C_k) is specified, for a given DDT from origin, as follows:

$$C_k(\text{DDT}) = \beta_{a/e}t_{a/e} + \beta_b t_{b,k} + \beta_t t_{t,k} + \beta_c p_k + \beta_{\text{EDT}}\text{EDT}_k + \beta_{\text{LDT}}\text{LDT}_k \quad (16)$$

and for a given DAT at destination:

$$C_k(\text{DDT}) = \beta_{a/e}t_{a/e} + \beta_b t_{b,k} + \beta_t t_{t,k} + \beta_c p_k + \beta_{\text{EAT}}\text{EAT}_k + \beta_{\text{LAT}}\text{LAT}_k \quad (17)$$

with schedule penalties given by:

- EDT_k and LDT_k, Early Departure Time and Late Departure Time (difference between DDT and the scheduled departure time);
- EAT_k and LAT_k, Early Arrival Time and Late Arrival Time (difference between DAT and the scheduled arrival time).

Obviously, EDT_k and LDT_k attributes in (16) and EAT_k and LAT_k in (17) are mutually exclusive. Maximum earliness and lateness is 60 minutes for a given DDT/DAT both for work and study purposes. Fare for each path k (p_k) depends on the simulated linear fare systems (Eq. 13 and 14). It is obtained as sum of:

- access fare component (p_0), associated to the boarding link;
- variable on-board fare component (p_v), associated to the (or some) on-board links.

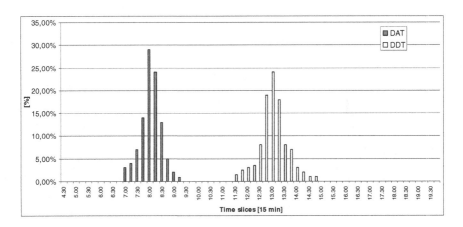

Fig. 6. DAT and DDT distributions for study purpose

Fare is, then, converted into generalized cost by means of β_c, which is represented by the inverse of value of time for users.

The values of parameters β for path attributes are presented in Table 8. They are obtained from values presented in [16], relating parameters of different time attributes to the on-board time parameter ($\beta_b = 1.00$). In order to penalize the number of transfers, parameter β_t is equal to 10.00, estimated from an aggregate calibration. The value of parameter β_c is equal to 8.00 (dis/€) for work purpose and to 13.80 (dis/€) for study purpose.

Path choice model is deterministic and fully pre-trip and provides the minimum shortest space-time path on the run-based network from each origin to each destination.

Table 8. Parameters for path attributes

Parameters		Work	Study
$\beta_{a/e}$	[dis/min]	3.05	0.38
β_b	[dis/min]	1.00	1.00
β_t	[dis/min]	10.00	10.00
β_{EDT},β_{EAT}	[dis/min]	1.64, 1.70	2.92, 3.60
β_{LDT},β_{LAT}	[dis/min]	1.64, 1.70	2.92, 3.60
β_c	[dis/€]	8.00	13.80

dis = disutility

3.4 Transit Assignment

Transit assignment is based on a schedule-based approach on not congested run-based network. It is performed with the support EMME/2© software [24], which considers a minimum cost path algorithm on a space-time network. As space-time network can become very large, the algorithm generates dynamically the part of the network that is actually needed for the computations (instead of explicitly building the whole network). The algorithm computes paths either forward (starting from the origin), for trips with a desired departure time, or backward (starting from the destination) for trips with a desired arrival time, implicitly generating the minimum cost space-time path for each origin-destination couple ([24] and [19]).

3.5   Revenues Estimation

Revenues are calculated for each simulated fare scenario and each mode (bus, rail and mixed), considering as inputs loads on runs from transit assignment and fares from fare models (13) and (14).

Revenues are divided considering the access fare component (p_0) and the variable on-board fare (p_v). This last one is calculated according to model (13) for the not integrated distance fare system (with parameters from Table 3) and according to model (14) for the integrated zone fare system (with parameters from Table 4 for each simulated scenario).

Table 9 shows the results of revenues estimation. Total daily revenues for the not integrated distance fare system is equal to 26960 €/day. Total daily revenues for scenario 1 ($p_0 = 0.35€$; $\beta = 0.40€/n$) of integrated zone linear fare system is 19107 €/day, with a reduction of 29.10% related to the current fare system. Concerning scenarios 2 ($p_0 = 0.35€$; $\beta = 0.40€/n$) and 3 ($p_0 = 0.55€$; $\beta = 0.60€/n$), revenues are respectively 24550 €/day (-8.90%) and 28849 €/day ($+7.00\%$). Revenues for mixed mode are always negligible, as expected.

Table 9. Daily revenues estimation for the simulated fare systems

Fare system		Revenues [€/day]			
		Bus	Rail	Mixed	Total
Distance	Access	3375	2719	109	6202
	On-board	14095	6458	204	20758
	Total	17470	9177	313	26960
Zone 1	Access	5307	2696	89	8002
	On-board	8069	3036	132	11105
	Total	13375	5732	221	19107
	Δ%	−23.40	−37.50	−29.40	−29.10
2	Access	6762	3476	144	10238
	On-board	10015	4298	171	14313
	Total	16777	7773	315	24550
	Δ%	−4.00	−15.30	+0.60	−8.90
3	Access	8203	4212	231	12415
	On-board	11978	4456	199	16434
	Total	20181	8668	430	28849
	Δ%	+15.50	−5.50	+37.40	+7.00

Δ% = percentage variation related to current fare system.

Figure 7 shows the percentage differences in daily revenues ($\Delta\%$ revenues) and in daily transit demand ($\Delta\%$ demand) between the three scenarios of the integrated zone linear fare system and the not integrated distance fare system one. Concerning scenario 1, revenues reduction is -29.10%, while transit demand increment is $+0.50\%$ ($+119$ pax/day). In scenario 2 there is a reduction of revenues of -8.9%, while transit demand is not changed. Finally, in scenario 3 there is an increment of $+7.00\%$ in revenues versus a reduction in transit demand of -0.8% (-182 pax/day).

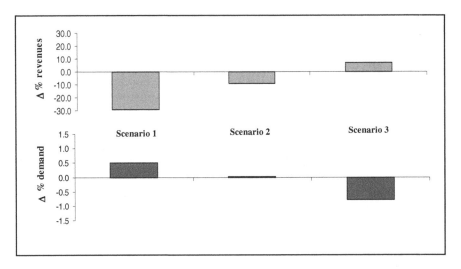

Fig. 7. Percentage differences in revenues and demand among the three simulated zone fare systems and the distance one

4   Conclusions and Research Perspectives

The paper presents a general procedure for the estimation of effects on demand and on transit revenues due to the definition of an integrated fare system in regional transit services. The necessary input data for a modeling approach are pointed out and some fare system models are specified. Finally, the procedure is applied to the Province of Reggio Calabria, Italy. Two fare systems are simulated and the results are highlighted in terms demand and revenues.

The procedure allows to simulate demand elasticity at mode and path choice dimensions, to simulate the effects of integrated linear fare systems and to explicitly simulate the mixed mode, composed by two transit modes, through the definition of intermodal transfer nodes. Non-linear fare systems can not actually be simulated, due to their non-additive nature. The simulation will be possible, using algorithms which are able to explicitly enumerate all paths on the network, which is now very costly, both in terms of CPU time and memory space.

Future research will concern the analysis of effects of integrated fare systems on other classes of users (occasional), a deeper analysis on demand with the support of stated preferences investigations, and a more disaggregate analysis of service revenues for each transit company. Moreover, it will be investigated the possibility to simulate integrated non-linear fare systems, by means of algorithms which are able to explicitate paths on the network.

References

1. Montella, B., Gallo, M., D'Acierno, L.: Stato dell'arte sulle strutture tariffarie nelle aziende di TPL. In: Russo, F. (ed.) Modelli e metodi per la programmazione dei servizi di trasporto pubblico locale: uno stato dell'arte, FrancoAngeli, Milan, pp. 101–116 (2002)
2. Gattuso, D., Carbone, G., Chindemi, A.: Integrazione modale ed unificazione tariffaria del TPL. Esperienze recenti a scala sovracomunale. In: Russo, F. (ed.) Modelli e metodi per la programmazione dei servizi di trasporto pubblico locale: uno stato dell'arte, FrancoAngeli, Milan, pp. 84–100 (2002)
3. Gattuso, D., Musolino, G.: Integrazione tariffaria a scala regionale nel TPL. Analisi comparata di cinque esperienze europee avanzate. Rivista Trasporti e Territorio (April 2004)
4. Italian Railways: Integrazione tariffaria. Orientamenti tecnico operativi per la progettazione di un sistema tariffario integrato nell'ambito delle grandi aree metropolitane. Internal Report (1995)
5. Gattuso, D., Carbonea, G., Chindemi, A.: A methodology for fare integration in transit systems: application to an Italian extra-urban area. In: Proceedings of AET Conference 2002 Cambridge (2002)
6. Gattuso, D., Carbone, G., Chindemi, A.: Una metodologia per l'integrazione tariffaria nei sistemi di trasporto pubblico: applicazione ad un caso reale. In: Russo, F. (ed.) Modelli e metodi per la programmazione dei servizi di trasporto pubblico locale: applicazione a casi reali, FrancoAngeli, Milan, pp. 240–260 (2003)
7. A.A., V.V.: Linee Guida per la programmazione dei servizi di Trasporto Pubblico Locale. Ministero dell'Istruzione, dell'Università e della Ricerca. Laruffa Editore, Reggio Calabria (2002)
8. Schöbel, A.: Zone planning in public transportation. In: Bianco, L., Toth, M. (eds.) Advanced Methods in Transportation Analysis, pp. 117–134. Springer, Heidelberg (1996)
9. Schöbel, A., Schöbel, G.: Wabplan. A software tool for design and evaluation of tariff systems. Internal report (1999)
10. D'Acierno, L., Gallo, M., Montella, B.: Un modello per la determinazione delle tariffe ottimali per il Trasporto Pubblico Locale. In: Cantarella, G., Russo, F. (eds.) Metodi e Tecnologie dell'Ingegneria dei Trasporti, Seminario 2001, FrancoAngeli, Milano, pp. 256–272 (2001)
11. Pratelli, A., Schoen, F.: Un modello di zonizzazione ottimale per sistemi di trasporto pubblico integrati. In: Proceedings of Metodi e tecnologie dell'ingegneria dei trasporti 10-12 Dicembre 2002, Reggio Calabria. Forthcoming (2002)
12. Hamacher, H.W., Schöbel, A.: Design of zone tariff systems in public transportation. Operations Research 52(6), 897–908 (2004)
13. Babel, L., Kellerer, H.: Design of tariff zones in public transportation networks: theoretical results and heuristics. In: Mathematical Methods of Operations Research, pp. 1–16. Springer, Heidelberg (2003)

14. Anez, J., de la Barra, T., Perez, B.: Dual graph representation of transport networks. Transportation Research 30B, 209–216 (1996)
15. Florian, M.: Deterministic Time Table Transit Assignment Preprints of PTRC Seminar on National models, Stockholm (1998)
16. Nuzzolo, A., Russo, F., Crisalli, U.: Transit network modeling. The schedule-based dynamic approach. FrancoAngeli, Milan (2003)
17. Ben-Akiva, M., Lerman, S.R.: Discrete choice analysis. MIT Press, Cambridge, Mass (1985)
18. Cascetta, E.: Transportation systems engineering: theory and methods. Kluwer, Dordrecht (2001)
19. Florian, M.: Finding shortest time-dependent paths in Schedule-Based transit networks: a Label Setting algorithm. In: Wilson, R., Nuzzolo, A. (eds.) Schedule-Based Dynamic Transit Modeling. Theory and Applications, pp. 96–112. Kluwer Academic Publishers, Dordrecht (2003)
20. Nielsen, O.A., Jovicic, G.: A large scale stochastic timetable-based transit assignment model fro route and sub-mode choices. In: Proceedings of 27th European Transportation Forum, Seminar F, pp. 169–184. Cambridge, England (1999)
21. Nguyen, S., Pallottino, S., Malucelli, F.: A modeling framework for the passenger assignment on transportation network with time-tables. Transportation Science 35, 238–249 (2000)
22. Postorino, M.N., Musolino, G., Velonà, P.: A methodology for demand evaluation by traffic counts in transit systems: application to an Italian extra-urban area. In: Proceedings of AET Conference 2002, Cambridge (2002)
23. NetEngineering, Il sistema ferroviario metropolitano regionale dell'area centrale veneta. Grafica Atestina-Este, Padova (2001)
24. INRO, EMME/2© user's manual (release 9.0). Montreal, Canada (2002)

Intelligent Train Scheduling on a High-Loaded Railway Network

Antonio Lova[1], Pilar Tormos[1], Federico Barber[2], Laura Ingolotti[2],
Miguel A. Salido[2], and Monsterrat Abril[2]

[1] DEIOAC, Universidad Politecnica de Valencia, Spain
{ptormos,allova}@eio.upv.es
[2] DSIC, Universidad Politecnica de Valencia, Spain
{fbarber,lingolotti,msalido,mabril}@dsic.upv.es

Abstract. We present an interactive application to assist planners in adding new trains on a complex railway network. It includes many trains with different characteristics, whose timetables cannot be modified because they are already in circulation. The application builds the timetable for new trains linking the available time slots to trains to be scheduled. A very flexible interface allows the user to specify the parameters of the problem. The resulting problem is formulated as a CSP and efficiently solved. The solving method carries out the search assigning values to variables in a given order verifying the satisfaction of constraints where these are involved. When a constraint is not satisfied, a guided backtracking is done. Finally, the resulting timetable is delivered to the user who can interact with it, guaranteeing the traffic constraint satisfaction.

1 Introduction

The problem considered by our application is to obtain a valid and quality scheduling for new trains on a railway network, which may or may not be occupied by other trains. The timetables for the new trains are obtained in a search space that is limited by traffic constraints, user requirements, railway infrastructure and network occupation. The system displays the resulting schedules graphically and provides an interactive interface so that the user can modify them. Planning rail traffic problems are basically optimization problems which are computationally difficult to solve. These problems belong to the NP-hard class of problems. Several models and methods have been analyzed to solve these problems [1], [3], etc. The majority of the papers published in the area of periodic timetabling in the last decade are based on the Periodic Event Scheduling Problem (PESP) introduced by Serafini and Ukovich [8]. Specifically, an efficient model that uses the PESP and the concept of symmetry is proposed in [6]. However, we cannot use the concept of symmetry because: (i) we allow different types of trains; this doesn't guarantee the necessary symmetry to be able to use these

F. Geraets et al. (Eds.): Railway Optimization 2004, LNCS 4359, pp. 219–232, 2007.

models, and (ii) the use of the infrastructure may not be symmetrical. There are works that are related to railway problems such as: the allocation of n trains in a station minimizing the number of used tracks and allowing the departure of trains in the correct order [4], the allocation of new stations through the railway network to increase the number of users [7], etc. There exist tools to solve certain kinds of problems such as the Rescheduling tool [2] or the TUFF scheduler [5]. The Rescheduling tool allows to modify a timetable when trains in a section of track cannot run according to the infrastructure, ensuring that the scheduling rules are not violated. The TUFF scheduler describes a constraint model and solver for scheduling trains on a network of single tracks used in both directions. Computer-aided systems for railways problems have become a very useful tool in assisting planners to obtain an efficient use of railway infrastructures.

2   Problem Description

We propose to add new trains on an heterogeneous high-loaded railway network, minimizing the traversal time of each new train. The timetables for the new trains will be obtained in a search space that is limited by traffic constraints, user requirements, railway infrastructure and network occupation. The problem specification does not demand that all considered trains visit the same sequence of locations. It may there be many types of trains implying different velocities, security margins, commercial stops and journeys. Our method takes into account the following scenario to generate the timetables corresponding to the new trains:

1. two sets of ordered locations $L_D = \{l_k, l_{k+1}, ..., l_{k+m}\}$ and $L_U = \{l_h, l_{h+1}, ..., l_{h+n}\}$, such that $\{\exists i, j \backslash l_i \in L_D \wedge l_{i+1} \in L_D \wedge l_j \in L_U \wedge l_{j+1} \in L_U \wedge l_i = l_{j+1} \wedge l_{i+1} = l_j\}$. A pair of adjacent locations can be joined by a single or double track section. L_D and L_U correspond to the journey of trains going in down and up direction, respectively; and they are part of a railway line $L = \{l_0, ..., l_z\}$, $(L_D \subset L, L_U \subset L)$.

2. a set of trains for each direction. $T_D = \{t_0, t_2, ..., t_d\}$ is the set of trains that visit the locations in L_D in the same order given by this sequence and we said that these trains go in *down direction*. $T_U = \{t_1, t_3, ..., t_u\}$ is the set of trains that visit the locations in L_U in the same order given by this sequence and we said that these trains go in *up direction*. The subscript i in the variable t_i indicates the departure order among the new trains going in the same direction.

3. a journey for each set of trains (T_U and T_D) specifies the traversal time for each section of track in L_D and in L_U ($R_{i \rightarrow i+1}$), and the minimum stop time (S_i) for commercial purposes in each l_i.

Considering $t_y \mathrm{dep\_}l_x$ and $t_y \mathrm{arriv\_}l_x$ as the departure and arrival times of train t_y from/at location l_x, the problem consists in finding the running map that minimizes the average traversal time, satisfying all the following constraints:

- *Initial departure time.* The first train must leave from the first station of its journey within a given time interval ($[min_D, max_D]$ for trains going in down direction and $[min_U, max_U]$ for trains going in up direction).

$$min_D \leq t_0 dep\_l_k \leq max_D \wedge min_U \leq t_1 dep\_l_h \leq max_U. \tag{1}$$

- *Frequency of Departure.* It specifies the period (F_U/F_D) between departures of two consecutive trains in each direction from the same location,

$$\{\forall t_i, l_j \backslash t_i \in \{T_D - \{t_d\}\} \wedge l_j \in \{L - \{l_{k+m}\}\}\}, t_{i+2} dep\_l_j = t_i dep\_l_j + F_D. \tag{2}$$

$$\{\forall t_i, l_j \backslash t_i \in \{T_U - \{t_u\}\} \wedge l_j \in \{L - \{l_{h+n}\}\}\}, t_{i+2} dep\_l_j = t_i dep\_l_j + F_U. \tag{3}$$

- *Minimum stops.* A train must stay in a location l_j at least S_j time units,

$$\{\forall t_i, l_j \backslash t_i \in \{T_D \cup T_U\} \wedge l_j \in \{L_D \cup L_U - \{l_k, l_h, l_{k+m}, l_{h+n}\}\}\},$$

$$t_i dep\_l_j \geq t_i arriv\_l_j + S_j. \tag{4}$$

- *Exclusiveness.* A single track section must be occupied by only one train at the same time.

$$\{\forall t_j, t_i, l_x, l_y / t_j \in T_D \wedge t_i \in T_U \wedge l_x \in L_D - \{l_{k+m}\} \wedge$$
$$l_y \in L_U - \{l_h\} \wedge l_x = l_{y+1} \wedge l_{x+1} = l_y\},$$

$$t_i dep\_l_y \geq t_j arriv\_l_y \vee t_j dep\_l_x \geq t_i arriv\_l_x. \tag{5}$$

- *Reception Time.* At least are required R_x time units at location l_x between the arrival times of two trains going in the opposite direction (Figure 1).

$$\{\forall t_j, t_i, l_x / t_j \in T_D \wedge t_i \in T_U \wedge l_x \in \{L_D - \{l_k\} \cap L_U - \{l_h\}\}\},$$

$$t_j arriv\_l_x \geq t_i arriv\_l_x + R_x \vee t_i arriv\_l_x \geq t_j arriv\_l_x + R_x. \tag{6}$$

- *Expedition Time.* At least are required E_x time units at location l_x between the arrival and departure times of two trains going in the opposite direction (Figure 1).

$$\{\forall t_j, t_i, l_x / t_j \in T_D \wedge t_i \in T_U \wedge l_x \in \{L_D - \{l_{k+m}\} \cap L_U - \{l_{h+n}\}\}\},$$

$$t_j dep\_l_x \geq t_i arriv\_l_x + E_x \vee t_i dep\_l_x \geq t_j arriv\_l_x + E_x . \tag{7}$$

- *Precedence Constraint.* Each train employs a given time interval ($R_{x \to x+1}$) to traverse each section of track ($l_x \to l_{x+1}$)in each direction.

$$\{\forall t_i, t_j, l_x, l_y / t_i \in T_D \wedge t_j \in T_U \wedge l_x \in \{L_D - \{l_{k+m}\}\} \wedge l_y \in \{L_U - \{l_{h+n}\}\}\},$$

$$t_i arriv\_l_{x+1} = t_i dep\_l_x + R_{x \to x+1} \wedge t_j arriv\_l_{y+1} = t_j dep\_l_y + R_{y \to y+1}. \tag{8}$$

- *Capacity of each station.* The number of trains that may stay simultaneously in a station depends on the number of available tracks in it.
- *Closure times.* Traffic operations and/or passing of trains are not allowed at the closing times of a station.

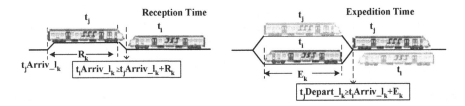

Fig. 1. Reception and Expedition time constraint

3   Solving Method: The Sequential Algorithm

In this section we explain the used algorithm to solve the specified problem in Section 2. We have named it *"Sequential"* because of the way that it generates the timetable for each new train (Figure 2). For each iteration, the Sequential

```
01    procedure Sequential_Algorithm(I,C)
02    begin
03       while(Enough_Time())
04          S=Generate_Set_Ref_Station()
05          ref_st=Get_Ref_Station(S)
06          init_dep_down=Get_Init_Dep_Time(min_D,max_D)
07          init_dep_up=Get_Init_Dep_Time(min_U,max_U)
08          sched1=Get_Partial_Sched(init_dep_down,l_k,ref_st,T_D)
09          sched2=Get_Partial_Sched(init_dep_up,l_h,ref_st,T_U)
10          init_dep_down=t_0arriv_l_ref_st+S_ref_st
11          init_dep_up=t_1arriv_l_ref_st+S_ref_st
12          sched3=Get_Partial_Sched(init_dep_down,ref_st,l_{k+m},T_D)
13          sched4=Get_Partial_Sched(init_dep_up,ref_st,l_{h+n},T_U)
14          new_sched=sched1+sched2+sched3+sched4
15          if(Is_Better(new_sched,best_sched))
16             best_sched=new_sched
17       end while
18       Show(best_sched)
19    end
```

Fig. 2. Sequential Algorithm

algorithm constitutes a subset of the whole search space where it searches the values for the problem variables that satisfy all the problem constraints (Section 2). The assignment of valid values to the problem variables generates (if there is a feasible solution in the subset) a timetable for each new train (line 08-14 in Figure 2). The elements of the reduced search space depend on three values, which are:

1. *Initial departure time* for the first train going in down direction (init_dep_down in Figure 2). In each iteration is chosen randomly a value that belongs to the time interval $[min_D, max_D]$ (Constraint 1 in Section 2). This time interval is part of the input parameters I (one of the input parameters of the Sequential Algorithm in Figure 2) given by the final user.

2. *Initial departure time* for the first train going in up direction (init_dep_up in Figure 2). In each iteration is chosen randomly a value that belongs to the time interval $[min_U, max_U]$ (Constraint 1 in Section 2). This time interval is part of the input parameters I (one of the input parameters of the Sequential Algorithm in Figure 2) given by the final user.

3. *Reference Station* (ref_st in Figure 2). When the assigned value to a problem variable, that represents the departure time of a train, violates the Constraint 5 defined in Section 2 (avoid crossing between trains going in opposite direction), the process must decide which of the two trains will have to wait for the section track release. The decision taken by the process will state a priority order between the trains competing by the same resource, the single track section.

When the two trains competing by a single track section are: a train in circulation and a new train, that should be added to the railway network, the priority order always will be the same. The new train will have to wait until the train in circulation releases the single track section.

When the two trains competing by a single track section are new trains, the priority order is decided according to the position of the single track section with respect to one station, which we name the *reference station*. The *reference station* divides the journey of each new train in two parts. The first part: from the initial station of the journey, to the *reference station*. The second part: from the *reference station*, to the last station of the journey. Each train will have priority on the single track sections that belong to the first part of its journey. In Figure 3, S_2 is the reference station and it divides the journey of each new train in two parts. For the trains going in down direction (t_0, t_2 and t_4) the first part of theirs journey is composed of the track sections (S_0-S_1) and (S_1-S_2), the second part is composed of (S_2-S_3) and (S_3-S_4). For the trains going in up direction (t_1, t_3 and t_5) the first part of theirs journey is composed of (S_4-S_3) and (S_3-S_2), the second part is composed of (S_2-S_1) and (S_1-S_0). In Figure 3 is pointed out by a circle the possible crossing between the trains t_0 and t_1 in case that t_0 left from S_2 as soon as possible. The single track section (S_2-S_3) belongs to the first part of t_1 journey and therefore this train has greater priority than t_0 on this track section. Then, t_0 will have to wait until the single track section (S_2-S_3) is released by the train t_1. The pointed lines represent the position of the train t_0 if it had left from the station S_2 as soon as possible. The continued lines represent the real departure time of the train t_0 after the single track section had been released by t_1. The same reasoning is applied by the other cases pointed out by circles in Figure 3.

The sequential algorithm iterates until the time given by the user has been spent completely or until the user interrupts the execution (line 03 in Figure 2). For each iteration the sequential algorithm compares the obtained scheduling with the best timetable obtained until that time. The scheduling that produces the least average traversal time for each new train is the best scheduling. The function $Is\_Better(new\_timetable, best\_timetable)$ returns TRUE if the new

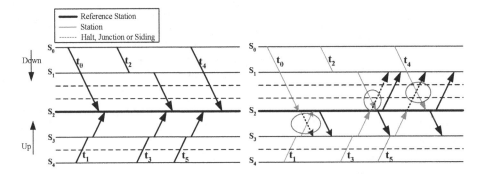

Fig. 3. Priority Order between new trains, defined by a Reference Station

scheduling (*new_sched* in Figure 2), obtained in the current iteration, is better than the best scheduling (*best_sched* in Figure 2), obtained until that time. Finally, the sequential algorithm returns the best scheduling that has been obtained during the running time.

3.1   Description of an Iteration of the Sequential Algorithm

In each iteration, the sequential algorithm generates a complete scheduling for each group of trains (T_D and T_U) in two steps. In Figure 2, *sched1* and *sched2*

```
01    procedure Get_Partial_Sched(init_dep,first_st,last_st,T)
02    begin
03       st=first_st
04       dep_time=init_dep
05       while(st != last_st)
06          next_st=Get_Next_St(st)
07          i=Get_First_Train(T)
08          while(t_i !=NULL)
09             error=Verify_Constraints(st,next_st,dep_time,t_i)
10             if(error>0)
11                if(Is_Required_Frequency())
12                   i=Get_First_Train(T)
13                   t_i dep_l_st=t_i dep_l_st+error
14                end if
15             else
16                if(Is_Required_Frequency())
17                   dep_time=t_i dep_l_st+F_T
18                else
19                   dep_time=t_{i+2} arriv_l_st+S_T
20                end if
21                i=i+2
22             end if
23          end while
24          st=next_st
25       end while
26    end
```

Fig. 4. Partial Scheduling

```
01    int function Verify_Constraints(st,next_st,dep_time,t_i)
02    begin
03       k=st
04       m=next_st
05       trav_time=Get_Traversal_Time(st,next_st)
06       t_i dep_l_k=dep_time
07       t_i arriv_l_m=dep_time+trav_time
08       error=Verify_Crossing(t_i dep_l_k,t_i arriv_l_h)
09       if(error=0)
10          error=Verify_Overtaking(t_i dep_l_k,t_i arriv_l_h)
11          if(error=0)
12             error=Verify_Availability_Tracks(t_i arriv_l_h)
13             if(error=0)
14                error=Verify_Closure_Time(t_i arriv_l_h)
15                if(error=0)
16                   Set_Timetable(st,next_st)
17                end if
18             end if
19          end if
20       return error
21    end
```

Fig. 5. Constraints Verification and Timetable Assignment

correspond to the generated scheduling for the first part of the new trains journey going in down and up direction respectively. In Figure 2, *sched3* and *sched4* correspond to the generated scheduling for the second part of the new trains journey going in down and up direction respectively. The complete scheduling is generated in this way in order to state greater priority to each new train on the first part of its journey. In case of existing a crossing possibility in a track section, the greater priority will be given to the new train whose timetable had been assigned previously on that section track. The first and second part of a journey are stated by the chosen reference station at the current iteration. Figure 4 shows how is generated the scheduling for one part of the whole journey. $Verify\_Constraints(st, next\_st, dep\_time, t_i)$ verifies that all problem constraints are satisfied by the train t_i in the track section limited by the stations st and $next\_st$. This function returns the time that must be added to the departure time of t_i in order to satisfy the violated constraint. If the function returns 0, then none constraint has been violated. Considering the set $L' = \{l_x, l_{x+1}, ..., l_{x+p}\}$ as the ordered set of locations visited by the train t_i, from $l_x = st$ to $l_{x+p} = next\_st$. This function assigns values to the variables $ti\_dep\_l_j$ and $ti\_arriv\_l_h$ such that $x <= j < x+p$ and $x < h <= x+p$ (Figure 5). Figure 6 shows an example of how is verified the constraint that avoids a crossing between trains going in opposite directions. Consider that train t_2 is the train t_i, the train whose timetable is being created, and t' is the train whose timetable has been created previously and cannot be modified. The initial value assigned as departure time from st=l_x to t_2 causes a crossing of t' with t_2 according to the picture on the left of Figure 6. The function $Verify\_Constraints$ computes the time necessary that must be added to dep_time in order to avoid this crossing and this value is returned by the function. The picture on the right of Figure 6 shows the necessary delay in the departure time of t_2 from S_2.

The precedence constraint (Constraint 8 in Section 2) is used to propagate the values among the variables($trav\_time$ in Figure 5 is the spent time to go from st to $next\_st$). Minimum stops constraint(Constraint 4) is used to compute the departure time of a train from its arrival time at the same location(S_T in Figure 4). $Get\_Next\_Station(st, T)$ returns the next station to st in the journey corresponding to trains belonging to T. $Get\_First\_Train(T)$ returns the first train that starts the journey corresponding to trains belonging to T. $Is\_Required\_Frequency(st)$ returns TRUE if all the trains must keep the same departure frequency in the station st.

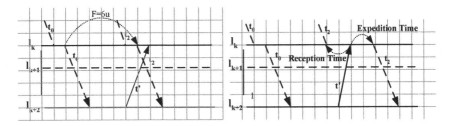

Fig. 6. A crossing detected in a section of track

This solving procedure is the core of a system to assist rail operators in the decision making process. The general architecture of that system is described in the following section.

4   Using the System

We have developed an "Aid System for Train Scheduling" (ASTS) to assist rail operators in the planning and use of railways. ASTS is an interactive application

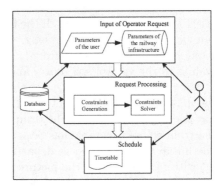

Fig. 7. General System Architecture

that is implemented in C++. It connects to a database to store requests and solutions and to keep all information that is related to railway infrastructure, journeys, type of trains, stations, etc. The main modules of our system are described in Figure 7. Following, we explain them in more detail.

4.1   Input of Operator Request

The system returns a valid scheduling for each operator requirement. The main input parameters are: network occupation, number and type of trains, interval of allowed frequencies and interval of allowed initial departure times. Figure 8 and Figure 9 show an example where an operator specifies a request to allocate new trains on a railway network. In the example, the operator needs new trains of the type "A COR_SANTI_L1" for trains going in the down direction and "SANTI_A COR_L1" for trains going in the up direction. For each type, the operator specifies a frequency (01:11:00 and 01:20:00), a number of trains (13 and 13) and a time interval to start the corresponding journeys ([05:00:00-08:30] and [05:00:00-08:15:00]).

The user uses the interface shown in Figure 9 to specify the network occupation; that is, which trains are in circulation and must be taken into account during the solving process. Each train in circulation may belong to a different type of train. In Figure 9, window W1 shows all the types of trains that are in

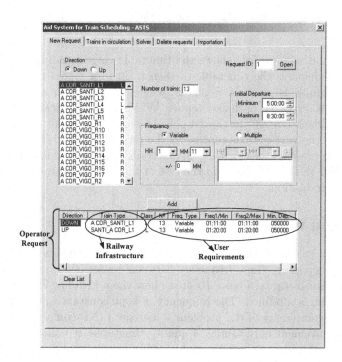

Fig. 8. Interface to specify a scheduling request for new trains

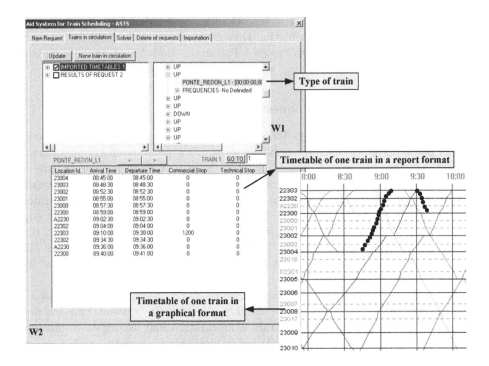

Fig. 9. An interface to specify the network occupation

circulation, and window W2 shows the timetable corresponding to each one of them. ASTS displays the timetable of each train in a report or in a graphical form. An example is given in Figure 9, where the timetable corresponding to "TRAIN 1" is displayed both as a report (W2) and in graphical form (line highlighted with black circles). In the graphical form, the horizontal axis represents the time and the vertical axis represents the positions of each train. Thus, each oblique line represents the timetable of one train. The thickest lines represent the timetables for the new trains, and the rest of the lines correspond to timetables of trains in circulation.

4.2   Request Processing

As mentioned in section 3, the timetable of one train depends mainly on three parameters: departure time, frequency between consecutive trains and priority assignment. For instance, Figure 10 shows how the scheduling changes when the departure time is modified. The frequency is kept constant in the two configurations. The complexity of the problem is increased (NP-complete to NP- hard) with the requirement of obtaining a quality timetable. It is not enough for the final user to obtain a correct timetable. This requirement implies finding the best combination of the three cited parameters.

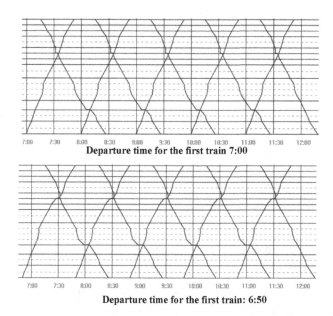

Departure time for the first train 7:00

Departure time for the first train: 6:50

Fig. 10. Two different scheduling for two different departure times for the same group of trains

The system performs the search within the time given by the operator. When this time ends, it stores to a database the best scheduling obtained according to the quality criteria given by the operator.

4.3   Scheduling Delivery

Once the running time given by the operator is completed, the system interrupts its process and stores the best scheduling obtained in this time interval in a database. The operator is sent a report about the results obtained (Figure 11).

Report	×
State:	Feasible Solution
Objective Function:	163700
Average Time:	01:44:56
Execution Time:	00:00:24
% Exc Down:	50.382
% Exc Up:	40.184
	OK

Fig. 11. Summary about the resulting schedule

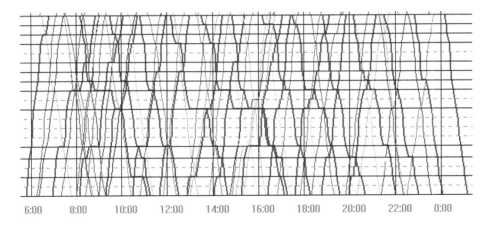

6:00   8:00   10:00   12:00   14:00   16:00   18:00   20:00   22:00   0:00

Fig. 12. A graphical example of a resulting schedule

Later, at any time, the operator can retrieve the resulting timetables from the database in a graphical form (Figure 12). The timetables can be modified manually by the operator. These modifications are checked by the system and only those that do not violate any constraint are allowed. The changes made by the user to timetables can be stored in the database.

The resulting scheduling for the new trains is shown in Figure 12. The thickest lines represent the new scheduling. The rest of the lines correspond to timetables of other trains, which in turn, represent the occupation of the railway network, on which the new trains had to be allocated.

5   Results

Several parameters may determine the quality of a scheduling for new trains in a railway company. Our system considers as quality criterion:

- Average traversal time for each new train.

In addition to this criterion other quality parameters are provided to the user:

- Number of conflicts managed among new trains as well as between new trains and those in circulation. Solving conflicts implies technical stops and then a less attractive timetable from the customer point of view.
- Divergence, defined as the difference between the average delay (measured as a percentage) of trains in down direction and the average delay of trains in up direction. Higher divergence could imply worse schedules.

Another factor that affects the quality of the schedules obtained by our system is the running time that the operator gives to solve the problem. Considering the input parameters of Table 1, the system provides three different outputs for an execution time of 5, 10 and 40 seconds respectively (Table 2).

Table 1. Input parameters

Dir	New Trains	Locat.	Train in Circulation	Freq.	Departure Time Interval
Down	13	20	Yes	[01:10:00-01:20:00]	[05:00:00-08:30:00]
Up	13	20	Yes	[02:00:00-2:20:00]	[05:30:00 08:15:00]

Table 2. Three different outputs

Execution Time(sec)	No of Conflicts	Divergence	Average Traversal Time
5	Average 3	54%-40%	01:51:00
10	Average 3	53%-38%	01:47:27
40	Average 2	50%-40%	01:45:23

6 Conclusions

Our system assigns to each train a departure and arrival time for each location of its journey, taking into account the operator requirements and the described traffic constraints. The system accepts a wide range of scenarios on which it is possible allocate new trains correctly and efficiently. One of these scenarios can be a high-loaded railway network with different types of trains. The system responds with different timetable configurations depending on the operator request and the execution time. The operator can make decisions easily and quickly, modifying the input parameters by comparing the different scheduling obtained. This comparison is very complex and very time consuming to do manually in a complex network. As all optimization processes for NP-hard problems, the heuristics can be improved in order to obtain better solutions in less time. Currently, ASTS is in an evaluation phase. We are considering adding other quality criteria in order to increase the scope. We also want to integrate ASTS with other tools proper of the railway company.

Acknowledgments

This work has been partially supported by the research projects TIN2004-06354-C02-01 (Min. de Educ. y Ciencia, Spain-FEDER), FOM-70022/T05 (Min. de Fomento, Spain), GV/2007/274 (G. Valenciana), and FP6-021235-2, IST-STRIP (U.E.). Thanks to ADIF and to J. Estrada for his continuous support.

References

1. Bussieck, M.R., Winter, T., Zimmermann, U.T.: Discrete optimization in public rail transport. Math. Programming 79(1-3), 415–444 (1997)
2. Chiu, C.K., Chou, C.M., Lee, J.H.M., Leung, H.F., Leung, Y.W.: A constraint based interactive train rescheduling took. Constraints 7, 167–198 (2002)
3. Cordeau, J.F., Toth, P., Vigo, D.: A survey of optimization models for train routing and scheduling. Transportation Science 32(4), 380–404 (1998)
4. Koci, L., Di Stefano, G.: A graph theoretical approach to shunting problems. In: Proceedings of ATMOS 2003 Algorithmic Methods and Models for Optimization of Railways, Electronic Notes in Theoretical Computer Science, Elsevier 92, vol. 1 (2003)
5. Kreuger, P., Carlsson, J.O., Sjoland, T., Astrom, E.: The tuff train scheduler. In: Puebla, G. (ed.) Proceedings of the workshop on Tools and Environments for (Constraints) Logic Programming at the International Logic Programming Symposium ILPS'97 (1997)
6. Liebchen, C.: Symmetry for periodic railway timetables. In: Proceedings of ATMOS 2003 Algorithmic Methods and Models for Optimization of Railways, Electronic Notes in Theoretical Computer Science 92, vol. 1 (2003)
7. Mammana, M.F., Mecke, S., Wagner, D.: The station location problem on two intersecting lines. In: Proceedings of ATMOS 2003 Algorithmic Methods and Models for Optimization of Railways, Electronic Notes in Theoretical Computer Science, Elsevier 92, vol. 1 (2003)
8. Serafini, P., Ukovich, W.: A mathematical model for periodic scheduling problems. SIAM Journal on Discrete Mathematics 2(4), 550–581 (1989)

Platform Assignment

Sabine Cornelsen[1,*] and Gabriele Di Stefano[2]

[1] Universität Konstanz, Fachbereich Informatik & Informationswissenschaft
cornelse@inf.uni-konstanz.de
[2] Università dell'Aquila, Dipartimento di Ingegneria Elettrica
gabriele@ing.univaq.it

Abstract. We consider a station in which several trains might stop at the same platform at the same time. The trains might enter and leave the station to both sides, but the arrival and departure times and directions are fixed according to a given time table. The problem is to assign platforms to the trains such that they can enter and leave the station in time without being blocked by any other train. We consider some variation of the problem on linear time tables as well as on cyclic time tables and show how to solve them as a graph coloring problem on special graph classes. One of these classes are the so called circular arc containment graphs for which we give an $\mathcal{O}(n \log n)$ coloring algorithm.

1 Introduction

We consider the following problem. We are given a set of trains that has to enter a station with Marshalling topology. Each train might enter the station either from the left hand side or from the right hand side and might leave the station to the left hand side or to the right hand side. The direction from which it enters the station and in which it leaves the station, however, is fixed. Also the arrival time and departure time is fixed. Hence, each train is labeled with the time interval $[t_{\mathrm{arr}}, t_{\mathrm{dep}}]$ in which it stays in the station, and an entering and leaving direction d_{arr} and d_{dep}, respectively. For example, a train labeled [24,25]RL is a train that enters the station at 24 from the right hand side and leaves the station at 25 to the left hand side. In the meanwhile, it stops on one of several parallel platforms (tracks) in the station. There might be several trains waiting at the same platform. We further assume that the platforms have infinite length. The problem is to assign each train to a platform such that it can leave the station in time without being blocked by other trains. E.g., the assignment in Fig. 1a is not feasible, since train [-2,2]RL would be blocked by train [-1,4]LL. The assignments in Fig. 1b and c are both feasible. However, the interesting assignment is the one in Fig. 1c, that uses the minimum number of platforms.

We consider as well linear time tables as cyclic time tables. In cyclic time tables, the situation repeats itself each time period T, i.e., a label $[t_{\mathrm{arr}}, t_{\mathrm{dep}}]d_{\mathrm{arr}}d_{\mathrm{dep}}$

* Work partially done while the author was visiting the University of L'Aquila, supported by the Human Potential Program of the EU under contract no HPRN-CT-1999-00104 (AMORE Project).

F. Geraets et al. (Eds.): Railway Optimization 2004, LNCS 4359, pp. 233–245, 2007.

Fig. 1. a) A non-feasible platform assignment. b) A feasible platform assignment that is achieved with the first-fit algorithm. c) A feasible platform assignment that uses the minimum number of platforms.

represents a series of trains $(z_i)_{i \in \mathbb{Z}}$. Train z_i arrives at time $t_{arr} + iT$ from direction d_{arr} and leaves at time $t_{dep} + iT$ in direction d_{dep}. We further assume that no train stays longer than the time period T in the station, i.e., that $t_{dep} - t_{arr} < T$. The goal is a platform assignment that repeats itself every time period T, i.e., to assign each train of a series belonging to the same label to the same platform. E.g., we are not interested in an assignment as indicated in Fig. 2a, but rather in an assignment as indicated in Fig. 2b.

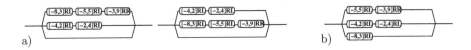

Fig. 2. Cyclic platform assignment with time period $T = 16$. a) A platform assignment that repeats itself every second time period. b) A cyclic platform assignment that repeats itself every time period.

As an additional constraint, we will sometimes consider the so called *midnight constraint*, i.e., the condition that all trains enter the station before the first train leaves the station. The name of this constraint is motivated by shunting problems in which all trains enter a shunting depot in the evening and leave it in the morning. But the midnight constraint even seems to be a reasonable constraint for modeling the situation in some stations where first all trains enter the station such that passengers can change trains and afterwards all trains leave the station. A train is a *turning back train*, if it leaves the station in the direction from where it entered it, i.e., if $d_{arr} = d_{dep}$.

The platform assignment problem is closely related to the shunting problem some variation of which have been studied in the following work [1,3,5,6,8,13]. We will consider the platform assignment problem as a graph coloring problem. Given a platform assignment problem, we define the following *constraint graph*. The vertices of the graph are the trains (or the series of trains if cyclic time tables are considered). Two trains are adjacent if they cannot be put on the same platform. The platform assignment problem then corresponds to coloring the constraint graph. For some variation of the shunting problem, this approach has been studied, e.g., in [4,12].

In this paper, we show the following results for the platform assignment problem. In the cases of linear time tables with midnight constraint or without turning back trains, we show that the constraint graph is a permutation graph and hence it can be colored in $\mathcal{O}(n \log n)$ time. However, the algorithm does not consider the trains in the order in which they enter the station. We further give a 2-competitive online-algorithm to solve this problem. This algorithm also runs in $\mathcal{O}(n \log n)$ time.

In the case of cyclic time tables with midnight constraint, the constraint graph is not necessarily perfect. For this problem, we give a 3-approximation algorithm that runs in quadratic time. Cyclic time tables without turning back trains yield a constraint graph that is a comparability graph and, hence, can be colored in $\mathcal{O}(n^2)$ time. If we further restrict the input to trains all entering from the right hand side and leaving to the lefthand side the constraint graph is a circular arc containment graph. We show how to color such graphs optimally in $\mathcal{O}(n \log n)$ time. The results are summarized in the following table.

	linear time table	cyclic time table
with midnight constraint	$\mathcal{O}(n \log n)$	3-approx within $\mathcal{O}(n^2)$
without midnight constraint without turning back trains	$\mathcal{O}(n \log n)$	$\mathcal{O}(n^2)$
without midnight constraint only type RL		$\mathcal{O}(n \log n)$

The distribution of this paper is as follows. In Sect. 2, we give a formal definition of the platform assignment problem and show that it is equivalent to coloring the conflict graph. In Sect. 3, we consider some variations of the linear platform assignment problem and in Sect. 4 variations of the cyclic platform assignment problem. As a byproduct, we give our algorithm for coloring circular arc containment graphs in Sect. 4.2. Finally, in Sect. 5, we discuss some open problems.

2 The Conflict Graph

In this section, we give a formal definition of the platform assignment problem with linear or cyclic timetable, define the conflict graph and show that the platform assignment problem is equivalent to the coloring problem on the conflict graph.

LINEAR PLATFORM ASSIGNMENT PROBLEM (LPA)

Given a set
$$Z = \{[t_{\text{arr}}^{(1)}, t_{\text{dep}}^{(1)}] d_{\text{arr}}^{(1)} d_{\text{dep}}^{(1)}, \ldots, [t_{\text{arr}}^{(n)}, t_{\text{dep}}^{(n)}] d_{\text{arr}}^{(n)} d_{\text{dep}}^{(n)}\}$$

of train labels ($t_{\text{arr}}^{(i)}, t_{\text{dep}}^{(i)} \in \mathbb{R}, t_{\text{arr}}^{(i)} < t_{\text{dep}}^{(i)}; d_{\text{arr}}^{(i)}, d_{\text{dep}}^{(i)} \in \{R, L\}$) with the property that all arrival and departure times are distinct and given a number k, is there an assignment $p : Z \to \{1, \ldots, k\}$ such that the following *train operation procedure* does not return false?

1. Sort arrival and departure times increasingly.
2. For all arrival and departure events in the sorted list
   (a) If the next event is an arriving train z. If z arrives from the right, append z to the righthand side of list $L_{p(z)}$ else append z to the lefthand side of $L_{p(z)}$.
   (b) If the next event is a departing train z. If z departs to the righthand side and is on the righthand side of list $L_{p(z)}$, or if z departs to the lefthand side and is on the lefthand side of list $L_{p(z)}$ then delete z from $L_{p(z)}$, else return false.

For a train label $z = [t_{\text{arr}}, t_{\text{dep}}]d_{\text{arr}}d_{\text{dep}}$, an integer i and a time period T set $z + iT = [t_{\text{arr}} + iT, t_{\text{dep}} + iT]d_{\text{arr}}d_{\text{dep}}$.

CYCLIC PLATFORM ASSIGNMENT PROBLEM (CPA)

Given a set Z of train labels (with the property that all arrival and departure times are distinct), a time period T (such that $t_{\text{dep}} - t_{\text{arr}} < T$ for all $[t_{\text{arr}}, t_{\text{dep}}]d_{\text{arr}}d_{\text{dep}} \in Z$), and a number k, is there an assignment $p : Z \to \{1, \ldots, k\}$ such that the above train operation procedure applied to $\{z + iT; z \in Z, i \in \mathbb{Z}\}$ (with $p(z + iT) = p(z)$) would never return false?

The *conflict graph* for a platform assignment problem (Z, k) or (Z, T, k), respectively, is defined as follows. The vertices are the train labels. Two vertices z_1 and z_2 are adjacent if and only if the platform assignment problem $(\{z_1, z_2\}, 1)$ or $(\{z_1, z_2\}, T, 1)$ is a false-instance.

GRAPH COLORING PROBLEM

Given a graph $G = (V, E)$ and a number k. Is there an assignment $c : V \to \{1 \ldots, k\}$ such that $c(v) \neq c(u)$ for $\{u, v\} \in E$?

Theorem 1. *Let G be the conflict graph of a platform assignment problem (Z, k) or (Z, T, k), respectively. Then the solutions of the platform assignment problem and the solutions of the graph coloring problem correspond.*

Proof. Let $p : Z \to \{1, \ldots, k\}$ be a solution of the platform assignment problem. Let $p(z_1) = p(z_2)$. Assume without loss of generality that z_1 has to leave the station before z_2 and that z_1 departs to the right hand side. If $\{z_1, z_2\}$ where an edge of G then, by definition of the conflict graph, z_2 would be on the right hand side of z_1 at the time that z_1 had to depart. But then p would not have been a valid assignment. Hence setting $c(z) = p(z), z \in Z$ yields a solution for the graph coloring problem.

Now, let $c : Z \to \{1, \ldots, k\}$ be a solution of the graph coloring problem. Set $c(z) = p(z), z \in Z$. Suppose that at some point the train operation procedure returns false. So suppose without loss of generality that at some point z_1 has to leave to the right hand side, but that there is another train z_2 in the list $L_{p(i)}$ on the right hand side of z_1. But then $\{z_1, z_2\}$ would be an edge in G – contradicting that c is a valid coloring. □

A *transitive orientation* of an undirected graph $G = (V, E)$ is an orientation of its edges with the following property.

$$(u, v) \in E \text{ and } (v, w) \in E \Longrightarrow (u, w) \in E$$

A *comparability graph* is a graph that has a transitive orientation. A *permutation graph* is a graph G that is associated with a permutation π of the set $\{1, \ldots, n\}$. The set of vertices of G is $\{1, \ldots, n\}$ and two vertices i and j with $i < j$ are adjacent if and only if $\pi(i) > \pi(j)$. A graph G is a permutation graph if and only if G and its complement are comparability graphs (see e.g. [7, p. 158]). The *chromatic number* of a graph G is the minimum number k such that (G, k) is a yes-instance of the graph coloring problem. A graph G is *perfect* if for all induced subgraphs H of G the size of a maximum clique of H equals the chromatic number of H. Comparability graphs and, hence, permutation graphs are perfect [7, p. 133].

The *first-fit algorithm* is a heuristic to solve the graph coloring problem. It works as follows. Start with an ordering $v_1, \ldots, v_n$ on the vertices. Color the vertices in this order. Assign vertex v_i the first color that was not assigned to an adjacent vertex of v_i. Chvátal [2] characterized all orderings that are such that the first-fit algorithm solves the graph coloring problem on the graph and all its induced subgraphs optimally. This is for example true, if the vertices are ordered according to a transitive orientation. In general, the first-fit algorithm runs in $\mathcal{O}(n^2)$ time. For permutation graphs, it can be implemented to run in $\mathcal{O}(n \log n)$ time (see e.g. [7, p. 168]). Note however that for arbitrary vertex-orderings the first-fit algorithm can behave arbitrarily bad even on permutation graphs [10].

3   Linear Timetables

In this section, we consider three variations of the LPA. In Sect. 3.1, we assume that the midnight constraint is given. Else even the restricted problem in which every train enters from and leaves to the right hand side is $\mathcal{NP}$-complete: it corresponds to coloring circle graphs (see, e.g., [4] for a review). In Sect. 3.2, we consider an online solution for the LPA with midnight constraint. Finally, in Sect. 3.3, the midnight constraint need not be true, but there may not be any turning back trains. In the following let n be the number of trains in an instance of an LPA.

3.1   With Midnight Assumption

In this section, we assume that we are given an input of the LPA with the midnight constraint, i.e., that the time intervals intersect in at least one point. For example, in Fig. 1, 0 is a common point in all time intervals. Let $G_{\text{lin}}^{\text{mid}}$ be the conflict graph for this problem. A schematic image of $G_{\text{lin}}^{\text{mid}}$ is indicated in Fig. 3. Each rectangle represents all trains that have the same arrival and departure direction. E.g., the rectangle labeled RL represents all trains with $d_{\text{arr}} = R$ and $d_{\text{dep}} = L$. We will refer to this as the *type* of a train. Within one

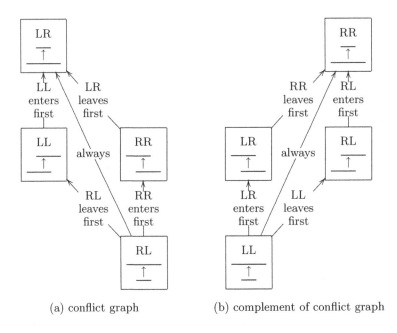

(a) conflict graph   (b) complement of conflict graph

Fig. 3. A transitive orientation of the conflict graph and its complement

rectangle, ⊥ and ⊥ indicate that there is an edge between two trains if the time intervals overlap, but neither is contained in the other. ⊥ and ⊥ indicate that there is an edge between two trains if the time interval of one train is contained in the time interval of the other train. The edges between the rectangles indicate when there is an edge between two trains of different type. An orientation of the edges is visualized by upward pointing arrows.

Theorem 2. *The linear platform assignment problem with midnight constraint can be solved in $\mathcal{O}(n \log n)$ time.*

Proof. We show that $G_{\text{lin}}^{\text{mid}}$ is a permutation graph and that a representation as permutation can be found in $\mathcal{O}(n \log n)$ time. Hence, since permutation graphs can be colored in $\mathcal{O}(n \log n)$ time, the LPA with midnight constraint can be solved in the same asymptotic running time.

To show that $G_{\text{lin}}^{\text{mid}}$ is a permutation graph, we have to show that it is a comparability graph and that its complement is also a comparability graph. A transitive orientation of the conflict graph is indicated in Fig. 3. Exchanging LL with RL and LR with RR, one can see that the complement of the conflict graph has the same structure as the conflict graph itself. Hence it is also a comparability graph.

According to the general rule [7, p. 158], the permutation that corresponds to the conflict graph can be constructed as follows. First the trains are labeled according to the ordering that results from the transitive orientation of the conflict graph and its complement. Then the trains are permuted according to the reversed transitive orientation of the conflict graph and the original transitive orientation of its complement. Hence, from the schematic representation in Fig. 3, we can immediately see, that for two trains z_1, z_2 and for each of the two orderings, we can decide in $\mathcal{O}(1)$ if z_1 is before z_2. Hence, since ordering can be done in $\mathcal{O}(n \log n)$ time, the representation of $G_{\text{lin}}^{\text{mid}}$ as a permutation can found in $\mathcal{O}(n \log n)$ time. $\qquad\square$

The two orderings constructed in the proof of Theorem 2 for constructing the representation as a permutation for $G_{\text{lin}}^{\text{mid}}$ have a quite intuitive meaning. The trains departing to the left are labeled in increasing order of their departure time followed by the trains departing to the right in decreasing order of their departure time. This is exactly the order in which the trains can be positioned on one platform such that they can leave in time. For the permutation, the trains arriving from the left are ordered in decreasing order of their arrival time followed by the trains arriving from the right in increasing order of their arrival time. This corresponds exactly to the order in which the trains would be positioned at midnight if only one platform was given. E.g., the representation as permutation for the conflict graph of the LPA in Fig. 1 is constructed as follows. First the trains are labeled according to there departure direction and time.

$$1 : \text{[-4,1]RL}, \ 2 : \text{[-2,2]RL}, \ 3 : \text{[-1,4]LL}, \ 4 : \text{[-3,3]RR}$$

Then they are permuted according to there arrival direction and time.

$$3 : \text{[-1,4]LL}, \ 1 : \text{[-4,1]RL}, \ 4 : \text{[-3,3]RR}, \ 2 : \text{[-2,2]RL}$$

Hence the resulting permutation is $\begin{pmatrix} 1 \ 2 \ 3 \ 4 \\ 3 \ 1 \ 4 \ 2 \end{pmatrix}$.

3.2 Online Solution

The first-fit algorithm for coloring permutation graphs, colors the vertices in the order in which they occur in the permutation. I.e., in the solution of the LPA with midnight constraint given in Sect. 3.1, we first have to decide how to color the trains that enter last from the left hand side. Hence, it is not an *online-algorithm*, i.e., the trains are not assigned a platform in the order in which they enter the station. In fact, Fig. 1a showed that the first-fit algorithm that assigns platforms to trains in the order they enter the station would not necessarily yield the minimum number of platforms. The next theorem states that there is an algorithm that assigns platforms to trains in the order in which they enter the station which uses at most twice the number of platforms that the optimal offline-algorithm in Sect. 3.1 would use.

Theorem 3. *There is a 2-competitive online-algorithm for the linear platform assignment problem which runs in $\mathcal{O}(n \log n)$ time.*

Proof. First, we divide the input into two subproblems – the trains entering from the left hand side and the trains entering from the right hand side. The conflict graph for both subproblems is a permutation graph. For the trains entering from the right hand side, we use the representation as a permutation as it was computed in Sect. 3.1. For the trains entering from the left hand side, we reverse the orientation of the conflict graph and its complement given in Fig. 3. Then we apply the standard procedure described in the the proof of Theorem 2 on these two transitive orientations to obtain a representation as a permutation. Representing the conflict graphs of the two subproblems like that, the first-fit algorithm for coloring them is an online algorithm to solve the two subproblems optimally in $\mathcal{O}(n \log n)$ time. Hence, applying the first-fit algorithm with the additional constraint that a train that enters form the left hand side and a train that enters from the right hand side may not be put on the same platform yields a 2-competitive online-algorithm for solving the whole problem in $\mathcal{O}(n \log n)$ time. $\square$

3.3   Without Turning Back Trains

In this section, we do not require the midnight constraint, but we assume that we have an instance of the linear platform assignment problem without turning back trains. Hence, we consider the quite typical case that a train coming from one side will continue its trip to the other side of the station. Fig. 4 shows that also in this case the conflict graph is a permutation graph. $\underline{\ .\uparrow.\ }$ indicates that there is an edge between two trains if neither of the time intervals is contained in the other. Hence, we can conclude the following theorem.

Theorem 4. *The linear platform assignment problem without turning back trains can be solved in $\mathcal{O}(n \log n)$ time.*

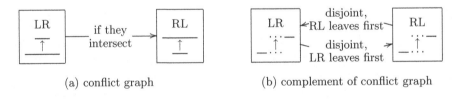

(a) conflict graph   (b) complement of conflict graph

Fig. 4. Transitive orientations of the conflict graph and its complement for the linear platform assignment problem without turning back trains

4   Cyclic Time Tables

In this section, let T denote the time period of a CPA and let n be the number of series of trains in the input. We represent an interval $[t_{\mathrm{arr}}, t_{\mathrm{dep}}]$ in the input as a circular arc $[t_{\mathrm{arr}} \frac{2\pi}{T}, t_{\mathrm{dep}} \frac{2\pi}{T}]$. The thus represented input for the example in Fig. 2 is shown in Fig. 6. In this section, we consider three subproblems of the CPA. In Sect. 4.1, we give a 3-approximation algorithm for, the CPA with midnight constraint. In Sect. 4.2, we first briefly consider the CPA without turning back trains in general. Then we solve the special case that there are only trains of type RL or only trains of type LR by giving an $\mathcal{O}(n \log n)$ algorithm for coloring the class of graphs called circular arc containment graphs.

4.1 With Midnight Assumption

In this section, we assume again that the midnight-assumption holds. Hence, without loss of generality, we may assume that 0 is contained in the intersection of all intervals of the input. If there is only one type of trains (among LL, LR, RL, RR) the midnight-assumption implies that cyclic time tables can be handled in the same way as simple intervals. This changes, however, if there are trains that differ in either the incoming direction or the outgoing direction (but not both) such that the intersection of its time arcs is not connected, i.e. if the difference between the departure time of the train that leaves last and the arrival time of the train that arrives first is greater or equal T. In fact, in a solution that repeats every time period, two such trains can never be put on the same platform. The conflict graph $G_{\text{cyc}}^{\text{mid}}$ of the CPA with midnight constraint is illustrated in Fig. 5.

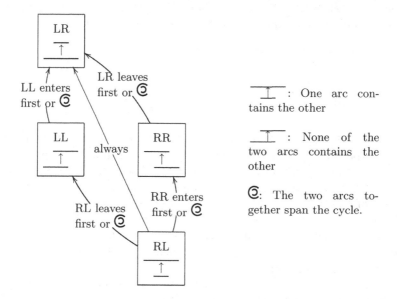

Fig. 5. The conflict graph for the cyclic platform assignment problem with midnight constraint

If there are only trains of type LL and RR involved, $G_{\text{cyc}}^{\text{mid}}$ is always a permutation graph. The case in which the input consists only of trains of type LR and RL will be consider more generally (without midnight constraint) in the next subsection. In general, $G_{\text{cyc}}^{\text{mid}}$ need not be perfect, even if only trains of type RR/LR, RR/RL, LR/LL, or RL/LL are involved. See Fig. 6 for an example. The time complexity of the CPA with midnight constraint in these cases is sofar open.

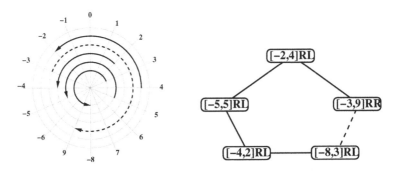

Fig. 6. The cyclic time table considered in Fig. 2 and its conflict graph which is a C_5 and hence not perfect. Clockwise arcs indicate trains leaving to the right and counter clockwise arcs indicate trains leaving to the left. Dashed arrows indicate turning-back trains. The dashed edge in the conflict graph is of type $\mathfrak{G}$.

Theorem 5. *There is a 3-approximation algorithm for the cyclic platform assignment problem with midnight constraint that works in $\mathcal{O}(n^2)$ time.*

Proof. We consider the acyclic orientation on the graph that is indicated in Fig. 5. The height function which is defined recursively by setting

$$h(v) = \begin{cases} 1 & \text{if v is a sink,} \\ 1 + \max\{h(w);\ (v,w) \in E\} & \text{otherwise} \end{cases}$$

is a proper coloring of $G_{\text{cyc}}^{\text{mid}}$ and the number of colors $\chi(h)$ that are used is equal to the number of vertices in the longest directed path in $G_{\text{cyc}}^{\text{mid}}$ [7, p. 132]. Note that the orientation within the four types RR, LL, RL, an LR is even transitive. Any directed path P in $G_{\text{cyc}}^{\text{mid}}$ contains trains of at most three of these types and all trains of one type are consecutive in P. Hence, the subgraph induced by P can be covered by three cliques – one for each type of train that is contained in P. Let ω be the size of a maximum clique in $G_{\text{cyc}}^{\text{mid}}$ and let χ be the chromatic number of $G_{\text{cyc}}^{\text{mid}}$. Then $\chi(h) \leq 3\omega \leq 3\chi$. Hence, the height function yields a 3-approximation algorithm.

The height function can be computed in time linear in the size of $G_{\text{cyc}}^{\text{mid}}$ by a DFS-algorithm [7, p. 46] and hence in $\mathcal{O}(n^2)$ time.   □

4.2 Without Turning Back Trains

In case only trains of type LR and RL are involved, the conflict graph remains a comparability graph, even without the midnight constraint. See Fig. 7a for the transitive orientation. Hence, the conflict graph can be colored in $\mathcal{O}(n^2)$ time. Fig. 7b demonstrates, however, that in that case the conflict graph does not have to be a permutation graph anymore.

In the rest of this section, we now consider the case that there are only trains of type RL or only trains of type LR. In this case, the conflict graph is a circular arc

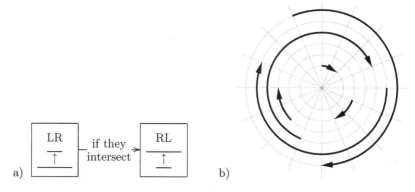

a)   b)

Fig. 7. Trains of type LR and RL only, without requiring the midnight constraint. a) Transitive orientation of the conflict graph b) Example for which the conflict graph is a C_6 and, hence, not a permutation graph – even if only trains of type LR are involved.

containment graph (CACG) [12]. A CACG is a graph whose vertices are circular arcs and in which two circular arcs are adjacent if and only if one of them is contained in the other. In what follows, let n be the number of vertices and m the number of edges of a CACG. A maximum clique of a CACG and, hence (since CACGs are comparability graphs and, hence, perfect), the chromatic number of a CACG can be determined in $\mathcal{O}(n \log \log n)$ time [9]. Nirkhe [11] showed that a coloring for a CACG can be found in $\mathcal{O}(n \log n + m)$ time. We will show how to modify the first-fit algorithm for permutation graphs such that it can be applied to color circular arc containment graphs in $\mathcal{O}(n \log n)$ time.

We will refer to a *circular arc* by an interval $[a, b]$ with $-2\pi < a < 2\pi, ; 0 < b \leq 2\pi$, $0 < b - a < 2\pi$. A circular arc $[a, b]$ is contained in a circular arc $[c, d]$ if

1. $c < 0$ and $a > 2\pi + c$ or if
2. $c < a$ and $d > b$.

Let G be the CACG on the set $V = \{[a_1, b_1], \ldots, [a_n, b_n]\}$ of n circular arcs. We may assume that all $a_1, \ldots, a_n, b_1, \ldots, b_n$ are distinct [11].

Since ordering can be done in $\mathcal{O}(n \log n)$ time, we assume that the arcs in V are ordered as follows.

$$i < j \text{ iff } a_i < a_j$$

Then, clearly, this induces a transitive orientation on G. We apply the first-fit algorithm. To obtain the required running time, we apply the algorithm below, maintaining the following arrays.

COLOR: COLOR(i) is the list of vertices that is colored i.
FIRST: FIRST$(i) = 2\pi$ if so far no arc has been colored with color i. Else, let $[a, b]$ be the first arc that is colored with color i. Then FIRST$(i) = a + 2\pi$.
LAST: LAST$(i) = 0$ if so far no arc has been colored with color i. Else, let $[a, b]$ be the last arc that is colored with color i. Then LAST$(i) = b$.

1. **for** $j = 1, \ldots, n$
2.   Find smallest k such that FIRST$(k) > a_j$
3.   Find smallest $i \geq k$ such that LAST$(i) < b_j$
4.   **if** COLOR(i) is empty
5.   Set FIRST$(i) = a_j + 2\pi$
6.   Append $[a_j, b_j]$ to COLOR(i)
7.   Set LAST$(i) = b_j$

Lemma 1. *The above algorithm solves the graph coloring problem on the class of circular arc containment graphs in $\mathcal{O}(n \log n)$ time.*

Proof. First, we show that the algorithm is the first-fit algorithm and, hence, solves the problem optimally. (Recall that the arcs are ordered according to a transitive orientation.) Note that no not-yet colored arc contains an already colored arc. There are the following two cases in which arc $[a_j, b_j]$ is contained in the arc $[a, b]$ colored by color i.

1. $a_j > a + 2\pi$. By construction, $a + 2\pi \geq$ FIRST(i). Hence, $a_j >$ FIRST(i).
2. $b_j < b$. Hence, since by construction $b \leq$ LAST(i), it follows $b_j <$ LAST(i).

Hence, Line 2 and 3 of the algorithm guarantee that $[a_j, b_j]$ is colored with the first color that was not used to color an adjacent arc of $[a_j, b_j]$.

For the running time observe the following. FIRST(i) is monotonely increasing, i.e. FIRST$(i) \leq$ FIRST(j) for $i < j$. The value k that is chosen in Line 2 of the algorithm is monotonely increasing with every step. LAST(i) is monotonely decreasing for $i \geq k$, i.e. LAST$(i) \geq$ LAST(j) for $k \leq i < j$. Hence, Line 2 and 3 can both be performed in logarithmic time by binary search. It follows that the over all running time is in $\mathcal{O}(n \log n)$. $\qquad \square$

5   Open Problems

We have considered some variations of the platform assignment problem with linear and cyclic time tables. In the following, we list some interesting questions concerning this topic that sofar remained open.

- How bad can the first-fit algorithm behave for solving the linear platform assignment problem with midnight constraint as an online problem? Recall that the first-fit algorithm on arbitrary vertex-orderings of a permutation graph can behave arbitrarily bad [10].
- Is the cyclic platform assignment problem with midnight constraint $\mathcal{NP}$-complete?
- A solution for the cyclic platform assignment problem in which trains from a series do not have to be assigned to the same platform could use less platforms than a solution that has to be the same every time period. See Fig. 2 for an example. How big can the difference be?

References

1. Blasum, U., Bussieck, M.R., Hochstättler, W., Moll, C., Scheel, H.-H., Winter, T.: Scheduling trams in the morning. Mathematical Methods of Operations Research 49(1), 137–148 (1999)
2. Chvátal, V.: Perfectly ordered graphs. In: Topics on Perfect Graphs. Annals of Discrete Mathematics, vol. 21, pp. 63–65. North-Holland, Amsterdam (1984)
3. Dahlhaus, E., Horak, P., Miller, M., Ryan, J.F.: The train marshalling problem. Discrete Applied Mathematics 103(1–3), 41–54 (2000)
4. Di Stefano, G., Koci, M.L.: A graph theoretical approach to the shunting problem. In: Gerards, B. (ed.) Proceedings of the Workshop on Algorithmic Methods and Models for Optimization of Railways (ATMOS 2003), Electronic Notes in Theoretical Computer Science, vol. 92 (2004)
5. Freling, R., Lentink, R.M., Kroon, L.G., Huisman, D.: Shunting of passenger train units in a railway station. Technical Report EI2002-26, Econometric Institute, Erasmus University Rotterdam. To appear in Transportation Science (2002), http://www.eur.nl/WebDOC/doc/econometrie/feweco20020917130601.pdf
6. Gallo, G., Di Miele, F.: Dispatching buses in parking depots. Transportation Science 35(3), 322–330 (2001)
7. Golumbic, M.C.: Algorithmic Graph Theory and Perfect Graphs. Computer Science and Applied Mathematics. Academic Press, San Diego (1980)
8. He, S., Song, R., Chaudhry, S.S.: Fuzzy dispatching model and genetic algorithms for railyards operations. European Journal of Operational Research 124(2), 307–331 (2000)
9. Lou, R.D., Sarrafzadeh, M.: Circular permutation graph family with applications. Discrete Applied Mathematics 40, 433–457 (1992)
10. Nikolopoulos, S.D., Papadopoulos, C.: On the performance of the first-fit coloring algorithm on permutation graphs. Information Processing Letters 75, 265–273 (2000)
11. Nirkhe, M.V.: Efficient algorithms for circular-arc containment graphs. Master's thesis, University of Maryland (1987), http://techreports.isr.umd.edu/report/1987/MS_87-11.pdf
12. Rossi, A.: Il problema dell'ordinamento dei treni in un deposito: modellazione e soluzione algoritmica. Master's thesis, Università dell'Aquila (2003)
13. Winter, T., Zimmermann, U.T.: Real-time dispatch of trams in storage yards. Annals of Operations Research 96, 287–315 (2000)

Finding All Attractive Train Connections by Multi-criteria Pareto Search

Matthias Müller-Hannemann and Mathias Schnee

Darmstadt University of Technology, Department of Computer Science,
Hochschulstraße 10, 64289 Darmstadt, Germany
{muellerh,schnee}@algo.informatik.tu-darmstadt.de
http://www.algo.informatik.tu-darmstadt.de

Abstract. We consider efficient algorithms for timetable information in public transportation systems under multiple objectives like, for example, travel time, ticket costs, and number of interchanges between different means of transport. In this paper we focus on a fully realistic scenario in public railroad transport as it appears in practice while most previous work studied only simplified models.

Algorithmically this leads to multi-criteria shortest path problems in very large graphs. With several objectives the challenge is to find *all* connections which are potentially attractive for customers. To meet this informal goal we introduce the notion of *relaxed Pareto dominance*. Another difficulty arises from the fact that due to the complicated fare regulations even the single-criteria optimization problem of finding cheapest connections is intractable. Therefore, we have to work with fare estimations during the search for good connections.

In a cooperation with Deutsche Bahn Systems we realized this scenario in a prototypal implementation called PARETO based on a time-expanded graph model. Computational experiments with our PARETO server demonstrate that the current central server of Deutsche Bahn AG often fails to give optimal recommendations for different user groups. In contrast, an important feature of the PARETO server is its ability to provide many attractive alternatives.

1 Introduction

We consider efficient algorithms for timetable information in public transportation systems (with emphasis on public railroad systems) under multiple objectives. We concentrate on three optimization criteria: travel time, fare and number of train changes. However, it is easy to add further criteria like the possibility of seat reservation or safety margins for train changes in case of delays.

Previous work. In recent years several papers studied models for timetable information. These models are based on suitably constructed digraphs on which one can apply shortest path algorithms to answer queries. Two main approaches have been proposed: the *time-expanded* [13,17,8,7,18] and the *time-dependent*

F. Geraets et al. (Eds.): Railway Optimization 2004, LNCS 4359, pp. 246–263, 2007.

approach [11,12,10,1]. In the *time-expanded* digraph we have a node for each event (departure or arrival of a train) at a station. Basically, there are two kinds of edges: train edges connecting the departure of a train with its arrival at the next station and waiting edges within a station. Fixed weights are assigned to the edges like e.g. travel time. This construction usually creates very large but sparse graphs.

In the *time-dependent* approach [1] every node represents a station and two nodes a and b are connected by an edge if there is a train that departs at a and arrives at b without stopping in between. The key idea in a *time-dependent* digraph is that the time-delay of an edge depends on the point in time the edge is used. So the weights on the edges are computed "on-the-fly". This models the phenomenon, that the delay an edge induces depends on the path that is used to reach this edge.

Most of the cited work on public railroad information systems solely considers single-criteria optimization. Next we briefly sketch the previous work on multi-criteria shortest path problems. For a more complete overview, we refer to the section on shortest paths in the recent annotated bibliography on multi-objective combinatorial optimization [2,3]. The standard approaches to the case that *all* Pareto optima have to be computed are generalizations of the standard algorithms for the single-criterion case. Instead of one scalar distance label, each node $v \in V$ is assigned a number of k-dimensional vectors, which are the lengths of all Pareto optimal paths from s to v (clearly, for $k = 1$ the Pareto optima are exactly the distance labels). For the bicriteria case, generalizations of the standard label setting (Dijkstra's algorithm) [4] and label correcting [16] methods have been developed. In the monograph of Theune [19] algorithms for the multi-criteria case are described in detail in the general setting of cost structures over semi-rings. A *two-phase method* has been proposed by Mote et al. [5]. They use a simplex-type algorithm to find a subset of all Pareto optimal paths in the first place, and a label-correcting method to find all remaining Pareto optimal paths in the second phase.

The crucial parameter for the run time and the space consumption is the total number of Pareto optima over all visited nodes. The insight that this number is exponential in $|V|$ in the worst case has motivated the design of approximation algorithms. Hansen [4] and Warburton [20] both present a fully polynomial approximation scheme (FPAS) for finding a set of paths which are approximately Pareto optima for the bicriteria shortest-path problem. The *(resource)-constrained* or *weight-restricted shortest-path problem* [9] is a simplifying (yet still $\mathcal{NP}$-hard) variation of the bicriteria case. Here only one Pareto optimal path is to be computed, namely the one that optimizes the first criterion subject to the condition that the second criterion does not exceed a given threshold value. A theoretical study on the size of the Pareto set in practical applications appeared in [8].

Our contribution. As mentioned in the beginning, the challenge is to find all reasonable train connections meeting a query. The traditional concept of Pareto optimality bears two problems: First, many attractive alternatives will be sorted out if classical Pareto dominance is applied. Second, some Pareto-optimal solutions may be practically useless. The latter problem can easily be overcome: we just eliminate practically useless connections in a postprocessing step. To cope with the first problem, we introduce in Section 2 the concept of *relaxed Pareto dominance* which allows us to identify also "near-optimal" solutions.

This work arose in cooperation with Deutsche Bahn Systems. Within this project we built an information server called PARETO. As described above, there has been a number of papers on time-table queries in recent years, but most of them considered only very simplified scenarios. In contrast, our aim was to incorporate all necessary details and side-constraints such that the output of our server could in principle be used even for purposes like ticketing.

In [7] we already argued that the time-expanded model might be more appropriate to model such complex scenarios than the time-dependent approach. The recent studies of Pyrga et al. [14,15] supported this claim and also indicated that the better computational efficiency of the time-dependent model diminishes greatly if the model becomes more realistic. Therefore, we use a time-expanded graph model to capture all practical requirements. Unfortunately, due to complicated fare regulations (in particular, fares are not additive on the edges in the underlying graphs) even the single-criteria optimization problem of finding cheapest connections is intractable. Therefore, we have to work with fare estimations during the search for good connections.

In principle, our system uses a generalized version of Dijkstra's algorithm for finding the Pareto-optimal set of solutions. This system has been adapted to our relaxed notion of Pareto-optimality and to the approximation of fares during the search. As speed-up techniques we use goal-directed search and improved dominance tests with the terminal station of a query. The goal-directed search is made effective by using lower bounds based on an auxiliary *station graph* (which will be defined in Section 5.1).

Finally, we provide results from a computational study. It turns out that the current version of the Deutsche Bahn server often fails to find optimal solutions for several different user groups. Moreover, our PARETO server has the advantage to offer many more attractive alternatives which the Deutsche Bahn server does not find. This is an important feature for marketing reasons to improve customer satisfaction. Namely, with the larger pool of attractive connections that PARETO delivers, the railway company may use the timetable information system to balance the usage of trains more evenly and to improve the availability of seat reservations. If there are enough good alternatives then connections with trains working to capacity may simply be suppressed in the interface of the information system.

Overview. The rest of the paper is organized as follows. First, in Section 2, we provide further background information about railway timetable information systems. In particular, we explain in detail the specification of queries and side constraints for feasible answers. We discuss different possibilities to evaluate and to compare train connections with respect to several objectives simultaneously, and introduce our concept of relaxed Pareto dominance. In Section 3, we briefly sketch our time-expanded graph model used for timetable information and discuss it in comparison with a time-dependent graph model. In Section 4, we describe our information server PARETO. Specific speed-up and space-saving techniques are discussed in Section 5. Then, in Section 6, we present our computational results and compare them with results of the Deutsche Bahn server. Finally, we conclude with a summary and possible future extensions.

2   Railway Timetable Information Systems

2.1   Specification of Queries

A *query* to a timetable information system usually includes:
   The (start or) *source station* of the connection, the *target station* and an *interval* in time in which either the departure or the arrival of the connection has to be, depending on the *search direction*, the user's choice whether to provide the interval for departure ("forward search") or arrival ("backward search").[1] Additional query options include:

Vias and duration of stay. A query may contain one (or more) so called *vias*, stations the connection has to visit and where at least the specified amount of time can be spent, e.g. from Cologne to Munich via Frankfurt with a stay of at least two hours for shopping in Frankfurt.

Train class restrictions. Each train has a specific *train class* assigned to it. These classes are high-speed trains such as the German ICE and French TGV; ICs and ECs; Interregios and the like; local trains, "S-Bahn" and subway; busses and trams. The *query* may be restricted to a subset of all *train classes*. By excluding high speed trains one might be able to find cheaper connections.

Attribute requirements. Trains have *attributes* describing additional services they provide. Such attributes are for example: "bike transportation possible", "sleeping car", "board restaurant available". A user can specify attributes a connection has to satisfy or is not allowed to have. We allow Boolean operators for specifying *attribute requirements* like: (a restaurant OR a bistro) AND bike transportation.

2.2   Connections Matching a Query

A connection needs to be feasible and has to satisfy all requirements of the query specification to match the query. Some additional feasibility requirements are:

Meta Stations and Source-/Target-Equivalents. For a passenger it might be unimportant to start at a specific stations, as long as these stations are relatively close

---

[1] Note that the specification of an interval is crucial for typical pre-trip queries although previous work often assumes single point intervals.

together. Virtual *meta stations* group such stations together (like the railway station and bus stops that can be found right next to each other at the central station of any city). *Source/target-equivalents* group stations together in a similar fashion, but not as a new virtual station: Every *source/target-equivalent* consists of a station and its possible replacements.

A meta-station or source/target-equivalent may replace the source and target station as well as any via in a query.

Special attributes: NotIn / NotOut. There are some train and station related attributes that do have a special meaning for the stops of a train. Although a train stops at a station, boarding or leaving the train or both may not be allowed. Especially for night and high-speed trains there are some stations near the origin of the train where one is only allowed to enter the train and some stations near the end where one is only allowed to leave it. In a night train passengers should not be disturbed by too much "traffic" inside the train. In both cases the trains should not be used only for a short transfer. Passengers are encouraged to rather use local transportation instead.

Traffic days. Most trains do not operate on a daily basis. There is a lot of change during the year. Some trains only operate on workdays, others only on Sundays. National and local holidays affect the days of operation as well as school holidays.

Interchanges. The time table data provides rules for a lower bound on the time between the arrival of a train and the departure of its connection if a change of trains occurs. Arranged from most general to most specific these are:

- *Interchange rules at stations.* Every station has two interchange times: one for interchanges between two faster (or higher valued) trains like the German ICE or French TGV and one for all other interchanges.
- *Transfers between transfer classes.* Each train is associated with a *transfer class*. The time needed for the train change depends on the transfer classes of the arriving train at arrival and the leaving train at departure. There are rules with and without dependence on the station of the train change.
- *Line to line transfers.* Similar to the *transfer classes* each train may be associated with a *line* it serves and specific rules for line changes.
- *Service to service transfers.* The most specific interchange rule gives interchange times between individual trains.

2.3   Measuring the Quality of Connections

Most timetable information systems only regard one criterion, namely *travel time*. As mentioned before we want to focus on the three criteria travel time, ticket costs, and number of interchanges. Simply minimizing any of these three independently (or all three separately) is obviously not the method of choice. In the *weighted multi–criteria* case an evaluation function c may look like:

$$c = \varphi \cdot \text{travel time} + \xi \cdot \text{number of interchanges} + \vartheta \cdot \text{ticket cost}.$$

Table 1. Example connections

Connection	Departure	Travel time (minutes)	Number of interchanges	Price (Euro)	Pareto optimal
c_1	7 : 30	110	2	75	
c_2	8 : 00	100	2	75	√
c_3	8 : 00	160	0	60	√
c_4	8 : 00	200	3	35	√
c_5	8 : 00	300	3	34	√
c_6	8 : 15	110	1	45	√

Different choices for the set of parameters $\{\varphi, \xi, \vartheta\}$ express the difference in importance of the three criteria. Users may never see some interesting alternatives (for them) if either they or a system/operator sets the wrong parameters.

To overcome this problem the concept of *Pareto-optimality* treats all criteria as equally important. For two given k-dimensional vectors $x = (x_1, \ldots, x_k)$ and $y = (y_1, \ldots, y_k)$, x *dominates* y if $x_i \leq y_i$ for $1 \leq i \leq k$ and $x_i < y_i$ for at least one $i \in \{1, \ldots, k\}$. Vector x is *Pareto optimal* in set X if there is no $y \in X$ that dominates x. Here, we assume for simplicity that all cost criteria shall be minimized. In our scenario we compare 3-dimensional vectors (travel time, ticket costs, number of interchanges) for our connections. Note that this approach is easily extendable to cover further criteria.

Consider the connections of Table 1: Connections c_2 to c_6 are Pareto optimal. The single-criterion and weighted-criteria approaches (for some parameters) both do not find c_6 which is probably the most promising of all connections for most people. Unfortunately, the classical Pareto approach has its drawbacks as well: Suppose connection c_6 does not exists in the list. Although connection c_1 is dominated by c_2 it still arrives earlier at its destination. A passenger using a timetable information system at the departure station might prefer c_1 as it leaves more time to get to his final destination from the target station instead of waiting 30 minutes at the departure station. In spite of being Pareto optimal connection c_5 is of no practical use at all. The almost as cheap alternative c_4 is much more attractive.

Relaxed Dominance

To tackle the drawbacks of the simple Pareto dominance approach we *relax* the dominance rule in the *relaxed Pareto dominance* case. This means that more pairs of connections become mutually incomparable. In addition to the cost criteria travel time, travel fare and number of train changes, further aspects are taken into account to define the smaller relation between connections.

Formally, we now consider n-dimensional (integral or real-valued) vectors $x = (x_1, \ldots, x_k, x_{k+1}, \ldots, x_n) \in S$ where the first k components are cost criteria and the remaining $n - k$ components encode additional data (like departure and arrival time, highest used train class). Furthermore, for each cost criterion we have a non-negative *relaxation function* $f_i : S \times S \mapsto \mathbb{R}_0^+ \cup \{+\infty\}$. For any two $x, y \in S$ we now define that x *dominates* y *(in the relaxed sense)* if $x_i + f_i(x, y) \leq y_i$ for all $1 \leq i \leq k$ and $x_i + f_i(x, y) < y_i$ for at least one $i \in \{1, \ldots, k\}$. Note, that in order to be able to apply relaxed Pareto dominance during search and/or for final filtering of the connections, it is essential that dominance is a transitive relation. This restricts the set of reasonable relaxation functions. Next we give examples how to specify suitable relaxation functions f_i.

- The travel time spent for getting less expensive connections has to yield a fair *hourly wage*, say of Δ Euros per hour. (In the examples of Table 1 an hourly wage of less than one Euro is not enough to make connection c_5 worth considering.) This can be modeled as follows. Suppose we want to compare connections A and B with associated costs c_A, c_B in Euros and travel times t_A, t_B in minutes, respectively. Then connection A dominates B with respect to the cost criterion only if

$$c_A + \frac{\max\{t_A - t_B, 0\}}{60} \cdot \Delta < c_B.$$

- The larger the time difference between the departure and arrival times of two connections is, the less these connections should influence each other.
  Suppose we want to compare connections A and B which have departure times d_A, d_B, arrival times a_A, a_B and travel times t_A, t_B (all data given in minutes), respectively. Then connection A dominates B with respect to the criterion travel time if

$$t_A + \alpha(t_A) \cdot \Delta(A, B) + \beta(t_A) < t_B,$$

where, e.g., we may choose $\alpha(t_A) := t_A/360$ and $\beta(t_A) := 5 + \sqrt{t_A}/4$, and define

$$\Delta(A, B) = \begin{cases} 0 & \text{if } d_A > d_B \text{ and } a_A < a_B, \\ \min\{|d_A - d_B|, |a_A - a_B|\}, & \text{otherwise .} \end{cases}$$

- Different kinds of connections shall not dominate each other (e.g., connections using an event train (e.g. a special train to a sports event) or night trains). Using night trains one does not want to arrive as fast (and/or cheap) as possible but has the chance to arrive relaxed and even save a night's stay at a hotel. Both these alternatives should not be dominated by connections using other kinds of transportation. This can be modeled by defining a relaxation function to be $+\infty$ if the encoding of the train class attributes forbids a mutual domination.

Incomparable connections do not dominate each other, thus attractive alternatives are not suppressed. It is easy to check that all the proposed relaxation functions preserve the desired transitivity of our Pareto dominance relation. In Section 4.2 it will turn out that this concept can also be used to handle special tariffs of pricing systems.

Our overall goal is to determine the complete set of connections not dominated by relaxed dominance. However, some other aspects are still not covered, like the stability of a connection, i.e. how high is the possibility of getting all interchanges, the aim to use a sleeping cart as long as possible during the night, the maximization of a stay at "nicer" locations, lovely panorama etc.

3   Modeling Timetable Information in Graphs

3.1   A Time-Expanded Graph Model for a Realistic Scenario

Schulz et al. [17] introduced the time-expanded model to compute earliest arrival paths which assumed zero interchange times. Modifications to allow the counting of train interchanges with $\{0, 1\}$-weights in a Dijkstra-search have been proposed in [8,7,15]. The extension of Pyrga et al. [15] models also some basic interchange rules with the concept of change nodes: In their time-expanded digraph there is one node for each departure and arrival event of every train. The departure of a train at one station and its arrival at the next station are connected with a *train edge*. The so-called *change nodes* are copies of all arrival and departure events. A *waiting edge* is introduced between each change node and the next change node in time at the same station (introducing a waiting arc over midnight between the last and first change node at that station as time is taken modulo a single day). Between the arrival of a train and its departure at the same station there is a *stay in train edge* in the graph.

We further extended the model to cover all interchange rules and the special attributes NotIn/NotOut (cf. Section 2.2): If boarding is permitted we have an *entering edge* from the change node copy to the original departure node. If leaving a train is possible, we have one *leaving edge* connecting the arrival with the change node at the point in time from which on all other events are reachable, i.e. the time difference of this node to the arrival is the maximum over the interchange times required by all change rules concerning this train at this station. For all trains reachable before this point in time we have *special interchange edges* from the arrival to the departure nodes of the corresponding trains. The first two types of edges can also be found in the model by Pyrga et al. but with different semantics.

It is easy to see that we have indeed covered all interchange rules and the attributes *NotIn/NotOut*. Take a look at Fig. 1 (left) for an example. Arriving with train t we here can either stay in train t (use stay in train edge e) or change to t^* what is possible due to some interchange rule e.g. service to service transfer (special interchange edge f). However, we can not take t' (for example, if the minimum interchange time at the station does not allow this). Therefore, we needed the special interchange edge to reach t^* and not to reach t' from t although entering t' is allowed from the change level. Every event from time b on is again reachable (leaving edge g to change node at time b), e.g., we can take train t'' (via entering edge h). Train t' stops at the station only for boarding (no leaving edges for the arrival at time a).

Traffic days, possible attribute requirements and train class restrictions with respect to a given query can be handled quite easily. We simply mark train edges as *invisible* for the search if they do not meet all requirements of the given query.

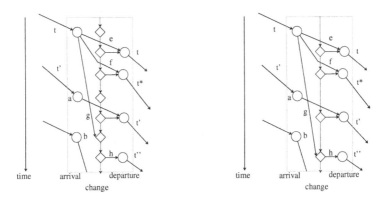

Fig. 1. Time expanded model without special interchange edges (left) and an extension to skip arrival change nodes in forward search (right) cf. Section 5.4

With respect to this visibility of edges, there is a one-to-one correspondence between feasible connections and paths in the graph.

We associate component-wise non-negative cost vectors to the edges. Here we describe only the choice for forward search, the necessary modifications for backward search should be obvious. For the cost criterion travel time, the cost for edge $e = (v, w)$ is the difference between the timestamps of the nodes w and v. For the cost criterion number of train changes, all entering edges and all special interchange edges get a cost value of 1, and all other edges a value of 0. (More precisely, such an assignment counts the number of used trains.) Ticket costs are more difficult to handle. We come back to this issue in Section 4.2.

3.2   Discussion: Time-Expanded vs. Time-Dependent Models

For single–criteria shortest path search, the time-dependent model seems to be more attractive due to its smaller sized graph that drastically speeds up the search. This advantage does not hold for more realistic scenarios. In recent experiments for computing all Pareto optima for two criteria Pyrga et al. [14] did not show a big advantage for time-dependent models. Not only did the size of the graph significantly grow due to their modeling of constant transfer times (which are still far from reality), the computational time required for solving the problem on the time-expanded graph was only 58% higher than for the time-dependent graph. Their constructions in [14] show how difficult the extension to model more realistic scenarios is. The general interchange rules and the special attributes disallowing boarding and deboarding even violate assumptions made for the most realistic model of interchanges known for time-dependent graphs (i.e. taking a later connection might allow for taking an earlier connection at some subsequent station). The time-dependent approach is not as easily extendable to cover traffic days, attribute requirements and train class restrictions as the time-expanded approach as already observed in [7].

Considering time intervals instead of points in time for possible departures requires either lists of labels for each node or computing a solution for each node in the interval. The latter is surely not reasonable for larger intervals. In summary, the time-expanded model is much more flexible and extensible (for possible further extensions see Section 7). Therefore, we used the time-expanded model for our algorithm.

4   The Information Server PARETO

4.1   The Search Algorithm in PARETO

Our algorithm is a "Pareto version" of Dijkstra's algorithm using multi-dimensional labels. See Möhring [6] or Theune [19] for a general description and correctness proofs of this approach.

We keep the travel time, number of interchanges, ticket costs (cf. Section 4.2) and some additional information in the labels. For every node in the graph we maintain a list of labels that is not dominated by any other label at this node. Every time a node is extracted from the priority queue, its outgoing edges are scanned and (if they are not infeasible due to traffic days, attributes and train class restrictions etc.) labels for their head nodes are created. Such a new label is compared to all labels in the list at the head node and only inserted into that list and into the priority queue if it is not dominated by any other label in the list. On the other hand, labels dominated by the new label are removed. For changes to this basic algorithm see the rest of this Section (for modeling reasons) and Section 5 (for space and time consumption reasons).

4.2   Modeling Ticket Costs

Pricing systems of railway companies are very complex. Unfortunately, ticket costs are typically not proportional to the distance traveled. The cost of the distance traveled in one train depends not only on the number of kilometers but also on its train class and other train classes used in the connection. Currently there are different supplementary fares for the different higher speed train classes. Furthermore, the system undergoes rapid change. In building timetable information systems it should not be the task to rebuild pricing components.

To be resistant to the changes in the pricing system (to some degree) we have a black-box pricing component (BPC) that can be used to calculate the ticket cost for some connection. Unfortunately, one call to this black-box routine is very costly: The path information stored in labels has to be converted into structures for the BPC. Much additional information like attributes on the train edges has to be set to get a correct price. Therefore, it is not at all possible to calculate the correct price for every label and achieve a bearable running time.

As a consequence we use price estimates in the labels that are updated during the search. The distance between the two stations of a train edge is taken as the straight line distance obtained from the coordinates of the stations. For every train edge the price estimate is increased by the distance times a factor depending on the train class used. The supplementary fare is paid once and only for the highest train class involved.

This simplified model provides helpful estimates for the search. In order not to loose low cost connections due to this approximation we need a safety margin which is incorporated into the corresponding relaxation function for the relaxed Pareto dominance. Here another benefit of the Pareto relaxation (compare Section 2.3) comes into play, enabling us to model some exceptions: For example, there might be a special offer (like "Schönes Wochenende-Ticket" in Germany) for traveling on weekends for a fixed price independent of the distance but valid only on non-high speed trains. The relaxation allows us not to compare connections using no high speed trains to connections with high speed trains on weekends. After a search is completed, all connections are correctly priced by the BPC and relaxed Pareto dominance can be applied to true fares.

5   Speed-Up and Space-Saving Techniques

5.1   The Station Graph for Lower Bounds

For the strategies goal direction (Section 5.3) and to discard labels dominated by labels at the target station (Section 5.2) lower bounds for the distance from any node n to the target are required.

Let us consider lower bounds for the criterion travel time: Regarding space efficiency, it is not reasonable to store a precomputed lower bound for every pair of stations. Thus, these values must be computable "on the fly" during the search. One easy approach for calculating a lower bound on the remaining travel-time is to calculate the straight-line distance from the station $S(n)$ of node n to the target station Ω and divide this value by the fastest travel speed of all trains in the data, as used for example by Schulz, Wagner and Weihe [17]. Empirical testing revealed that this method leads only to a small speed-up.

Our idea, giving tighter lower bounds, uses the *station graph*. This graph consists of one node per station and we insert an arc from station A to B if there is a direct connection and take as the travel time the minimum among all such connections (not considering traffic days). If we simply reverse all edges in the *station graph* we may use one Dijkstra-search on the *station graph* starting at the target station Ω at the beginning of each search and get a lower bound on the travel time to Ω for *every* station or the information that no connection to Ω exists.

5.2   Domination by Labels at the Terminal

To reduce the number of labels investigated during the search, a simple heuristic improvement can be utilized that relies upon the simple fact, that if P is a Pareto optimal path, then any subpath P' of P must also be Pareto optimal.

We use the following lower bounds on the cost of a path from node v to station Ω for the three criteria:

- The value from the station graph as lower bounds on travel time.
- The trivial lower bound zero for the number of interchanges.
- The straight line distance from the station $S(v)$ of v to Ω multiplied by the cost for traveling in the lowest train class for the ticket cost.

We maintain a list of strict Pareto optimal labels at the terminal station Ω. Not all relaxed Pareto optimal labels are stored in that list to keep it short. Every new label is checked against each label in this list. If the values of the label plus the lower bounds are dominated by any label in the list there is no need to further regard the new one and it is not inserted into the priority queue.

5.3   Goal-Direction in PARETO

We use travel time as the criterion for goal direction and only fall back on the number of interchanges for breaking ties. The smaller relation for the priority queue orders the labels according to the sum of the time traveled so far, the lower bounds for the travel time to the target station (computed via the station graph (see Section 5.1)) plus γ times the number of interchanges for some constant $\gamma > 0$. Although the search cannot be terminated once the first label at the target station Ω is extracted from the priority queue using goal direction, labels at Ω are generated fairly early in the search process, thus improving the efficiency of the strategy from the previous section.

5.4   Skipping Arrival Events

Having a arrival and d departure nodes on the change level results in $2 \cdot (a + b)$ waiting edges ($a + d$ for forward and backward search each). In forward search there is no need to consider arrival events except the ones that are extracted from the priority queue (inserted as the target of a feasible train edge). Thus we may arrange the change nodes in two cycles, one by linking the departure change nodes with waiting edges according to increasing time values (for forward search) and the other by linking arrival change nodes ordered by decreasing time value (see Figure 1 right). Applying this construction we only need d waiting edges for forward search and a for backward search. Thus we save half of the edges and operations on the change level.

5.5   The Impact of Speed-Up Techniques

We ran our algorithm on over 5000 queries stemming from an internet server of Deutsche Bahn AG (original customer queries). They include forward and backward searches with and without train class restrictions and attribute requirements but no vias. The time table data was prepared for one week and all queries were shifted to the corresponding weekday in that week. The resulting graph has about 1 million original arrival and departure nodes and half a million of train edges and about 1.8 million additional edges. Up to now the sole purpose of our implementation was to countercheck the results of the server

Table 2. The impact of various speed-up techniques. The percentages (right columns) give the increase compared to the run with all speed-up techniques activated.

activated Speed-Ups	all	noDom		noGoal		noSkip		noSkip + noGoal	
given in	k	k	%	k	%	k	%	k	%
PQMinOps	309	2302	526	459	49	526	70	776	144
LabelsUsed	599	4544	659	884	48	802	34	1145	91
LabMaxAct	223	949	326	316	42	376	69	523	135
DomTarget	81	0	-100	77	-5	80	-2	60	-26
StationsHit	1.7	5.5	215	2.1	21	1.7	0	2.1	19
PQ Before	17	17	0	169	920	26	62	266	1507

currently used by Deutsche Bahn AG. Speed considerations were only secondary goals. The algorithm solves queries with an average computational time of less than 5 seconds on an Athlon XP 2100+ PC with 1 Gigabyte of RAM both under Unix using the GNU Compiler version 2.95.3 and under Microsoft Windows using MS Visual Studio 6.0.

As we did not fine tune the computational efficiency of our prototype, we consider in the following only operation counts to study the impact of the proposed heuristics. Computational results are shown in Table 2. The column headings describe the activated speed-up techniques for the corresponding column:

all. All speed-up techniques activated.

noDom. Domination by a label at the terminal station (see Section 5.2) is deactivated.

noGoal. Goal-direction (see Section 5.3) is deactivated.

noSkip. The modification of the graph in the skipping arrival events heuristic (see Section 5.4) is not done.

The rows in Table 2 give the average numbers (left) or the increase (in percent on the right) compared to the run with all speed-up techniques turned on. We used the following key values as operation counts:

- The number of extractMin()-operations on the priority queue for the whole search (*PQMinOps*) and before creation of the first label at the target station (*PQBefore*).
- The number of labels used for the search (*LabelsUsed*).
- The maximum number of labels active at the same time (*LabMaxAct*). This is the minimum number of labels that must fit into the main storage.
- The number of labels dominated by other labels at the target station (*DomTarget*).
- The number of stations that were hit by the search (*StationsHit*).

We now take a look at the three main speed-up techniques.

Dominated by labels at terminal. This technique is powerful in bounding the number of operations and labels needed for the search. When combined with goal direction it makes the search much faster and significantly reduces the memory consumption. Deactivating this strategy slows the search down by a factor of over 5. It leads to the exploration of nearly 3 times as many stations and requires about 4 times as many labels in memory at the same time.

Goal direction. This technique produces labels at the target station fairly early in the search and is instrumental to significantly improving the performance of strategy "dominated by labels at terminal". Deactivating goal direction results in an increase of nearly 50% of the first four key values. After deactivation about 10 times as many operations as with goal direction are needed to reach the terminal station.

Skip arrival events. This strategy reduces the number of extractMin()-operations on the priority queue. Unnecessary labels for waiting are neither created nor inserted into nor extracted from the priority queue.

6   Comparison to the Deutsche Bahn Server

In this section we want to present a comparison between the quality of results computed by PARETO and the server currently used by Deutsche Bahn AG. The comparison was performed by Deutsche Bahn Systems personnel due to licensing reasons concerning the Deutsche Bahn server. We will present a measurement of quality based on three costumer groups as utilized internally by Deutsche Bahn AG. PARETO was delivered to DB systems without knowing the utility functions beforehand (i.e. we did *not* fine tune PARETO to look good with respect to these functions). Various quality distributions with respect to these groups are shown.

6.1   Utility Functions for Customer Groups

A weighted multi–criteria function is used to evaluate the quality of a connection c that looks like this (compare Section 2.3):

$$N(c) = \varphi \cdot \text{travel time} + \xi \cdot \text{number of interchanges} + \vartheta \cdot \text{ticket cost}.$$

There are three groups of costumers, each with another set of parameters $\mathcal{P} = (\varphi, \xi, \vartheta)$. These parameters were chosen by Deutsche Bahn Systems.

- The first group of costumers is mainly interested in travel time. A representative is a *businessman*. The parameters are $\mathcal{P} = (100, 1000, 1)$.
- The second group of costumers is mainly interested in convenience, i.e. the number of interchanges. Travel time is not unimportant, too. This group may contain a family with children, elderly people or a *handicapped person*. The parameters are $\mathcal{P}' = (12.5, 1500, 1)$ for this group.
- The third group of costumers is mainly interested in ticket cost. No other group is willing to spend so much time (hourly wage of 4.80 Euro for traveling longer) or to accept additional interchanges to save so little money. A representative is a *student*. The parameters are $\mathcal{P}^* = (8, 5, 1)$.

6.2   Quality Distribution for Costumer Groups

All connections are evaluated for all three costumer groups independently. The best connection for a group c_{opt} found by one of the two servers is used to normalize the value. The *quality points* for connection c are positive integers not greater than 100: $Q(c) = \lfloor N(c_{opt})/N(c) \cdot 100 \rfloor$.

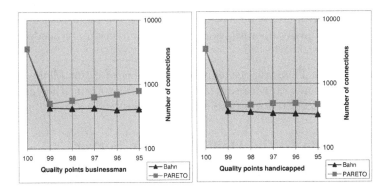

Fig. 2. Quality points for groups businessmen and handicapped. The number of connections is given in logarithmic scale.

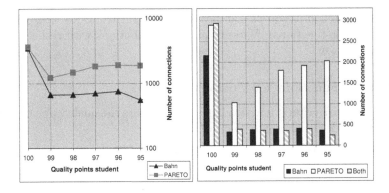

Fig. 3. Quality points for group student. On the right, we show a partition of the number of found connections with respect to the two servers.

For each user group, we evaluated the achieved quality points independently, see Figs. 2 and 3. Note that the number of connections is given in logarithmic scale. As we are only interested in "very good" connections we will not show connections with a value below 95. In this region of highest quality between 95 and 100 quality points, PARETO manages to outperform the Deutsche Bahn server significantly for every user group. (The gap becomes even larger for ranges below 95.) A deeper look into the results further revealed that the connections found by the two servers differ from each other surprisingly often. In Fig. 3 (right part), we display for the students group the partition of all connections into those found solely by the Deutsche Bahn server or by PARETO, respectively, and those found by both of them. Again, PARETO clearly delivers many more alternatives of very high quality above 94 points. The figures for the other user groups look rather similar.

7 Conclusion and Outlook

In this paper we have presented the concept of our timetable information server PARETO. Its major goal is to search for all "reasonably attractive" train connections for given customer queries and different groups of users. To formalize this goal and to avoid some weaknesses of ordinary Pareto dominance in multi-criteria optimization, we introduced the concept of *relaxed Pareto dominance*. Based on a fully-realistic time-expanded graph model, PARETO outperforms with respect to quality the current state of the art timetable information server of Deutsche Bahn AG. We briefly sketched some implementation issues especially concerning space and time consumption and experimental results.

There are many more interesting aspects in the field of timetable information. In the future, we plan to consider some of these and add extensions to the functionality of our tool. Such extensions are for example:

Secure interchanges. All connections where the time for a train change is not less than the minimum time needed for the change with respect to the interchange rules are valid. Connections providing a buffer time of k minutes for some $k > 0$ have a lower possibility of missing a train due to a delay of the arriving train. Such secure alternatives are especially noteworthy if a connection is the last connection of a day, i.e. a missed train would result in a night's stay to take the first train in the morning.

Guaranteed interchanges. The time buffer for an interchange is calculated as the difference between the arrival and departure of the trains involved minus the time needed for the interchange. Some connections have *guaranteed interchange times.* For example, ICE A waits for ICE B at station S at least 20 minutes. Although the interchange would be considered insecure for a calculated buffer of less than 2 minutes, the real buffer now is 20 minutes and that is pretty safe. Similar rules exist for the last busses in evenings that wait for the last trains to arrive.

Avoiding highly frequented trains. Trains often working to capacity currently have a static special attribute in the timetable data indicating high usage. If trains with known potential of working to capacity appear among the best connections, a timetable information system should be able to produce alternatives with less frequently used trains.

Seat reservability. Currently connections using identical trains in the same order are equivalent. If information about the reservation possibilities is available the location of train changes becomes important. The possibility of reservation throughout the whole connection might be possible for some orders of train changes and not for others.

Search for special offers. Every tariff that enables costumers to buy a ticket for a price below the actual standard ticket cost is considered a special offer (e.g. the already mentioned "Schönes Wochenende" ticket for non high speed connections). Basically there are two categories of special offers, one with and the other without quotas limiting the availability of the offer. Offers subject to quotas require very short updating periods for the data concerning the current quotas. Active search for such special offers (with or without quotas) is an important extension to finding cheaper connections.

Search for event trains. Special event trains are introduced for the purpose to bring participants jointly to events like soccer matches, the "love parade" in Berlin or the "Oktoberfest" in Munich. Such trains should be offered to the target group. Hence, it is necessary to be able to inform participants of such events explicitly about these trains even if they are not as good as others by standard criteria.

Acknowledgments

This work was done in cooperation with Deutsche Bahn Systems, Frankfurt am Main, Germany, and datagon GmbH, Waldems, Germany. In particular, we wish to thank Wolfgang Sprick for fruitful discussions and close collaboration in the development of PARETO.

References

1. Brodal, G.S., Jacob, R.: Time-dependent networks as models to achieve fast exact time-table queries. In: Proceedings 3rd Workshop on Algorithmic Methods and Models for Optimization of Railways (ATMOS 2003). Electronic Notes in Theoretical Computer Science, vol. 92, pp. 3–15. Elsevier, Amsterdam (2003)
2. Ehrgott, M., Gandibleux, X.: An annotated biliography of multiobjective combinatorial optimization. OR Spektrum 22, 425–460 (2000)
3. Ehrgott, M., Gandibleux, X. (eds.): Multiple Criteria Optimization: State of the Art Annotated Bibliographic Survey. Kluwer Academic Publishers, Boston (2002)
4. Hansen, P.: Bicriteria path problems. In: Fandel, G., Gal, T. (eds.) Multiple Criteria Decision Making Theory and Applications. Lecture Notes in Economics and Mathematical Systems, vol. 177, pp. 109–127. Springer, Berlin (1979)
5. Mote, J., Murthy, I., Olson, D.L.: A parametric approach to solving bicriterion shortest path problems. European Journal of Operations Research 53, 81–92 (1991)
6. Möhring, R.H.: Verteilte Verbindungssuche im öffentlichen Personenverkehr: Graphentheoretische Modelle und Algorithmen. In: Angewandte Mathematik - insbesondere Informatik, Vieweg, pp. 192–220 (1999)

7. Müller-Hannemann, M., Schnee, M., Weihe, K.: Getting train timetables into the main storage. In: Proceedings 3rd Workshop on Algorithmic Methods and Models for Optimization of Railways (ATMOS 2002). Electronic Notes in Theoretical Computer Science, vol. 66, Elsevier, Amsterdam (2002)

8. Müller-Hannemann, M., Weihe, K.: Pareto shortest paths is often feasible in practice. In: Brodal, G.S., Frigioni, D., Marchetti-Spaccamela, A. (eds.) WAE 2001. LNCS, vol. 2141, pp. 185–198. Springer, Heidelberg (2001)

9. Mehlhorn, K., Ziegelmann, M.: Resource constrained shortest paths. In: Paterson, M.S. (ed.) ESA 2000. LNCS, vol. 1879, pp. 326–337. Springer, Heidelberg (2000)

10. Nachtigal, K.: Time depending shortest-path problems with applications to railway networks. European Journal of Operations Research 83, 154–166 (1995)

11. Orda, A., Rom, R.: Shortest paths and minimum-delay algorithms in networks with time-dependent edge-length. Journal of the ACM 37(3), 607–625 (1990)

12. Orda, A., Rom, R.: Minimum weight paths in time dependent networks. Networks 21, 295–319 (1991)

13. Pallottino, S., Scutellà, M.G.: Equilibrium and advanced transportation modelling, ch. 11. Kluwer Academic Publishers, Dordrecht (1998)

14. Pyrga, E., Schulz, F., Wagner, D., Zaroliagis, C.: Towards realistic modeling of time-table information through the time-dependent approach. In: Proceedings 3rd Workshop on Algorithmic Methods and Models for Optimization of Railways (ATMOS 2003). Electronic Notes in Theoretical Computer Science, vol. 92, pp. 85–103. Elsevier, Amsterdam (2003)

15. Pyrga, E., Schulz, F., Wagner, D., Zaroliagis, C.: Experimental comparison of shortest path approaches for timetable information. In: Proceedings 6th Workshop on Algorithm Engineering and Experiments and the 1st Workshop on Analytic Algorithmics and Combinatorics, SIAM, pp. 88–99 (2004)

16. Skriver, A.J.V., Andersen, K.A.: A label correcting approach for solving bicriterion shortest path problems. Computers and Operations Research 27, 507–524 (2000)

17. Schulz, F., Wagner, D., Weihe, K.: Dijkstra's algorithm on-line: An empirical case study from public railroad transport. ACM Journal of Experimental Algorithmics, 5(12) (2000)

18. Schulz, F., Wagner, D., Zaroliagis, C.: Using multilevel graphs for timetable information in railway systems. In: Mount, D.M., Stein, C. (eds.) ALENEX 2002. LNCS, vol. 2409, pp. 43–59. Springer, Heidelberg (2002)

19. Theune, D.: Robuste und effiziente Methoden zur Lösung von Wegproblemen. Teubner Verlag, Stuttgart (1995)

20. Warburton, A.: Approximation of pareto optima in multiple-objective shortest path problems. Operations Research 35, 70–79 (1987)

The Railway Traveling Salesman Problem[*]

Georgia Hadjicharalambous[1], Petrica Pop[1], Evangelia Pyrga[1,2],
George Tsaggouris[1,2], and Christos Zaroliagis[1,2]

[1] Computer Technology Institute, P.O. Box 1122, 26110 Patras, Greece
[2] Dept of Computer Eng. & Informatics, University of Patras, 26500 Patras, Greece
{hadjicha,ppop,pirga,tsaggour,zaro}@ceid.upatras.gr

Abstract. We consider the Railway Traveling Salesman Problem (RTSP)
in which a salesman using the railway network wishes to visit a certain
number of cities to carry out his/her business, starting and ending at
the same city, and having as goal to minimize the overall time of the
journey. RTSP is an $\mathcal{NP}$-hard problem. Although it is related to the
Generalized Asymmetric Traveling Salesman Problem, in this paper we
follow a direct approach and present a modelling of RTSP as an integer
linear program based on the directed graph resulted from the timetable
information. Since this graph can be very large, we also show how to
reduce its size without sacrificing correctness. Finally, we conduct an
experimental study with real-world and synthetic data that demonstrates
the superiority of the size reduction approach.

1 Introduction

We consider a problem of central interest in railway optimization. We assume
that we are given a set of stations, a timetable regarding trains connecting these
stations, an initial station, a subset $\mathcal{B}$ of the stations, and a starting time. A
salesman wants to travel from the initial station, starting not earlier than the
designated time, to every station in $\mathcal{B}$ and finally return back to the initial
station, subject to the constraint that s/he spends the necessary amount of time
in each station of $\mathcal{B}$ to carry out his/her business. The goal is to find a set of
train connections such that the overall time of the journey is minimized. We call
this the *Railway Traveling Salesman Problem* (RTSP).

RTSP is related to the Generalized Asymmetric Traveling Salesman Problem
(GATSP) [2,3,4]. In that problem, a weighted complete directed graph is given
whose nodes are partitioned into clusters $V_1, V_2, \ldots, V_k$. The goal is to find a
minimum-weighted tour containing at least one node from each cluster. By virtue
of TSP, GATSP is an $\mathcal{NP}$-hard problem. It can be easily seen that TSP is
polynomially time reducible to RTSP as well: for each pair of cities, for which

---

[*] This work was partially supported by the Human Potential Programme of EU under
contract no. HPRN-CT-1999-00104 (AMORE), by the IST Programme (6th FP)
of EC under contract No. IST-2002-001907 (integrated project DELIS), and by the
Action PYTHAGORAS funded by the European Social Fund (Operational Program
for Educational and Vocational Training II) and the Greek Ministry of Education.

F. Geraets et al. (Eds.): Railway Optimization 2004, LNCS 4359, pp. 264–275, 2007.

there exists a connection, consider a train leaving from the first one to the second with travel time equal to the cost of the corresponding connection in TSP. This reduction sets RTSP in the class of $\mathcal{NP}$-hard problems.

Consider the so-called time-expanded digraph G constructed from the timetable information [5]. In that graph, there is a node for every time event (departure or arrival) at a station, and edges between nodes represent either elementary connections between the two events (i.e., served by a train that does not stop in-between), or waiting within a station. The weight of an edge is the time difference between the time events associated with its endpoints. Now, roughly speaking, and considering each set of nodes belonging to a specific station as a node cluster, RTSP reduces in finding a minimum-weight tour that starts at a specific node of a specific cluster, visits at least one node of each cluster in $\mathcal{B}$, and ends at a node of the initial cluster.

Despite their superficial similarity, RTSP differs from GATSP in at least three respects:

(i) GATSP is typically solved by transforming it to an instance of the Asymmetric TSP. The transformation is done by modifying the weights such that inter-cluster edges are "penalized" by adding a large weight to them and consequently the tour has to visit all the nodes within a cluster before moving to the next one. It is not clear whether such an approach can be directly applied to RTSP, where we have to take into account that the salesman must spend a minimum amount of time at each station (city).

(ii) RTSP starts from a specific node within a specific cluster, while GATSP starts from any node within any cluster.

(iii) GATSP requires to visit all clusters, while RTSP asks for visiting only a subset of them.

Consequently, we feel that a different approach is worth pursuing for the solution of RTSP; that is, an approach that does not follow the classical pattern of reducing RTSP into an instance of GATSP or to an instance of the Asymmetric TSP.

In this paper, we follow such a direct approach and present a modelling of RTSP as an integer linear program based on the time-expanded graph. Since this graph can be very large, we also show how to reduce its size without sacrificing correctness. This turns out to be rather beneficial, especially in the case where the number of stations $|\mathcal{B}|$ the salesman wants to visit is much smaller than the total number of stations. Finally, we conduct an experimental study with real-world and synthetic data that demonstrates the superiority of the size reduction approach.

2   Preliminaries

In this section, we describe the input of an RTSP instance and we provide some definitions. In the following we assume timetable information in a railway system, but the modelling and the solution approaches can be applied to any other public transportation system provided that it has the same characteristics.

A *timetable* consists of data concerning: *stations* (or bus stops, ports, etc), *trains* (or busses, ferries, etc), connecting stations, *departure* and *arrival times* of trains at stations. More formally, we are given a set of *trains* $\mathcal{Z}$, a set of *stations* $\mathcal{S}$, and a set of *elementary connections* $\mathcal{C}$ whose elements are 5-tuples of the form $(z, \sigma_1, \sigma_2, t_d, t_a)$. Such a tuple (elementary connection) is interpreted as train z leaves station σ_1 at time t_d, and the *immediately next* stop of train z is station σ_2 at time t_a. The *departure* and *arrival times* t_d and t_a are integers in the interval $T_{day} = [0, 1439]$ representing time in minutes after midnight.

Given two time values t and t', $t \leq t'$, their *cycle-difference*(t, t') is the smallest nonnegative integer l such that $l \equiv t' - t \pmod{1440}$.

We are also given a starting station $\sigma_o \in \mathcal{S}$, a time value $t_o \in T_{day}$ denoting the earliest possible departure time from σ_o, and a set of stations $\mathcal{B} \subseteq \mathcal{S} - \sigma_o$, which represents the set of the stations (cities) that the salesman should visit. A function $f_\mathcal{B} : \mathcal{B} \to T_{day}$ is used to model the time that the salesman must spend at each city $b \in \mathcal{B}$, i.e., the salesman must stay in station $b \in \mathcal{B}$ at least $f_\mathcal{B}(b)$ minutes.

We naturally assume that the salesman does not travel continuously (i.e., through overnight connections) and that if s/he arrives too late in some station, then s/he has to rest and spend the night there. Moreover, the salesman's business for the next day may not require taking the first possible connection from that station. Consequently, we assume that the salesman never uses a train that leaves too late in the night or too early in the morning.

3   The Time-Expanded Graph

The formulation of RSTP is based on the so-called *time-expanded* digraph [5]. Such a graph $G = (V, E)$ is constructed using the provided timetable information as follows. There is a node for every time *event* (departure or arrival) at a station, and there are four types of edges. For every elementary connection $(z, \sigma_1, \sigma_2, t_d, t_a)$ in the timetable, there is a *train-edge* in the graph connecting a *departure node*, belonging to station σ_1 and associated with time t_d, with an *arrival node*, belonging to station σ_2 and associated with time t_a. In other words, the endpoints of the train-edges induce the set of nodes of the graph. For each station $\sigma \in \mathcal{S}$, all departure nodes belonging to σ are ordered according to their time values. Let $v_1, \ldots, v_k$ be the nodes of σ in that order. Then, there is a set of *stay-edges*, denoted by $Stay(\sigma)$, (v_i, v_{i+1}), $1 \leq i \leq k-1$, and (v_k, v_1) connecting the time events within a station and representing waiting within that station. Additionally, for each arrival node in a station there is an *arrival-edge* to the immediately next (w.r.t. their time values) departure node of the same station. The cost of an edge (u, v) is *cycle-difference*(t_u, t_v), where t_u and t_v are the time values associated with u and v, respectively.

To formulate RTSP, we introduce the following modifications to the time-expanded digraph.

First, we do not include any elementary connections that have departure times greater than the latest possible departure time, or smaller than the earliest.

Second, we explicitly model the fact that the salesman has to wait at least $f_{\mathcal{B}}(b)$ time in each station $b \in \mathcal{B}$ by introducing a set of *busy-edges*, denoted by $Busy(b)$. We introduce a busy-edge from each arrival node in a station $b \in \mathcal{B}$, to the first possible departure node of the same station that differs in the time value by at least $f_{\mathcal{B}}(b)$.

Third, to model the fact that the salesman starts his journey at some station σ_o and at time t_o, we introduce a *source node* s_o in station σ_o with time value t_o. Node s_o is connected to the first departure node d_o of σ_o that has a time value greater than or equal to t_o, using an edge (called *source edge*) with cost equal to $cycle\text{-}difference(t_o, t_{d_o})$. In addition, we introduce a *sink node* s_f in the same station and we connect each arrival node of σ_o with a zero-cost edge (called a *sink edge*) to s_f.

Figure 1 gives an example of two stations in the time-expanded graph that illustrates our construction.

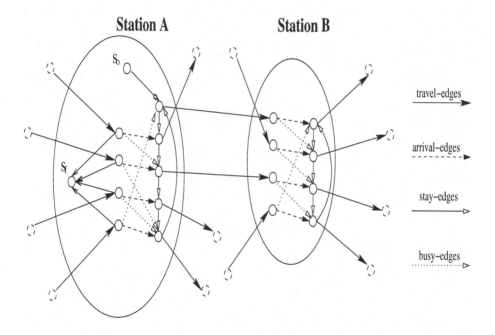

Fig. 1. An example of two stations in the time-expanded graph, where A is the starting station

4   Integer Linear Programming Formulation

The objective of RTSP is to find a tour from node s_o to node s_f that passes through each station $b \in \mathcal{B}$, with minimum total cost.

In order to model the RTSP problem as an Integer Linear Program (ILP), we introduce for each edge (u,v) a variable $x_{(u,v)} \in \mathbb{N}_0$ that indicates the number of times the salesman uses the edge (u,v). If $c(u,v)$ denotes the cost of edge (u,v), then the ILP becomes:

$$\min \sum_{(u,v)\in E} c(u,v) x_{(u,v)} \tag{1}$$

$$\text{s.t.} \sum_{(v,u)\in E} x_{(v,u)} - \sum_{(u,v)\in E} x_{(u,v)} = 0, \quad \forall u \in V - \{s_o, s_f\} \tag{2}$$

$$x_{(s_o,d_o)} = \sum_{(v,s_f)\in E} x_{(v,s_f)} = 1 \tag{3}$$

$$\sum_{e\in Busy(b)} x_e \geq 1, \quad \forall b \in \mathcal{B} \tag{4}$$

$$x_e \in \mathbb{N}_0, \quad \forall e \in E \tag{5}$$

Constraints (2) & (3) are the flow conservation constraints that form a path from node s_o to node s_f, while constraints (4) ensure that the salesman spends the required time in each station $b \in \mathcal{B}$.

The problem with the formulation so far is that it permits feasible solutions that contain cycles, disjoint from the rest of the path (*subtours*). To deal with this problem, we have to add some more constraints that will force the solution not to use any subtours.

Suppose that the selected stations are numbered from 0 to $|\mathcal{B}| - 1$. For each such station i, we introduce a new node f_i called the *sink node* of that station. Moreover, we create a copy $\bar{d}_k^i$ for the k-th departure node d_k^i of station i that has one or more incoming busy edges. We connect this new node with a zero cost edge to the original node. All busy edges now point to $\bar{d}_k^i$ instead of d_k^i, while all other edges remain unchanged. We also add an edge $(\bar{d}_k^i, f_i)$, for all k such that $\bar{d}_k^i \in V$. An example is given in Fig. 2.

We introduce now for each edge $e \in E$ a new set of variables y_e^i, $0 \leq i < |\mathcal{B}|$, and the following constraints:

$$\sum_{(v,u)\in E} y_{(v,u)}^i - \sum_{(u,v)\in E} y_{(u,v)}^i = 0, \quad \forall u \in V - \{s_o, f_i\} \tag{6}$$

$$y_{(s_o,d_o)}^i = \sum_{(v,f_i)\in E} y_{(v,f_i)}^i = 1 \tag{7}$$

$$y_e^i \leq x_e, \quad \forall e \in E - \{(u,f_i) \in E\} \tag{8}$$

The above constraints form a multicommodity flow problem, introducing a commodity for each selected station and asking for one unit of flow of each commodity $i \in [0, |B| - 1]$ to be routed from the source node s_o to the corresponding selected station's sink node f_i. An additional condition imposed on the y-variables is that the flow y_e^i of commodity i on any edge $e \in E$ cannot have a value greater than x_e. This means that if the x-flow is zero on some edge e,

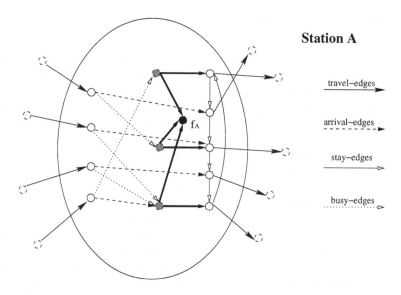

Station A

travel–edges

arrival–edges

stay–edges

busy–edges

Fig. 2. Station A of Fig. 1 after introducing the new nodes and edges used for the subtour elimination constraints. For simplicity, nodes s_o and s_f are not shown. The grey nodes are the copies of the departure nodes that had incoming busy edges, while the black node is the sink node of A. The thick black edges are the new edges that have been introduced to the graph.

then the y-flows will all be zero on that edge, too. Constraints (6) – (8) force the x variables to be assigned appropriately so that for each selected station there exists a path from s_o to some departure node in that station, using only edges for which the corresponding x variables are non-zero. Since now the only way to reach f_i is through the busy edges of station i, the flow modelled by the x variables is forced to use some busy edge for each selected station, which makes constraints (4) reludant.

4.1 Path Retrieval

The reason that the variables $x_{(u,v)}$, $(u,v) \in E$, are not restricted to be 0-1 variables is the fact that the salesman is allowed to pass through the same station σ more than one times, regardless of whether σ belongs to $\mathcal{S}$, or to $\mathcal{B}$, or it is the starting station σ_o.

Knowing the values of the variables $x_{(u,v)}$, we can easily retrieve the path, using the Flow Decomposition Theorem (see e.g., [1, Chap. 3]). Let n (resp. m) be the number of nodes (resp. edges) of the time-expanded graph G. We decompose the flows into at most m simple cycles and one simple path (from s_o to s_f) in time $O(nm)$. We can then construct the minimum cost tour as a linked list of edges as follows. We initially set the tour equal to the path, and then we iteratively expand it by merging it with a cycle that has a common edge with it. This can be done in time linear to the length of the final tour which is at most $O(nm)$.

4.2 Size Reduction Through Shortest Paths

The time-expanded graph can be rather large. In this section, we present a method that reduces the size of the graph, which in turn can be beneficial in the case where the salesman wishes to visit a relatively small number of stations compared to the total number of stations. We reduce the problem size by transforming the time-expanded graph G to a new graph G_{sh}, called the *reduced time-expanded graph*. Graph G_{sh} can be constructed by precomputing shortest paths among the stations that belong to $\mathcal{B}$, as follows.

Let again σ_{o} denote the start station. For each station σ in $\mathcal{B} \bigcup \{\sigma_{\mathrm{o}}\}$ we introduce a sink-node s_{σ} and we connect each arrival-node of σ with a zero cost edge to s_{σ}. Then, for each departure node d of $\sigma \in \mathcal{B} \bigcup \{\sigma_{\mathrm{o}}\}$, we compute a shortest path to the sink-nodes of every other station in $\mathcal{B} \bigcup \{\sigma_{\mathrm{o}}\}$. If such a shortest path does not pass through some other station in $\mathcal{B} \bigcup \{\sigma_{\mathrm{o}}\}$, then we insert a *shortest path edge* from d to the last arrival node of that path. The cost of this edge is equal to the cost of the corresponding shortest path.

We can now transform G in the following way. We first remove the sink-nodes (and their incoming edges) that were previously introduced. Next we delete from G all the nodes and edges that belong to stations that are not in $\mathcal{B} \bigcup \{\sigma_{\mathrm{o}}\}$. The remaining arrival nodes (that belong to some station $\sigma \in \mathcal{B} \bigcup \{\sigma_{\mathrm{o}}\}$) that were not used by any of the shortest paths that were previously computed are also deleted, as well as the remaining travel-edges. This procedure results in the reduced time-expanded graph G_{sh}.

5 Implementation Details

In order to speed-up the computations (trying to reduce the number of variables used by the integer programs) when we do not use the shortest path reduction, we try to eliminate the unnecessary nodes of the graph. To be more precise, we remove all arrival nodes of all stations $\sigma \notin \mathcal{B}$. These nodes have a single incoming edge (the one from the corresponding departure node), and a single out-going edge (to some departure node of the same station), while there are no busy edges starting from them. Therefore, we can eliminate these nodes (as well as their adjacent edges), and introduce a new virtual edge connecting the departure node that was the source of the incoming edge to the departure node that was the target of the outgoing edge, with cost the sum of the costs of the two eliminated edges. In this way, we can reduce the number of edges and nodes in the graph, and consequently the number of variables and constraints in the integer program.

6 Experiments

6.1 Data Sets

The construction of the graphs is based both on synthetic as well as on real-world data. For the synthetic case, we have considered grid graphs (with nodes

Table 1. Graph parameters for the original time-expanded graph in each data set. Note that the number of connections is equal to half the number of nodes, since two nodes form one connection, and the number of edges is twice the number of nodes. The number of edges does not include the incoming edges of s_f and the outgoing edge of s_o or any other artificial nodes or edges.

Data Set	Number of Stations $|\mathcal{S}|$	Number of Nodes	Number of Edges	Number of Connections	Conn/ Stations
nd_loc	23	778	1556	389	16.91
nd_ic	21	684	1368	342	16.29
synthetic 1	20	400	800	200	10
synthetic 2	30	600	1200	300	10
synthetic 3	40	800	1600	400	10
synthetic 4	50	1000	2000	500	10
synthetic 5	60	1200	2400	600	10
synthetic 6	70	1400	2800	700	10
synthetic 7	80	1600	3200	800	10

representing stations). Each node has connections (in both directions) with all of its neighbouring nodes, i.e., the stations that are located immediately next to it in its row or column in the grid.

The connections among the stations were placed at random among neighbouring stations, such that there is at least one connection in every pair of neighbouring stations (for both directions) and the average number of elementary connections for each station is 10. The time-differences between the departure and the arrival time of each elementary connection are independent uniform random variables, chosen in the interval $[20, 60]$ (representing minutes), while the departure times are random variables in the time interval between the earliest and the latest possible departure time. We have created graphs whose number of stations varies from 20 to 80.

The real-world data represent parts of the railroad network of the Netherlands. The first data set, called nd_ic, contains the Intercity train connections among the larger cities in the Netherlands, stopping only at the main train stations, and thus are considered faster than normal trains. These trains operate at least every half an hour, while the number of stations is equal to 21. The second real-world data set, nd_loc, contains the schedules of the trains that connect the cities in only one region, including some main stations, while trains stop at each intermediate station between two main ones. The total number of stations in this case is 23.

The characteristics of all the graphs that were used, for both real and synthetic data, are shown in Table 1.

6.2   Description of Experiments

For each data set, several problem instances were created, varying the number $|\mathcal{B}|$ of the *selected stations*, i.e., the set of stations that the salesman must visit.

For both graphs based on real and synthetic data, we have used two values for $|\mathcal{B}|$, namely 5 and 10. Note that $|\mathcal{B}|$ does not contain the starting station. Because of that, when $|\mathcal{B}|$ is set equal to the total number of stations, the actual value that will be used is $|\mathcal{B}| - 1$, since one station has to be the starting one.

For each combination of data set and a value of $\mathcal{B}$, we have selected the stations that belong to $\mathcal{B}$ randomly and independently from each other. The selection of stations has been repeated many times, and the mean values among all corresponding instances were computed. For each instance we have created, the corresponding integer program was given as input to GLPSOL v4.6 (GNU Linear Programming Kit LP/MIP Solver, Version 4.6). The time needed by GLPSOL to find the optimum solution for each case has been measured.

We have tested both the original (with our modifications) version of the time-expended graph, as well as the reduced version based on precomputed shortest paths described in Section 4.2. The time for this precomputation has also been measured.

6.3   Results and Discussion

The results of the experiments performed are reported in Tables 2 and 3 (a graphical comparison is illustrated in Fig 3).

The values measured are the average values for 50 different sets of the selected stations. The standard deviation of the running times provided by GLPSOL for instances of the same parameters was large, showing a great dependence on the graph structure.

It can easily be seen that the value of $|\mathcal{B}|$ has a great impact on the running time, when the original graphs were used. The larger the $|\mathcal{B}|$, the larger the running time.

Table 2. Graph parameters for the reduced graphs for all data sets (average values)

Real graphs					
		$|\mathcal{B}| = 5$		$|\mathcal{B}| = 10$	
Data Set	$|\mathcal{S}|$	Nodes	Edges	Nodes	Edges
nd_loc	23	209.4	415.2	343.9	698.1
nd_ic	21	181.6	385	303.6	708.7

Synthetic graphs					
		$|\mathcal{B}| = 5$		$|\mathcal{B}| = 10$	
Data Set	$|\mathcal{S}|$	Nodes	Edges	Nodes	Edges
1	20	95.75	209.35	197.7	420.3
2	30	91.3	204.7	181.1	430.4
3	40	95.7	213.0	188.3	437.8
4	50	92.7	205.2	184.6	440.3
5	60	94.8	216.2	181.9	456.8
6	70	93.0	215.4	184.0	491.7
7	80	91.2	207.2	174.6	457.7

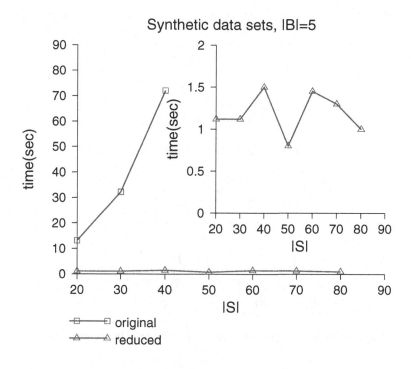

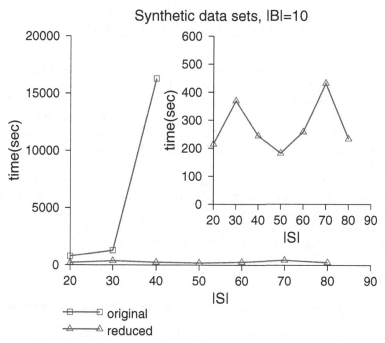

Fig. 3. Graphical comparison of running times for synthetic data sets

Table 3. Results for the synthetic data sets. Time is measured in seconds.

Running times for real data sets

Reduced Graphs		
$|\mathcal{B}|$	5	10
nd_loc	319.0	9111.9
nd_ic	29.1	6942.6

Running times for the synthetic data sets

Original Graphs			
$|\mathcal{S}|$	20	30	40
$|\mathcal{B}| = 5$	13.12	32.24	72.06
$|\mathcal{B}| = 10$	781.12	1287.00	16293.80

Reduced Graphs							
$|\mathcal{S}|$	20	30	40	50	60	70	80
$|\mathcal{B}| = 5$	1.12	1.12	1.50	0.80	1.45	1.30	1.00
$|\mathcal{B}| = 10$	214.76	369.59	244.18	181.85	257.96	431.80	233.26

Shortest path computation times for synthetic data sets

$|\mathcal{S}|$	20	30	40	50	60	70	80
$|\mathcal{B}| = 5$	0.13	0.18	0.24	0.29	0.35	0.41	0.46
$|\mathcal{B}| = 10$	0.28	0.38	0.48	0.59	0.69	0.79	0.88

The size reduction approach based on precomputed shortest paths results in much smaller graphs than the original. Table 2 shows the characteristics of the reduced graphs that have resulted by the use of the size reduction technique.

Also, it is clear from Table 3 that the size reduction approach results in a great speedup w.r.t. the time achieved by considering the original time-expanded graph to find an optimal (or close to optimal) solution. Moreover, the smaller the value of $|\mathcal{B}|$, the larger the speedup.

For a fixed size of $\mathcal{B}$ the running time seems to grow rapidly with the increase of $\mathcal{S}$ when the shortest path technique is not being used. On the contrary, this is not the case when this technique is used. In this case, when we consider synthetic data, the running times are fluctuating. Nevertheless, it is quite clear that the time depends mainly on the size of $\mathcal{B}$, rather than the size of $\mathcal{S}$.

References

1. Ahuja, R., Magnanti, T., Orlin, J.: Network Flows. Prentice-Hall, Englewood Cliffs (1993)
2. Fischetti, M., Salazar-Gonzalez, J., Toth, P.: The Generalized Traveling Salesman Problem and Orienteering Problem. In: Gutin, G., Punnen, A. (eds.) The Traveling Salesman Problem and its Variations, Kluwer, Dordrecht (2002)

3. Gutin, G.: Traveling Salesman and Related Problems. In: Gross, J., Yellen, J. (eds.) Handbook of Graph Theory, CRC Press, Boca Raton (2003)
4. Noon, C., Bean, J.: A Lagrangian Based Approach for the Asymmetric Generalized Traveling Salesman Problem. Operations Research 39, 623–632 (1991)
5. Schulz, F., Wagner, D., Weihe, K.: Dijkstra's Algorithm On-Line: An Empirical Case Study from Public Railroad Transport. ACM Journal of Experimental Algorithmics 5(12) (2000)

Rotation Planning of Locomotive and Carriage Groups with Shared Capacities

Taïeb Mellouli[1] and Leena Suhl[2]

[1] Department of Management Information Systems and Operations Research, Martin-Luther-University Halle-Wittenberg, Universitaetsring 3, 06108 Halle (Saale)
mellouli@wiwi.uni-halle.de
[2] Decision Support & OR Laboratory, University of Paderborn, Warburger Str. 100, 33098 Paderborn, Germany
suhl@uni-paderborn.de

Abstract. In a large railway passenger traffic network, a given set of trips or service blocks are to be serviced by equipment consisting of several groups of locomotives/carriages. The allowed groups per service block are predefined as patterns or multisets of locomotives and carriages. A given type of locomotive/carriage may occur with varying numbers in several groups. We search for a cost-minimal assignment of locomotive/carriage groups to rotations taking special restrictions into account, especially, we shall find the optimal mix of groups obeying given capacities on the level of locomotive and carriage units for each type.

Our solution approach is based on a multi-layer (multi-commodity) network flow model where each layer represents a locomotive/carriage group, and the requirement of servicing each trip exactly once is modeled by cover/partitioning constraints. In this paper, we concentrate on railway specific requirements and present special techniques to model and optimize locomotive and carriage groups with shared capacities. These techniques erable us to solve large-scale practical problem instances of German Railways into optimality.

1 Introduction

In railway passenger traffic, carriages and locomotives have to be assigned to trips in order to carry out a given schedule which has been published for passengers. In a large network this scheduling and routing task may be very complex, and until recently it was not generally possible to compute cost-minimal rotations for a given timetable with hundreds or thousands of trips when considering practical requirements such as maintenance rules or multiple types of carriages and locomotives.

We consider a railway network for passenger traffic, consisting of scheduled trips (service trips), each from a given departure station to a given end station. A trip may be divided into legs, also called *service blocks* during which coupling and uncoupling operations of train equipment are not allowed. Each service block has to be serviced with adequate equipment according to requirements considering

F. Geraets et al. (Eds.): Railway Optimization 2004, LNCS 4359, pp. 276–294, 2007.

technology, number of seats, comfort degree, and so on. The equipment consists of one or several groups of locomotives and/or carriages, which contain a given number of locomotives and/or carriages according to given group types. There may be several alternative groups to be used in putting together the equipment for a service block.

For example, the vehicle group types VG11, VG14 and WG15 may consist of carriage types ABn, ABnrz, BDnf, Bn, Bndf, Bnbdz, and Bnrz in the following way:

VG11: 1*ABnrz + 1*Bnbdz + 2*Bnrz VG14: 2*ABnrz + 2*Bn + 1*Bndf + 1*Bnrz VG15: 2*ABn + 1*BDnf + 3*Bn.

As an example, carriages of type ABnrz are needed for vehicle group types VG11 and VG14, carriages of type Bn for VG14 and VG15, and carriages of type BDnf only for VG15.

Thus, each locomotive and carriage belongs to a given equipment type, and the number of vehicles (locomotives/carriages) of each equipment type is limited.

Individual trains, however, consist of *groups* of locomotives and/or carriages, selected out of a given set of vehicle group types, each given group type being set together as a (multi)set of vehicles with a fixed number of members out of each vehicle type. Thus, we aim to develop an optimization model which minimizes the cost of equipment simultaneously satisfying the requirements on the vehicle group level and the given capacities on the locomotive and/or carriage vehicle type.

The total operational cost is to be minimized so that all requirements considering types of locomotives and carriages and the way they are assembled into train units are fulfilled.

Generally speaking, we mean by *train unit, train assembly* or *train consist* a group of compatible units of equipment that travel along some part of the physical rail network. A train assembly may include a given number of first class and second class carriages together with one or two locomotives. In most cases of railway applications, multiple types of locomotives and carriages are in use, and for each service unit a set of compatible types is given. In the following, we use the term *loco/car* or *vehicle* as an abbreviation and abstraction of a unit of locomotive, steering-wheel waggon, or rail carriage/car/waggon. A *vehicle type* or a *loco/car type* is the type of a locomotive or carriage unit.

In this paper, we address the rotation planning problem of the railway application area under these requirements. The task is to generate rotations for locomotives and carriages being of a given equipment type and simultaneously being part of one or more loco/car groups. A loco/car type may be involved in several groups and capacities of equipment types have to be taken into account. Thus, on one hand, we have to consider individual loco/cars in order to meet the capacity requirements, and on the other hand, (types of) loco/car groups, in order to take the type requirements into account. Capacities are shared in the sense that different groups share the same loco/car types. This approach is currently being used at German Railways (Deutsche Bahn), and we will present algorithms tested with their data.

The paper is organized as follows. In section 2, a literature review is provided together with details on our previous research work concerning both the railway application area and solution approaches of rotation planning problems. In section 3, the problem of rotation building for loco/car groups with shared loco/car capacities is formalized. In section 4, a mathematical model based on a multi-layer (multi-commodity) flow network is presented where each network layer represents a loco/car group, and the requirement of servicing each trip exactly once is modeled by cover/partitioning constraints. Especially, a special aggregation scheme of "equivalent" loco/car groups is applied in order to solve large-scale problems of German Railways directly by a standard mathematical optimizer. Finally, we present computational results in section 5 and discuss problems of practical relevance solved by exact optimization together with suitable decision support tools.

2   Previous Work and Solution Approach

2.1   Literature Review

Although the problem of simultaneously assigning locomotives and carriages to trips and building rotations is very important to railways and has to be solved on a regular basis in practice, there are relatively few contributions to it in the scientific literature. One of the first papers was Ramani/Mandal (1992) dealing separately with the assignment of locomotives and carriages, and using a local improvement procedure to improve the overall solution. Ben-Kheder et al. (1997) described a system for the simultaneous assignment of locomotives and carriages for passenger trains at SNCF. The system treats both types simultaneously but uses aggregated modules that are then assigned as a whole, thus not dealing explicitly with compatibility constraints. Zirati et al. (1997) consider the problem of assigning locomotives requiring inspection within a time limit of the considered one-week planning horizon.

Cordeau et al. (1998) give a survey on research until 1998. Since then, a few papers have been published. In Cordeau et al. (2000) an optimization model was developed which assigns both locomotives and carriages simultaneously, solving a tactical periodic problem as an integer programming problem based on a time-space network. The authors propose a multi-commodity network model which they solve using Benders decomposition. A second paper of the same authors extends this model in Cordeau et al. (2001) for the practical case where maintenance and equipment substitution is taken into account.

Brucker et al. (2003) formulate the railway carriage routing problem as an integer multi-commodity network flow problem with nonlinear objective function and present a local search solution approach for it.

A recent publication of Abbink et al. (2004) considers the tactical problem of finding the most efficient schedule of for a set of rolling stock to train series, so that as many people as possible can be transported with a seat, especially when there is little seating capacity available during rush hours.

2.2   Own Research Work and Solution Approach

The Decision Support and Operations Research Laboratory at the University of Paderborn, Germany, has been involved since 1996 in several projects concerning design and development of optimization models and decision support tools in public transport. In the railway domain, we have studied practical tasks within both planning and operations control phases, developed optimization models for maintenance routing problems, and designed dispatcher support tools with embedded simulation capabilities, cf. Suhl and Mellouli (1999), *"computer-aided scheduling of public transport"*, Suhl et al. (2001), and Mellouli (2001) CASPT'2000 Berlin.

Our research work on the development of rotation building models and software date back to the work of the second author (Suhl 1995) where an extension to time windows was developed and applied to airlines.

The first author thoroughly applied and extended time-space networks based on connection-lines to deal with various rotation building problems in public transport. In 1997, he developed a state-expanded time-space flow network to deal with maintenance routing problems for railways and airlines (Mellouli 2001), and in 1999 a new aggregation scheme for potential deadhead trips (empty movements) which is crucial to solve hard practical requirements directly by state-of-the-art optimization software and to derive new complexity results for the rotation building problem (Mellouli 2003). This aggregation scheme for potential deadhead trips is successfully applied in the bus transit domain to solve large-scale multiple-depot, multiple-vehicle-type problems (Kliewer, Mellouli, and Suhl (2002)), as well as in the railway domain.

In 2001, our laboratory developed a prototype for rotation building for German Railways with the best optimization results in a prior study. Based on this, a development project with German Railways was accomplished in 2002. This paper presents parts of research results achieved by our laboratory and tested within this project. We concentrate on railway specific requirements and present special techniques to model and optimize loco/car groups with shared capacities. In order to solve large-scale practical problem instances of German Railways into optimality, the aggregation scheme for arcs representing all possible empty train movements is also used to which we refer to our mentioned works.

3   Problem Formalization and Analysis

In the following, we formalize the problem of rotation building for loco/car groups with shared capacities introduced in section 1. For this problem we are given:

- A set of *service blocks SB*: Each service block is a trip or a maximal trip part in which coupling and uncoupling operations are not performed. Thus, a service trip may consist of one block or of a sequence of blocks with different requirements on used train parts.

- A set VG of (types of) *vehicle groups* (or *loco/car groups*): Each $vg \in VG$ defines a group of locomotive and carriages which can be used as *train part* for some service blocks.
- A set of vehicle types VT (or loco/car types): Thus, VT consists of the different types of locomotives and carriages available.
- A set of home bases HB: Home bases are stations where vehicles may be stationed. For each $vt \in VT$ and $hb \in HB$, let $capacity(vt, hb)$ be the number of vehicles of type vt stationed at homebase hb.

For each $vg \in VG$ and each $vt \in VT$, let $number(vg, vt)$ be the number of vehicles of type vt occurring in the vehicle group vg. For instance, if vehicle group vg_1 consists of 2 vehicles of type vt_6 and one vehicle from the types vt_{12}, vt_{13}, and vt_{14}, respectively, so we have $number(vg_1, vt_6) = 2$, $number(vg_1, vt_{12}) = 1$, $number(vg_1, vt_{13}) = 1$, $number(vg_1, vt_{14}) = 1$, and $number(vg_1, vt) = 0$ for all other vehicle types vt.

Restrictions on assignments of service blocks to vehicle groups are regulated as follows:

- *Assignments of service blocks to vehicle groups:* For each service block $sb \in SB$, there may be several assignments of vehicle groups (not necessarily of different types) for different positions in a train unit. These vehicle groups define the *train assembly* that serves the service block sb. Alternative types of vehicle groups for the same train position are given by means of *global* or *local replacements* of (types of) vehicle groups.
- *Global and local replacements of vehicle groups:* A *global replacement* of the form $vg_i \leftarrow vg_j$ is declared independently of service block assignments. For each service block and train position, if the (type of) vehicle group vg_i can be assigned, then the (type of) vehicle group vg_j can be assigned alternatively.

  A *local replacement* is defined for each specific service block (and train position) by listing the possible (types of) vehicle groups that are allowed for serving this specific service block.

The test data of German Railways that is related to this specific problem comprises 31 different types of vehicle groups, two home bases, and 7,500 assignments of service blocks to vehicle groups. Most of the vehicle groups consist of 6, 5, or 4 vehicles (only one vehicle group contains a single locomotive and two others contain only one carriage vehicle as reinforcement).

The difficulty of the problem is directly related to the possibility of global and local replacement of vehicle groups. In the following, we consider two variants, a simple problem version without, and a complex one, with such replacements:

The simple problem version: Having no replacements of vehicle groups, the vehicle group used for each service block and train position is unique. So the problem can be decomposed according to different vehicle groups, by considering subsets of service blocks uniquely assigned to different vehicle groups, respectively. Minimizing the number of vehicle groups used for each sub-problem is then equivalent to minimizing the number of vehicles (locomotives and carriages) used. For the sub-problems, a polynomial-time procedure for basic rotation building (or vehicle scheduling) problem can be applied.

Let $MFSZ_{vg}$ be the number of vehicle groups used (minimum fleet size) for the subset of service blocks SB_{vg} assigned to vg. Then for each vehicle type vt the total number of used vehicles from type vt is equal to:

$$\sum_{vg \in VG} (MFSZ_{vg} * number(vt, vg))$$

The complex problem version: For each service block and train position, there may be several types of vehicle groups that can be assigned. This results from the local and global replacements of vehicle groups given for this problem setting. Global and local replacements define sets of alternative vehicle groups corresponding in some sense to "groups" of bus types/depots in the multiple-vehicle-type, multiple-depot vehicle scheduling problem (MDVSP, cf. Löbel (1998) and Kliewer, Mellouli, and Suhl (2002)).

Note that the use of the term "group" is different: For the multiple vehicle type problem, a vehicle type group is a set of alternative types of vehicles that can be used to serve a specific trip. For the considered railway application, a vehicle group defines an assembled pattern of vehicles of predefined types and numbers which are required for a certain train position. Because of this, we can say that, for each service block and train position in our problem setting, *a type group of vehicle groups* is given, i.e., a (type of) vehicle group is to be selected out of several feasible alternatives for each service block.

The extra difficulty of the problem is that minimizing vehicle groups of different types does not necessarily use a convenient constellation of locomotive and carriage types according to their availability. In the next sections, we review in short the multi-layer (multi-commodity) flow network for multiple vehicle types problems and apply it to types of vehicle groups for our case study. Then, we extend this model in order to create a "link" between vehicle or loco/car capacities and number of used vehicle groups of different types. Furthermore, we present an aggregation of "equivalent" vehicle groups decreasing the complexity and solution times for large-scale problems.

4   Multi-layer Flow Model and Shared Vehicle Capacity

There are basically two types of networks for basic rotation building and vehicle scheduling problems with one vehicle type: a *trip-as-node network* and a *trip-as-arc network*, where the latter is used in this paper. In both types of network for basic rotation building, vehicles of homogeneous type are modeled by flow units that originate from a depot node and terminate there for bus transit. For railways, there is generally no need of fixed depots and the "vehicle flow" *originates* from a virtual *source node* indicating period start and terminate at a virtual *target node* indicating period end. For the case of building genuine rotations for *periodic* (daily or weekly) timetables, frequently used in railways, this "vehicle flow" *circulates* within the network over *wrap-around period change arcs*. The sum of flow values on these arcs modeling *periodicity* corresponds to the number of used vehicles (of the considered type), since each vehicle used at the end of the period must "flow" back using one of these arcs in order to be used at the beginning of the next period at its space and time of availability (prescribed by its latest served activity, being a timetable trip or a deadhead trip).

In 4.1, we recall basic properties of multi-layer (multi-commodity) flow networks for rotation building with several vehicle types and apply the idea to vehicle group(type)s. Details on the design of this flow network for rotation building with our optimality-preserving deadhead trip aggregation are provided in 4.2 together with the objective function and constraints of the resulting mathematical model in 4.3. This model is extended in 4.3 and 4.4 for some specifics of the railway application studied in this paper.

4.1   The Basic Multi-layer Flow Model for loco/car Groups

To model rotation building problems with several vehicle types, we have to ensure that *vehicles of different types are not merged within the model network*. This can be achieved by constructing a multi-layer network (cf. Figure 1), where different *flow commodities* circulate on different layers, and thus cannot be merged. A network layer or a commodity corresponds to a vehicle type. For our case study, we adapt this multi-layer network where a vehicle type is replaced by a loco/car group (an not by a loco/car type).

The multi-layer network flow model results in a mixed-integer (linear) program (MIP) which is computationally much more difficult to solve than pure minimum cost flow problems. Besides the flow conservation constraints (that are to be formulated separately for each network layer, see 4.2), there are *cover/partitioning constraints* of non-flow type that ensure that each trip is included in the solution flow of only one network layer. These cover/partitioning constraints involve flow variables of value 0 or 1 and make the resulting optimization model of mixed-integer type.

The computational burden for solving these multi-layer network flow problems is due to the fact that the number of variables and constraints are multiplied by the number of network layers. Using the classical trip-as-node flow network model, whose number of variables for possible connection arcs already grows quadratically for basic problems with a single commodity, the resulting mixed-integer models cannot be solved directly by standard optimizers for timetables with thousands of trips. Special solution techniques, such as column generation or branch&price with Lagrangean relaxation, have been applied in order to solve problems of practical size (cf. Löbel (1998)), though in some cases only fleet minimal.

Using our model based on trip-as-arc network, the resulting models are much smaller owing to our powerful aggregation of possible deadhead trips. For large-scale instances of the multiple-depot problem of bus transit, computational results based on direct use of mathematical optimization software are presented in Kliewer, Mellouli, and Suhl (2002).

For each service block $sb \in SB$, let VG_{sb} be the list of possible vehicle groups that can be used to serve this specific service block. This list VG_{sb} is built

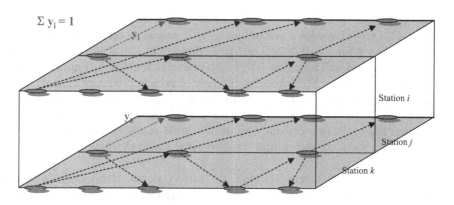

$\Sigma\, y_i = 1$

Fig. 1. Multi-layer flow network

by considering all vehicle groups included in local replacements for this specific service block sb and then adding each vg_j into VG_{sb} for each existing global replacement rule $vg_i \leftarrow vg_j$ where $vg_i \in VG_{sb}$ (transitive hull).

Let the 0/1-variable $Y_{sb,vg}$ denote the flow value on the arc modeling service block sb in the vg-layer of the network. The cover/partitioning constraints can be formulated as follows:

$$\sum_{vg \in VG_{sb}} Y_{sb,vg} = 1$$

for each service block $sb \in SB$.

As can observed in Figure 1, several Y variables exist for the same trip in different network layers. For basic problems with one commodity, all Y variables for trip arcs are set to 1, as only one network layer is involved. For the multi-layer network, the sum of flow variables $Y_{sb,vg}$ for a fixed service block sb over all network layers is equal to one (in order to guarantee that block sb is carried out by exactly one loco/car group). The optimizer will decide which of the 0/1-variables $Y_{sb,vg}$ (for a fixed service block sb) will be equal to 1. For a computed optimal solution, a value of $Y_{sb,vg} = 1$ means that service block sb is served by a loco/car group (with type) vg.

The overall model is a minimum cost multi-layer flow problem with side cover/partitioning constraints. To further reduce the size of network layer, for each commodity or loco/car group $vg \in VG$, we consider only the subset SB_{vg} of service blocks from SB that can be served by vehicle group vg. Thus, $SB_{vg} = \{sb \in SB \mid vg \in VG_{sb}\}$.

4.2 Trip-as-Arc Flow Network with Deadhead Trip Aggregation

Besides the partitioning/cover constraints, the mathematical model includes usual balance constraints on nodes for incoming and outgoing flow, separately for each vg network layer. Considering figure 1, the nodes in each network layer represent time-space points. These nodes are organized into connection lines

$CL(s, vg)$ to each station s, separately for each network layer vg. Each connection line $CL(s, vg)$ includes a line of nodes $N_i^{s,vg}$ modeling certain points in time at that station s for $i = 0, 1, 2, ..., n_{s,vg}$ (= number of nodes in $CL(s, vg)$), depending on vg and s. The nodes $N_i^{s,vg}$ of a connection $CL(s, vg)$ are connected by *waiting arcs*, where $X_i^{s,vg}$ denotes the flow of vehicle groups of type vg waiting from $N_{i-1}^{s,vg}$ to $N_i^{s,vg}$. Whereas the variables $X_i^{s,vg}$ connect nodes of the same connection line, the variables $Y_{sb,vg}$ for service blocks connect nodes of different connection lines – from a certain node of $CL(start\text{-}station(sb), vg)$ to a node of $CL(end\text{-}station(sb), vg)$ depending on start time and end time of service block sb, respectively.

Compatible service blocks sb_1 and sb_2 with $s = end\text{-}station(sb_1) = start\text{-}station(sb_2)$ and $end\text{-}time(sb_1) \leq start\text{-}time (sb_2)$ can be linked in the model through one (connection) node in $CL(s, vg)$ or through several consecutive (connection) nodes of this connection line over one or several waiting arcs. Thus connections of trips are not modeled explicitly but aggregated for several connections over the use of connection lines.

This aggregation is extended in Mellouli (2003) to the efficient representation of the quadratic number of deadhead connections needed to potentially connect all service blocks for the case where $s_1 = end\text{-}station(sb_1) \neq s_2 = start\text{-}station(sb_2)$. Our idea illustrated in Figure 2 is based on a two-stage aggregation for potential deadhead trips (matches).

The first stage aggregation is based on the matter of fact that each match is implicitly represented by taking the *first match* arc to the destination of the deadhead trip and then eventually going through waiting arcs of that destination station. The second stage aggregation is based on the observation that for a bundle of first match arcs connected to the same target service block f, only the *latest* one is needed, because we can go through waiting arcs of start station of deadhead trip until the *latest first match* in order to implicitly represent the connection of the omitted first matches.

Thus using waiting arcs of a trip-as-arc network only *latest first matches* (see Figure 2) are needed within the used connection line based network model in order to implicitly model all potential matches. The number of these latest first matches are considerably smaller than the number of service blocks multiplied by the number of stations. Since the number of stations is practically of factor 100 smaller for large models than the number of trips, a considerable reduction of arcs (and thus of model variables) is achieved. (In comparison, trip-as-node networks needs quadratic number of arcs relative to the number of service blocks). To get a figure on the impact of this aggregation, we could reduce in our railway case study with 7,500 service block arcs of 14 network layers, 134,643 matches to 5,861 first matches, and further to 1,661 latest first matches. This makes hard extensions of this network model solvable by direct use of mathematical optimization software for the large-scale railway application (cp. computational results in the next section). Results of the application of our model for multi-depot bus scheduling are presented in Kliewer, Mellouli, and Suhl (2002).

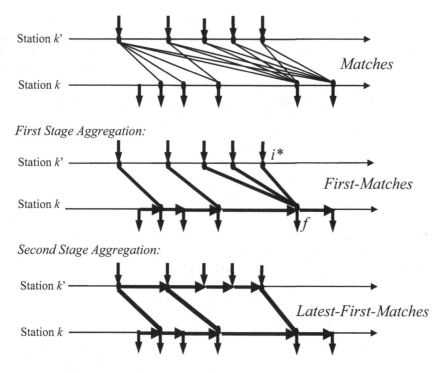

Fig. 2. Two-stage Aggregation for potential deadhead trips

4.3 Mathematical Model

The overall mathematical multi-layer flow model contains the partition/cover constraints given in 4.1 and flow balance constraints for the network described in the previous subsection based on connection lines and deadhead trip aggregation. For each connection node $N_i^{s,vg}$ in each connection line $CL(s,vg)$ (to each vehicle group vg and station s), let the set of:

- service block arcs incoming into this node be denoted by $E_i^{s,vg}$
- service block arcs outgoing from this node be denoted by $S_i^{s,vg}$
- latest first match arcs incoming into this node be denoted by $LE_i^{s,vg}$
- latest first match arcs outgoing from this node be denoted by $LS_i^{s,vg}$

Using these sets the flow balance constraints on connection line nodes $N_i^{s,vg}$ can be formulated as follows:

$\forall vg$, $\forall s$, and $\forall i = 0, 1, 2, ..., n_{s,vg}$ (number of nodes in $CL(s,vg)$):

$$X_i^{s,vg} + \sum_{sb \in E_i^{s,vg}} Y_{sb,vg} + \sum_{lfm \in LE_i^{s,vg}} Z_{lfm,vg}$$

$$= X_{i+1}^{s,vg} + \sum_{sb \in S_i^{s,vg}} Y_{sb,vg} + \sum_{lfm \in LS_i^{s,vg}} Z_{lfm,vg}$$

Here the Y- and X-variables represent the flow on service block arcs and wait-ing arcs respectively (as described in 4.1 and 4.2), the Z-variables the flow on latest first match (deadhead) arcs. Whereas the flow value for Y-variables is of 0/1-type, the X- and Z-variables are general integers, since several aggregated matches can use the same waiting or latest first match arc. One can show that the X-variables can be declared continuous, since they must take integer values anyway.

The partitioning/cover constraints for arcs of the same service block sb de-scribed in 4.1 are added into the model:

$$\forall sb \in SB \qquad \sum_{vg \in VG_{sb}} Y_{sb,vg} = 1$$

As described at the beginning of this section, periodicity of timetables is modeled by *wrap-around back arcs*. Especially. for the case of waiting arcs, the X-variable with last i-index in each connection line $CL(s, vg)$) is set equal to (or replaced by) that with 0-index of the same connection line.

$$\forall vg \forall s: \quad X^{s,vg}_{n_{s,vg}} = X^{s,vg}_0$$

Let the set of period change arcs for service blocks (e.g., starting on Sunday 23:00 and ending on Monday 2:00 for a weekly timetable) on vg-network layer be denoted by $PC(vg)$ and the set of period change arcs for latest first matches on vg-network layer be denoted by $PCL(vg)$, then the **fleet size constraints** (to each vg network layer) can be formulated as follows:

$$\forall vg: \quad FSZ(vg) = X^{s,vg}_0 + \sum_{sb \in PC(vg)} Y_{sb,vg} + \sum_{lfm \in PCL(vg)} Z_{lfm,vg}$$

Here, $FSZ(vg)$, the fleet size (= number of units) for vehicle group vg is set equal (as remarked at the beginning of this section) to the sum of flow values over wrap-around period change arcs for periodical timetables within the net-work layer for vg. It is important to see that not only waiting arcs can be period change arcs, but also arcs for service blocks and for latest first matches can be of this type.

The objective function of the overall model minimizes the overall fixed and variable costs of the rotation building problem. Fixed costs F_{vg} are those for used units of vehicle groups of type vg and the variable costs C_{lfm} for empty movements of vehicles incurred when using one of the latest first match arcs lfm of the network layer vg (set of all latest first matches being denoted by $LFM(vg)$. The objective function can now be stated easily:

$$minimize \quad \sum_{vg \in VG} FixCost_{vg} * FSZ(vg) + \sum_{lfm \in LFM(vg)} Cost_{lfm} * Z_{lfm,vg}$$

Note that $Cost_{lfm}$ can be made dependent on vehicle group vg proceeded empty from one station to another by setting $Cost_{lfm,vg}$ and that we can set small costs on waiting X-variables on connection lines of non-maintenance stations to favor standing times of vehicle groups at maintenance stations.

4.4   Modeling Shared loco/car Capacities Within the Flow Network

Recall that the problem is to build rotations for vehicle groups while assigning each service block to a vehicle group (type) and regarding the shared capacities on vehicle level, i.e., on loco/car level. To model the problem, we first take a network layer for each vehicle group and construct the multi-layer aggregated flow network as discussed in the previous subsections. The link between the number of used vehicle groups in the network layers and the available capacities of individual vehicles or loco/cars is reached by the following additional constraints (using the terminology in section 3):

Vehicle capacity constraint

$$\forall\, vt \in VT: \ \sum_{vg \in VG} FSZ(vg) * number(vg, vt) \leq capacity(vt)$$

where $capacity(vt)$ is the available number of vehicles or loco/cars of type vt over all homebases (sum of $capacity(vt, hb)$ over all $hb \in HB$) and $FSZ(vg)$ denotes the fleet size on the network layer for vehicle group vg.

Here, we have a direct relation to the **fleet size constraints** (to each vg network layer) formulated in the previous subsection.

If it is desired to consider different homebases for vehicles separately, we can write the vehicle capacity constraint as follows:

$$\forall\, vt \in VT \ \forall\, hb \in HB: \ \sum_{vg \in VG} FSZ(vg, hb) * number(vg, vt) \leq capacity(vt, hb)$$

As we have introduced a network layer to each (type of) vehicle group, the resulting model for the data of German Railways (see Section 3), including 32 network layers and 7,500 service blocks, risks to become computationally difficult and perhaps not directly solvable by optimization software. Therefore, we developed a technique to reduce the number of network layers in order to considerably reduce the model sizes. This refinement is discussed in the following subsection.

4.5   Model Refinement by Aggregating Vehicle Groups

As a motivation for this model refinement, we consider the characteristics of the test data of German Railways for this rotation problem for several vehicle group types. We have 32 vehicle groups denoted by $VG0, VG1, VG2, ..., VG31$. Most of the vehicle groups consist of 6, 5, or 4 vehicles. There is a vehicle group containing a locomotive (VG0) and two other vehicle groups that contain only one carriage vehicle as reinforcement.

Inspecting the 7500 assignments of service blocks (and train positions) to sets of feasible vehicle groups, we observed that relatively few different sets of feasible vehicle groups occur. These are the following 13 group sets:

$\{VG0\}$	$\{VG1, VG2, VG3\}$
$\{VG4, VG5, VG6, VG7\}$	$\{VG8, VG9\}$
$\{VG10, VG11\}$	$\{VG13, ..., VG20\}$
$\{VG21, ..., VG28\}$	$\{VG29, VG30\}$
$\{VG8\}$	$\{VG4\}$
$\{VG12\}$	$\{VG1\}$
$\{VG31\}$	

Now, we introduce the notion of *equivalent vehicle groups*. Two (types of) vehicle groups VG_i and VG_j are equivalent, if and only if they occur in the same sets of feasible vehicle groups (over the whole timetable). For example, $VG2$ and $VG3$ are equivalent, as both appear only once in the occurring set of alternative groups {VG1,VG2,VG3}. However, both $VG2$ and $VG3$ are not equivalent with $VG1$, since $VG1$ additionally appears in {VG1}.

Having an equivalence relation, we can build *equivalence classes* for vehicle groups, where each class includes a maximal set of equivalent vehicle groups. As vehicle groups are combined in classes, we call such an equivalence class a *combined vehicle group*. The following classes or combined vehicle groups are built for the above case:

$[VG0]$	$[VG1]$
$[VG2, VG3]$	$[VG4]$
$[VG5, VG6, VG7]$	$[VG8]$
$[VG9]$	$[VG10, VG11]$
$[VG12]$	$[VG13, ..., VG20]$
$[VG21, ..., VG28]$	$[VG29, VG30]$
$[VG31]$	

Vehicle groups within an equivalence class are interchangeable over the whole timetable. Having two equivalent vehicle groups VG_i and VG_j, each rotation that can be served by VG_i can be served by VG_j and vice versa. Now the idea is to generate a network layer to each combined vehicle group and not to each vehicle group. For the test data of German Rail, we get 14 instead of 32 network layers, as 14 combined vehicle groups are generated out of 32 vehicle groups. This makes large-scale instances of this problem type directly solvable by optimization software and enables integrating other requirements such as adding special constraints ensuring sufficient slots for servicing at maintenance bases with a even distribution in time.

Within a solution of the resulting flow model, the fleet size on a network layer for a combined vehicle group $CombVG$ specifies the number of vehicle groups required that can be freely chosen from the vehicle groups included in $CombVG$. For instance, if the fleet size for CombVG $[VG10, VG11]$ is 9, so several solutions are equivalent, namely ($7 * VG10$ and $2 * VG11$) or alternatively ($5 * VG10$ and $4 * VG11$), etc.

Observing this we can let the optimizer *split* the fleet size over all vehicle groups included in a $CombVG$ by including the following constraints:

Fleet size split constraints: For each combined vehicle group $CombVG$:

$$FSZ(CombVG) = \sum_{vg \in CombVG} FSZ(vg)$$

As above, additional fleet size constraints (to each layer) sets $FSZ(CombVG)$ equal to the sum of flow values on the wrap-around period change arcs within the network layer for $CombVG$.

If it is desired to consider different homebases for vehicles separately, we can write the fleet size split constraints as follows:

$$FSZ(CombVG) = \sum_{vg \in CombVG} \sum_{hb \in HB} FSZ(vg, hb)$$

Now, the vehicle capacity constraints of the last subsection connect the fleet size variables $FSZ(vg, hb)$ (or $FSZ(vg)$) for vehicle groups to the capacities of vehicles or loco/cars. Therefore, with both types of constraints, the optimizer will split the fleet size required for a $CombVG$ (network layer) among vehicle groups included in $CombVG$ while satisfying the capacities of vehicles used by the different vehicle groups.

5   Optimization Results and Decision Support Aspects

5.1   Computational Results

Applying the network flow approach and techniques presented in the last section, the resulting mathematical models for rotation building with the above requirements could be solved efficiently for the test data of German Railways (31 different types of vehicle groups, two home bases, and 7,500 assignments of service blocks to vehicle groups). The problem instances handle periodicity of the weekly timetable and empty train movements are allowed in order to reduce the fleet sizes used. Using our aggregation of empty movements as described in 4.2, 134,643 deadhead possibilities (matches) could be reduced in a first aggregation stage to 5,861 first matches and further in a second phase to 1,661 latest first matches.

The constructed 14-layer multi-commodity network flow model with special constraints for handling several vehicle groups with shared capacities results in a mathematical model with 206,000 variables and 46,000 constraints. Depending on the level of maintenance handling, the model is solved within 3 to 10 min by ILOG CPLEX on a 1 GHz Pentium III processor.

5.2   Decision Support Tools

Since railway operations are carried out in a complex dynamic environment with rapidly changing requirements, it is often necessary that human planners adjust the plans obtained by mathematical optimization techniques. In complex

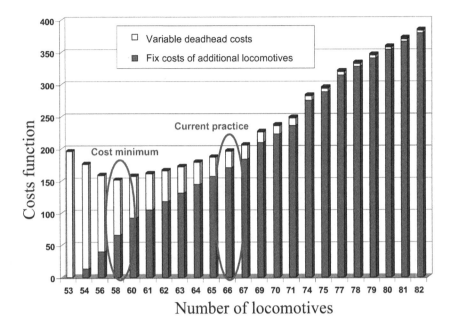

Fig. 3. Tradeoff fleet size versus empty movement costs

environments, this is only possible if optimization methods are embedded in a decision support system, providing a graphically interactive user interface which makes it easy to change the data and edit the results.

For example the following questions can be tackled within a what-if analysis: What is the practically best fleet size for a given timetable? What are the consequences of adding or deleting some trips of the timetable in terms of fleet size and operational costs? How sensitive are these costs against small changes in departure time of trips, in duration of scheduled or empty trips, or in minimum turn times required between consecutive trips? The first question assumes fixed input data (timetable, minimum turn time, empty trip duration), all others analyze changes in these input data.

Is it not sufficient, in order to answer the first question, to solve one problem instance, since the sum of fixed costs for vehicles and empty movement costs are minimized? Often, fixed costs of vehicles are set to a large value and not necessarily well and precisely scaled relatively to empty movement costs. A nice what-if analysis here is to analyze the trade-off between operative empty movement costs and the number of needed vehicles. Figure 3 shows the result of such a what-if-analysis for a timetable of German Railways with 1,098 (compound) trips and 77 terminal stations.

The analysis shows that the solution with minimum total costs is reached by a fleet of 58 locomotives. According to the used costs function, a considerable potential reduction of total fixed costs and empty movement costs is possible. An interesting aspect shown by this analysis is that empty movement costs increases non-linearly according to the number of saved vehicles. Nice to see is that these empty movement costs surmount the fixed locomotive costs for solutions with less than 58 locomotives. Thus, the minimum fleet size solution with 53 locomotives is not a cost-minimum solution.

Presenting a series of results with their properties, the decision maker chooses the best solution. Besides number of vehicles used and costs of empty movement trips incurred, other indices can be relevant such as the robustness of constructed rotations against delays or the number of induced opportunities for maintenance operations. Considering these aspects, a solution with 62-63 locomotives may be the best one from a practical point of view.

The questions related to a change in input data emphasizes the central position of timetable design in the production planning process and its interaction with rotation building. In fact, small changes in the input data may save considerable amounts in fixed costs of vehicles and empty movement costs. A useful what-if analysis here is to make some experiments changing the given minimum turn times (MTT). An analysis of this type for three fleets of German Railways is shown in Table 1:

Table 1. What-if analysis: Changing minimum turn times (MTT)

MTT (in minutes)	0	5	10	15	20	25	30	35	40
Fleet 1	34	34	34	35	35	37	37	37	38
Fleet 2	48	49	51	52	52	52	53	54	56
Fleet 3	53	54	57	61	66	68	73	75	76

Since the minimum turn time is handled as a "hard restriction" in rotation building (minimum duration between end time of one trip and start time of the next within a rotation), small changes may have considerable effect on the fleet size. Take two trips T1 from A to B and T2 from B to C. If T2 starts at 12:00, T1 arrives at 11:42 and the minimum turn-time is set to 20 minutes, then T2 is not a connection trip for T1 unless the minimum turn time is reduced to 18 minutes (or changing the duration or start times of T1 or T2). This situation is critical if no other suitable connection for T2 exists. In this case, a local change of turn-time and/or start time of T2 may save a vehicle or considerable amount of empty movements.

How to find critical locations where these savings are possible?

Finding critical locations can be supported by a chart plotting standing times (or availability) of vehicles at terminal stations. We realized a graphically interactive user-interface where these *vehicle availability charts* can be shown for all terminal stations (cf. Figure 4). An up-arrow indicates an arriving vehicle that becomes available at that station. A down-arrow indicates a departing vehicle from available ones at that point in time.

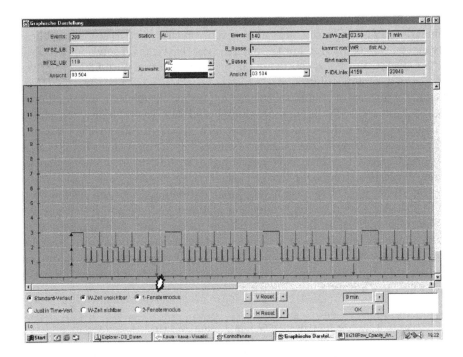

Fig. 4. Analysis of standing times

Generally long standing times within rotations which must appear in these vehicle availability charts (for some stations) constitute a strong argument that vehicles may not be utilized in an optimal way. This may appear within rotations of an optimal solution of rotation building and the cause of *practical non-optimality* may then lie in the input data themselves, and, thus concerns other planning phases, such as timetable design or trip scheduling.

In analyzing a German Railways sub-fleet, we encountered the situation given in Figure 4 at a station called AL. Among the three available vehicles at the start of the week (vertical axis), one vehicle (level 0 to level 1) is standing during the day. This vehicle is utilized at 23:01 and shortly thereafter (one minute later), another vehicle becomes available (after arrival).

Analyzing the situation at this station, we found out that two trips, say T1 and T2, arrive in AL at 22:55 and 22:58, respectively, and other two trips T3 and T4 start from AL at 23:01 and 23:10, respectively. The minimum turn time at station AL is set to 7 minutes. The vehicle serving T2 cannot serve T3 and must serve T4. The problem lies that only 6 minutes are available between arrival time of T1 and start time of T3. Since the minimum turn time is handled as a hard restriction by rotation building no connection from T1 to T3 is possible and the computed solution requires 3 instead of two vehicles. Discussing this with experts, they affirmed that local violations of the given minimum turn times are allowed in situations like this.

6   Conclusions

In this paper, we presented an exact optimization approach for rotation building for railways. In this domain, there are specific requirements related to the assembly of locomotives and carriages into train units. We discussed the case study appearing in some railways' fleet where service blocks are to be assigned to predefined groups of locomotives and carriages.

For this type of requirements, we developed new mathematical models based on time-space trip-as-arc network with aggregation schemes both for the basic and special problem. For the basic problem, an aggregation of all potential empty movements is applied which was published in former works. In this paper, we provided an extension of the multi-layer network flow model together with special constraints relating the fleet sizes expressed in number of used vehicle groups to the given capacities at locomotive and carriage level. For solving large-scale models, a special aggregation of "equivalent" vehicle groups is applied in order to reduce the number of involved network layers.

After applying the techniques described above, the resulting mixed-integer mathematical models showed a very small LP/IP gap. To our understanding, this behavior is due to the fact that the cover/partitioning constraints involved in the network flow model have much less non-zero elements than in standard set-partitioning and set-covering models. Large-scale instances of our models are solved directly using state-of-the-art mathematical optimization software.

Furthermore, we discussed some practically relevant questions related to the process of planning railway fleet and provided ways of integrating the optimization components with suitable decision support tools.

References

1. Abbink, E., van den Berg, B., Kroon, L., Salomon, M.: Allocation of railway rolling stock for passenger trains. Transportation Science 38(1), 33–41 (2004)
2. Ben-Kheder, N., Kintanar, J., Queille, C., Strainling, W.: Schedule optimization at SNCF: From conception to day of departure. Interfaces 28, 6–23 (1998)
3. Brucker, P., Hurink, J., Rolfes, Th.: Routing of Railway Carriages. Journal of global optimization 27, 313–332 (2003)
4. Cordeau, J.-F., Thoth, P., Vigo, D.: A survey of optimization models for train routing and scheduling. Transportation Science 32, 380–404 (1998)
5. Cordeau, J.-F., Soumis, F., Desrosiers, J.: A Benders decomposition approach for the locomotive and car assignment problem. Transportation Science 34(2), 133–149 (2000)
6. Cordeau, J.-F., Soumis, F., Desrosiers, J.: Simultaneous assignment of locomotives and cars to passenger trains. Operations Research 49(4), 531–548 (2001)
7. Kliewer, N., Mellouli, T., Suhl, L.: Multi-depot vehicle scheduling: a time-space network based exact optimization approach. In: Presented at the 9-th Meeting of the EURO Working Group on transportation. Bari. Italy (2002)
8. Löbel, A.: Optimal Vehicle Scheduling in Public Transit. PhD thesis, ZIB, Berlin. Shaker, Aachen (1998)

9. Mellouli, T.: A Network Flow Approach to Crew Scheduling based on an Analogy to a Train/Aircraft Maintenance Routing Problem. In: Voß, et al. (eds.) Computer-Aided Scheduling of Public Transport. LNEMS, vol. 505, pp. 91–120. Springer, Berlin (2001)

10. Mellouli, T.: Scheduling and Routing Processes in Public Transport Systems. Habilitation Thesis. University of Paderborn, Germany (2003)

11. Ramani, K.V., Mandal, B.K.: Operational planning of passenger trains in Indian Railways. Interfaces 22(2), 39–51 (1992)

12. Suhl, L.: Computer-Aided Scheduling: An Airline Perspective. Gabler–DUV, Wiesbaden (1995)

13. Suhl, L., Mellouli, T.: Requirements for, and Design of, an Operations Control System for Railways. In: Wilson, N. (ed.) Computer-Aided Transit Scheduling. LNEMS, vol. 471, pp. 371–390. Springer, Heidelberg (1999)

14. Suhl, L., Mellouli, T., Biederbick, C., Goecke, J.: Managing and preventing delays in railway traffic by simulation and optimization. In: Pursula, M., Niittymäki, J. (eds.) Mathematical methods on optimization in transportation systems. Applied Optimization, Ch. 1, vol. 48, pp. 3–16. Kluwer, Dordrecht (2001)

15. Zirati, K., Soumis, F., Desrosiers, J., Gélinas, S., Saintonge, A.: Locomotive assignment with heterogeneous consists at CN North America. European Journal of OR 97, 281–292 (1997)

An Estimate of the Punctuality Benefits of Automatic Operational Train Sequencing

Rien Gouweloos[1] and Maarten Bartholomeus[2]

[1] Atos Consulting, Papendorpseweg 93
3528BJ, Utrecht, The Netherlands
rien.gouweloos@atosorigin.com
http://www.atosorigin.com
[2] Holland Railconsult, Postbus 2855
3500GW, Utrecht, The Netherlands
mgpbartholomeus@hr.nl
http://www.hrn.nl

Abstract. In Dutch railway operations, most of the rescheduling decisions in the operational phase, following some disturbance, involve the resequencing of trains. These decisions are being taken using only approximate train information and operational rules. Improvements have been formulated but, since no insight in the potential gain in punctuality exists, lack a convincing business case. In this paper, using only elementary methods, we derive an estimate for this punctuality gain.

1 Introduction

Increasing the reliability of the Dutch Railways ranks high on the national political agenda and is therefore a high priority in the Railway sector. In fact the sectors strategy statement, called "Benutten en Bouwen" ("Utilize and Build"), makes this into its central tenet. This is a marked change from previous strategies, which concentrate on maximizing the volume of railway traffic per unit of infrastructure. Benutten en Bouwen takes a clear position: maximizing the utilisation of the infrastructure will be an illusionary goal, unless first reliability is improved drastically.

Given the many daily departures from the predefined timetable, it will be clear that the operational processes of rescheduling and dispatching train traffic are very busy indeed. Each day, thousands of minor and major rescheduling decisions are being taken. Obviously, the quality of these processes is of vital importance. Quantitatively however, very little is known about the influence of scheduling quality on reliability. Specifically, the question which part of the occurring unreliability is due to suboptimal rescheduling is unanswered. This paper addresses this question.

The dispatching process is far from being perfectly accurate. Train information, such as position (delay) and speed is only approximately available. Inaccuracy is further increased by the fact that, although traffic control in the Dutch

F. Geraets et al. (Eds.): Railway Optimization 2004, LNCS 4359, pp. 295–305, 2007.
© Springer-Verlag Berlin Heidelberg 2007

railways heavily uses information systems, train dispatching is still exclusively a human decision. A conscious decision has been made to limit the role of IT systems to presenting timetable and train status information to human decision makers. Efforts to assist these people by more advanced IT scheduling solutions are yet to leave the prototype stage. Although there are some valid reasons to leave dispatching in human hands, we have to realize that, compared to computers, human beings are not very good at making rapid, consistent and precise rule-based decisions. Thus, errors due to inaccurate train information and human operator errors combine into an imperfect decision making process. In this paper, we will confront this situation with a situation with perfect decision-making, referred to as "automatic sequencing".

In the Netherlands, the operational processes of rescheduling train traffic have been structured into a number of control layers. We will deal with the lowest layer, which we will call the dispatching function. In this paper we view dispatching as a series of decisions on the order of trains on given routes: sequencing. In practice, this covers the vast majority of all relevant dispatching decisions.

Detailed instructions have been prepared for the dispatchers when (not) to change the order of trains from the order as given by the original timetable. All situations with one delayed train have been exhaustively tabulated nationwide. Essentially, these so-called if-then sheets contain the following statements:

If train x has a delay between d_1 and d_2, **then** it should be given access to the infrastructure between trains y and z.

Our calculation builds on these if-then sheets. Through the if-then sheets, contact with actual data is being made. They are the reason we can get a result without the need for explicit modelling of infrastructure or timetable information.

The operation uses if-then sheet information for dispatching. Obviously, an operator needs information on delays in order to make the correct decisions. Unfortunately, this information is not perfectly available. We estimate that delays used in the decision process are distributed around the real delay values with a standard deviation of 2 minutes. The reliability of railway traffic is commonly operationalised in terms of the punctuality of passenger traffic. In the Netherlands, punctuality is defined as the percentage of trains with a delay of three minutes or less at major stations, as compared to the original timetable. We will derive a ballpark estimate for the amount of dispunctuality incurred from current operational procedures and systems.

This paper is organized as follows. In Section 2 we present out formalism. In Section 3 we discuss the various inputs to our calculation, such as if-then sheets, delay distributions and decision making errors. Section 4 presents the results. In Section 5 we present our conclusions and offer some suggestions for follow up.

This work has been done as part of a project under the jurisdiction of Railverkeers-leiding. The project called for the preparation of a business case for the development of improved control systems for the dispatching layer. Railverkeersleiding, part of the public domain ProRail organization, is responsible for operational capacity management of the Dutch Railway Infrastructure. Railver-keersleiding performs rescheduling of the infrastructure under continuous consultation with the transport companies.

2   Approach

2.1   Elements of the Formalism

Loss Function
Let us consider two trains A and B approaching an insertion point P. The optimal order must be decided upon. In the following, the optimal order will be the order in which the summed delays of A and B at some reference point will be minimal. This reference point is usually a point somewhere after P, where the two trains start having separate routes. If A and B have the same characteristics (speed) after P, any point after P will do and our optimal order rule reduces to a first come first serve rule.

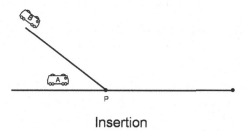

Insertion

Fig. 1. An insertion point

We simplify the dynamics of the problem by assuming that if train A (B) is given priority at the insertion point while hindering train B (A), train B (A) always leaves the insertion point at a fixed headway H after A (B). In this case, there always exists one relative delay of A and B d_c for which the optimal order changes from AB to BA. We call this the characteristic delay. The if-then sheets tabulate these characteristic delays for all train pairs at insertion points in the Netherlands.

We call the amount of extra summed delay for A and B at the reference point resulting from the choice for a suboptimal order at P, the loss L. We can plot L as a function of $v = d_A - d_B - d_c$: L is a piecewise linear function of v, symmetric in $v = 0$. In the diagram we also show our definition of four "regimes" 0 to III in the values of v.

It is of interest to note that an optimal decision in the summed delay is not necessarily optimal for the contributing trains taken separately. Therefore, in individual cases it is possible that the optimal decision decreases punctuality.

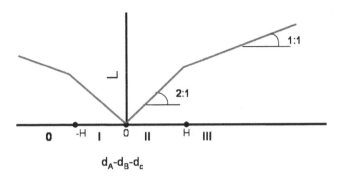

Fig. 2. Loss function

Delay Distribution

Trains approach the insertion point with a delay according to a density function $D(d)$. For this we use a conventional negative exponential fitted to the observed punctuality.

$$D(d) = 1/d_{av} \exp(-d/d_{av}) \quad (d > 0, 0 \text{ elsewhere}) \tag{1}$$

Note that in this we assume the same distribution will hold always and everywhere.

Control Error Distribution

Our purpose is to determine the punctuality effect of operator errors due to incorrect delay information. Let $F(d, d_g)$ be the probability density that a delay d_g will be assumed by the operator where actually a value d exists. For F we use a normal distribution

$$F(d, d_g) = F(d - d_g) = (1/(\sigma\sqrt{2\pi})) \exp(-(d - d_g)^2/2\sigma^2) . \tag{2}$$

Human operators tend to act conservatively, that is they tend to leave a planned train sequence in place, until it is very obvious that it should be changed. In terms of F, this behaviour means that F is no longer symmetric around $d - d_g$. In order to investigate these effects whilst leaving our formalism intact, we have introduced a parameter called the shift s:

$$F_s(d, d_g) = F_s(d - s - d_g) = (1/(\sigma\sqrt{2\pi})) \exp(-(d - s - d_g)^2/2\sigma^2) . \tag{3}$$

An erroneous decision will be made if a delay $d < d_c$ is assumed, while in fact $d_g > d_c$ and vice versa. For regime 0 (compare Figure 1) this chance is given by:

$$P_s(d - d_c) = \int_{d_c - d}^{\infty} F_s(d, d_g) dd_g = 1/2 \text{erfc}((d_c - d - s)/(\sigma\sqrt{2})) , \tag{4}$$

in which erfc is the complementary error function. Similar relations can be written for the case $d > d_c$ and $d_g < d_c$ and the other regimes.

2.2 Overall Punctuality Effect

We are interested in computing overall punctuality effects of suboptimal sequencing. Consider trains A and B with delays d_A and d_B respectively. Train A turns from being punctual to dispunctual when $d_A < 3$ while $L_A > 3 - d_A$. The reverse effect occurs if $d_A > 3$ while $L_A < 3 - d_A$.

Lets define the following probability distributions, all defined with respect to a sequencing point with characteristic delay d_c:

$P_1(d)$ The probability density for a train with delay d to be punctual,

$P_2(d_1, d_2)$ The probability density for trains with delays d_1 and d_2 to be incorrectly sequenced,

$P_3(d_1, d_2)$ The probability density that the extra delay for train 1 will exceed $3 - d_1$.

In order to keep this paragraph concise, we will only give explicit formulas for train A and regime 0 of the loss function. In this case:

$$P_1(d_A) = \Theta(3 - d_A) \tag{5}$$

$$P_2(d_A, d_B) = \text{erfc}(|(d_A - d_B) - d_c|/\sigma) \tag{6}$$

$$P_3(d_A, d_B) = \Theta(d_c + H - 3 + d_B) \tag{7}$$

where Θ is the Heaviside Step Function.

The probability train A turns from being punctual to dispunctual at a sequencing point with characteristic delay d_c is given by:

$$\int_{-\infty}^{\infty} dd_B V(d_B) \int_{-\infty}^{\infty} dd_A V(d_A) P_1(d_A) P_2(d_A, d_B) P_3(d_A, d_B) = \tag{8}$$

$$\int_{3-d_c-H}^{\infty} dd_B V(d_B) \int_0^{\min(3, d_c + d_B)} dd_A V(d_A) P(|d_A - d_B|) \ . \tag{9}$$

Similar expressions can be derived for

- the other regimes of the loss function;
- the probability that the train turns from being dispunctual to punctual;
- train B.

That is, all in all $4 * 2 * 2 = 16$ similarly structured double integrals have to be evaluated.

Lets call the grand total of these terms, the probability that an incorrect sequencing decision with characteristic time d_c turns some train into being dispunctual, $D(d_c)$.

It is straightforward to compute the chance of an incorrect decision E and the average value of the loss function L_{av}, for a decision with characteristic time d_c by:

$$E(d_c) = \int_{-\infty}^{\infty} dd_A \int_{-\infty}^{\infty} dd_B V(d_A)V(d_B)P(d_A - d_B - d_c) \tag{10}$$

$$L_{\mathrm{av}}(d_c) = \int_{-\infty}^{\infty} dd_A \int_{-\infty}^{\infty} dd_B V(d_A)V(d_B)P(d_A - d_B - d_c)L(d_A - d_B - d_c). \tag{11}$$

In all cases (9), (10) and (11) an overall effect is found by summing over the number of potential daily resequencing decisions $A(d_c)$ as given by the if-then sheets. After suitable normalizations for the total number of trains and punctuality measurement points, overall effects result.

2.3   Correction Terms

Some corrections to the formalism outlined above have been considered. These corrections are not expected to be very accurate, but serve to get some feeling for the reliability of the results.

Delay may be Nullified Before Measurement
Trains are planned with a driving time margin. That is, if the train has a delay d somewhere between punctuality measurement points (nodes), it will be able to reduce this delay somewhat before measurement. From the timetable, we have determined an average value dn for the amount a train can reduce its delay before the next node. We use separate values for resequencing points just after leaving a station (driving time margin for the distance between stations) and resequencing points underway between stations (on average, half the distance). This value is now used in the formulas above by simply adjusting the integration limits from "3" to "$3 + d_n$". This procedure is justified if the integrand does not vary strongly over the range of allowed values for the driving time margin. For the purpose of obtaining a rough estimate of the correction term, this is the case.

Delay may Persist After First Measurement
Our formalism measures the punctuality effects at the first measurement point (the next node). In fact, an effect may persist at later nodes, if the delay is not absorbed into the halting time at the node. This effect has been estimated at the second node by determining the average halting time margin hn from the timetables and adjusting the integration limits from "3" to "$3 + d_n^{(1)} + h_n + d_n^{(2)}$". Here $d_n^{(1)}$ is the average driving time margin from the sequencing point to the first node and $d_n^{(2)}$ the average driving time margin between nodes. They are added together with hn because at the second node, driving time margins of both the first and second stretch and one halting (at the first node) apply. Effects at later nodes can be computed in the same manner, but were found to be negligible.

Again, this procedure is justified if the integrand does not vary strongly over the range of allowed values for the halting time margin. Actually, this is not the case. Since, due to the exponential falloff of the delay distribution, trains with large halting time margins do not contribute strongly to the results, we have dealt with this problem by using an ad-hoc cut-off value on the halting time margins, considering only trains with small margins for our rough estimate of this correction term. Essentially, this means we consider trains passing through a station but not those at end nodes.

Delay may Cause Other Delays
We have not attempted to obtain a quantitative estimate for cascade effects. Operational data suggest strongly these effects are small.

3   Inputs to the Calculation

As mentioned earlier, contact with actual data is made through the if-then sheets. From these if-then sheets we obtain the number of potential resequencing decisions $A(d_c)$ on a day. Because of the corrections terms in Section 2.3, a distinction has to be made between resequencing decisions made between stations (open track) and at stations. Note: the majority of the resequencing decisions on stations concern departure situations.

Table 1. Number of potential resequencing decisions

Location \ d_c : 1	2	3	4	5	6	7	8	9	10	11	12	13
Open Track 87	336	497	594	752	454	539	695	461	306	260	181	160
Stations	250	714	1319	1070	1163	619	745	942	577	574	627	307	315

There are 5000 trains each day. The punctuality is measured at large stations (nodes), the average number of nodes for a train is 2.2, giving a total number of 11000 punctuality measurement points MS.

The value $d_{av} = 1.8$ minutes for the negative exponential delay distribution reproduces the 2002 punctuality of 82% for the Dutch railways. The conflict time or minimal headway (H) for insertion points is, according to headway calculations set to 90 seconds on the open track. For departure situations, values range from 97 to 133 seconds. We use 110 seconds.

The inaccuracy of the dispatching function in the sequence decision is set to $\sigma = 2$ minutes. This inaccuracy is composed of:

- rounding errors: in the decision process, dispatchers use three different delays, independently rounded to whole minutes;
- inaccuracy in the operating times themselves (30 - 60 sec);
- inaccuracy in the prediction of the delay at the insertion point;
- inaccuracy in the predicted travel time of the train after the insertion point;
- inaccuracy in the human dispatchers' decision making process.

Of these, the first two items are objectively known, for the others we have to rely on expert estimates. We are confident that the true standard deviation of the distribution will have a value between 1.5 and 3 minutes. More accurate determinations are feasible but have not been part of this investigation. We have used a shift (Formula 3) of 1 minute.

The parameters in the correction terms mentioned in Section 2.3 have been set as follows. For d_n, the driving time margin, i.e., the amount a train on average can reduce its delay between nodes, 1.7 minutes is used. The average halting time margin hn, after applying a cut-off value of 4 minutes to the data, was found to be 0.7 minutes. These values have been extracted from the timetable planning in the Netherlands.

In summary, unless stated otherwise we have used the following values for the parameters in our calculation:

Table 2. Default parameter values

Parameter	Symbol	Default value
Average delay	d_{ev}	1.8 minutes
Headway	H	90 seconds (open track)
		110 seconds (departures)
Dispatching error	Σ	2 minutes
Number of potential resequencing decisions	$A(d_c)$	Refer to Table 1
Average node-node driving time margin	d_n	1.7 minutes
Average halting time margin after cut-off	h_n	0.7 minutes
Cut-off value	h_{cut}	4 minutes
Shift	s	1 minute

4 Results

With the formulas from Section 2 and the default parameter values from Table 2 the effects of inaccuracy in the resequencing decisions can be computed. Table 3 gives the overall punctuality effect, the chance of an incorrect decision per train and the average loss per train, both with and without the correction terms mentioned in Section 2.3.

From the discussion so far, it should be obvious that our calculations depend upon a number of parameters of which the values are not accurately known. We have therefore experimented extensively with our formalism, in order to get a feeling for the range of possible results. As an illustration of these efforts, in Figure 3 we present the dispunctionality effect in cases where all parameters but one have been set to their default values (refer to Table 2) and one is varied. On

Table 3. Overall results

	Without correction	With correction
Punctuality effect	4.16%	3.62%
Chance of error	15%	
Average loss	0.5 minutes	

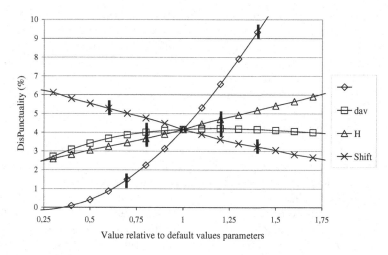

Fig. 3. Sensitivity analysis dispunctuality on input parameters

the horizontal axis the parameter value is given as a fraction of the default value. On the vertical axis, the resulting punctuality effect without correction terms is given. The vertical bars on the curves picture the range of plausible values for the parameter. Variations on the parameter values for d_n, h_n are similar to the variations on parameter s and not depicted in Figure 3. As to be expected, sensitivity is particularly high for the value of σ. Experimental determination of the control error distribution is highly desirable to narrow down our estimate. We will discuss the consequences of this (and similar) figure in the next section.

5   Discussion and Conclusions

The numbers given in Table 3 represent our best estimate of the effect of dispatching errors on punctuality. The numbers are unexpectedly large: around 4% of the observed dispunctuality stems from suboptimal resequencing. This is not in line with common opinion, which holds that suboptimal resequencing only contributes marginally to dispunctuality. Still, though many ingredients of our calculations are less than certain, our basic reasoning is so simple that it is hard to refute: the resequencing decision has an intrinsic accuracy of at least 1.5 minutes, there are many trains with delays between 1 and 3 minutes and many potential resequencing decisions (if-then rules) in this domain. Large effects are unavoidable. Note that the fact that the Dutch railway network is heavily used is crucial in this reasoning, since this ensures that many resequencing decisions have to be taken at delay values around 3 minutes, influencing punctuality. Our results are being corroborated by recent results of simulation study using the Combine Traffic Management System, which confront an "ideal" traffic management system (i.e. Combine) with data from practice at a specific location and find punctuality gains in the order of 5%.

As stated, our calculations were made within the context of an effort to construct a business case for investing in the dispatching control systems. For this purpose, we really need to know only that effects are large, say more than 0.5%. As our sensitivity analysis shows, varying the input parameters within the range of plausible values does not change this conclusion. The sensitivity analysis also makes us confident that our formalism, which relies on the use of average distributions, is reasonable.

Some other objections are possible.

- *Some inaccuracy is intrinsic to the process, therefore the "ideal decision making process" has no practical relevance.* Indeed, some errors are unavoidable. Especially the halting process at stations is unpredictable at the level of some tens of seconds. However, as Figure 3 shows, errors under 0.5 minutes do not contribute strongly to the overall result, while effects grow roughly linearly from $\sigma = 1.5$ to $\sigma = 3$ minutes. Therefore our conclusion holds.
- *Real sequencing is not done according to a simple summed delay criterion, other arguments are taken into account.* We have examined the if-then sheets, which have been prepared by hand by local experts for all single-delay situations, and found that our simple rule reproduces the if-then rule in very nearly all cases. Also the number of cases where a third train is relevant to the decision is not large. In an evaluation of a pilot of a conflict detection and resolution system based upon pairwise decisions, only very few exceptions were found.
- *The delay distribution (1) includes the effects, therefore cannot be input to the calculation.* This argument only holds merit if we find large effects, in which case our purpose is already met. Apart from that, we are very insensitive to the value of d_{av}. We have explicitly tested this point by varying dav to reproduce the actual punctuality including our effects. The predicted dispunctuality was almost constant.

The results are significantly larger than process experts expected. Why is that? We feel that for a long time the attention of human operators has been to maintain stable patterns (train sequences) close to the predefined timetable in order to maintain a workable mental image of the process. Optimizing at the level of plus or minus one/two minutes was neither possible, nor aimed for. As a result, the overall external goal of the railway sector, set at very definite punctuality levels in a very specific definition, a goal for which these relatively small errors are highly relevant, has only been aimed for in an indirect way and slipped out of focus.

In conclusion: our calculations were made within the context of an effort to construct a business case for investing in the dispatching control systems. It appears this investment would be an order of magnitude more effective than other punctuality oriented investments currently being suggested. In fact, even if our calculation overestimates the real effect by a factor of 2, which is within the realm of possibilities, the business case is still easily made.

Compared to the strategic (infrastructure planning) and tactical (timetable design) phase of railway operations, the operational phase has received relatively little attention from the mathematical community. Most work for the operational phase focuses on constructing real-time scheduling engines such as Marco and Combine. Experience suggests that for processes with are not under very precise control, as railway operations in the Netherlands neither are nor will be in the foreseeable future, an approach using very simple operational rules, which are well understood and easily amended by the operators, is to be preferred. Our calculations show that in the Netherlands, accurate implementation of one such rule may unlock significant gains in punctuality. We are aware that our formalism is quite basic. We challenge the mathematical community to improve upon our ideas, or indeed come up with an alternative, to get an approach to determine the effectiveness of specific simple dispatching rules.

Online Delay Management
on a Single Train Line

Michael Gatto, Riko Jacob, Leon Peeters, and Peter Widmayer

Institute of Theoretical Computer Science, ETH Zurich
{gattom,rjacob,leon.peeters,widmayer}@inf.ethz.ch

Abstract. We provide competitive analyses for the online delay management problem on a single train line. The passengers that want to connect to the train line might arrive delayed at the connecting stations, and these delays happen in an online setting. Our objective is to minimize the total passenger delay on the train line.

We relate this problem to the Ski-Rental problem and present a family of 2-competitive online algorithms. Further, we show that no online algorithm for this problem can be better than Golden Ratio competitive, and that no online algorithm can be competitive if the objective accounts only for the optimizable passenger delay.

1 Introduction

In the everyday operation of a railroad, it is unfortunately not uncommon for a train to arrive at a station with a delay. In such a situation, some of the train's passengers may miss a connecting train, resulting in an even larger delay for them since they have to wait for the next train. If, on the other hand, the connecting train waits, then it is delayed itself, and so are all the passengers it is carrying. Delay management consists of deciding which connecting trains should wait for what delayed feeder trains, usually with the objective of minimizing the overall discomfort faced by the passengers. Although railway optimization and scheduling problems have been studied quite intensively the last decade, the management of delayed trains has received much less attention.

Various approaches to delay management have been considered the last years, such as simulations, Linear and Integer Programming, or complexity and algorithmic analyses (Section 1.1 provides an overview of related research). However, except for very few exceptions, these papers consider delay management as an offline problem, where the delays are known a priori, and for which a global optimal solution is sought. But delays in a railway system are by nature not known a priori (even though they may be correlated). Therefore, delay management is inherently an online problem.

This paper studies delay management as an online problem, with a focus on competitive analyses. Given the lack of research on online delay management, we consider the basic case of a single train operating on a train line consisting of several consecutive legs. Moreover, previous research on offline delay management problems indicates that railway networks with a path topology are easier

F. Geraets et al. (Eds.): Railway Optimization 2004, LNCS 4359, pp. 306–320, 2007.

to analyze. At each intermediate stop of the train line, some passengers wish to board the train. Each passenger has a destination, and possibly an initial delay, meaning that she arrives at the transfer station with a delayed feeder train. Should a passenger miss her connection at the transfer station, then she has to wait for the next train. Further, we assume a timetable without buffer times, so a train cannot catch up on any of its delay.

The problem is to decide, for each intermediate station, whether the train waits for delayed passengers or not. If it waits, then all its passengers face an arrival delay, including the ones that were so far on time. The same holds for all passengers boarding the train at subsequent stations, since the train cannot catch up on its delay. If the train departs on time, then all connecting passengers will miss their train and have to wait for the next train. A waiting policy specifies at which intermediate stations the train waits for delayed passengers. The goal of the online delay management problem is to find a waiting policy that minimizes the total passenger delay, without knowing beforehand whether the connecting passengers will arrive with a delay at the subsequent intermediate stations. Although our descriptions are in terms of railways, we remark that our model and results are also applicable to other modes of scheduled public transportation, such as bus or metro.

1.1 Related Research

To the best of our knowledge, the only other theoretical online analysis of a delay management problem is by [1]. They consider a bus station with buses arriving at regular time intervals, and passengers arriving with a fixed arrival rate. For this problem, the objective is to decide which buses should wait for how long at the bus station such that the overall passenger waiting time is minimized.

A fair amount of research has been done on offline delay management. Several network-based Mixed Integer Programming (MIP) formulations were introduced [8,9], both for single criteria and bi-criterial objective functions. In particular, some formulations allow special cases to be solved to optimality efficiently, and the model's structure can be used to derive an appropriate Branch-and-Bound algorithm for solving the delay management problem to optimality.

Recently, [3] described polynomial time algorithms for special cases of the delay management problem, such as a limited number of transfers, or a railway network with a path topology. In a follow-up paper [4], a more general variant of the delay management problem was shown to be NP-complete both with and without slack times (or buffer times) in the timetable.

Another approach, based on simulation, applies deterministic waiting rules [2]. The delays are introduced on trains randomly over time with an exponential distribution, and the quality of the waiting rule is derived with an agent-based simulation tool.

Finally, [5,7] considered simulation studies for delay management, which are less related to this paper because we are interested in strategies with a theoretically guaranteed performance.

1.2 Results

We consider the above described setting of a simple railway network with a path topology, which is a natural next step after the station analysis by [1]. We relate this setting of the online delay management problem to a variant of the well-known ski rental problem [6]. Based on this relation, we propose a family of 2-competitive online algorithms, and show that the competitive analysis for this family of algorithms is tight. Further, we prove that no online algorithm for this setting can be better than Φ-competitive, where $\Phi \approx 1.618$ is the golden ratio. Finally, we consider the slightly different objective function of minimizing only the additionally faced delays of the passenger paths. For this objective, we show that no deterministic online algorithm can have bounded competitive ratio. Remarkably, the only strategy for this case not having infinite competitive ratio is the trivial strategy of waiting for any delayed passengers, and departing on time otherwise.

1.3 Outline

The next section defines the problem statement and our assumptions on the problem. Section 3 first introduces some notation, and discusses the relation between the online delay management problem and the so-called Discounted Ski Rental problem. Next, we present the family of 2-competitive online algorithms, as well as the tightness of the competitive analysis. Section 4 contains the two competitive ratio bounds for any online algorithm for the single train line, after which Section 5 concludes the paper and suggest some topics for further research.

2 Problem Statement

We consider a directed graph $G = (V, E)$, with the vertices $V = \{v_1, \ldots, v_n\}$. Let $E = \{(v_i, v_{i+1}) | i \in \{1, \ldots, n-1\}\}$, i.e. the graph forms a simple directed path from v_1 to v_n. The vertices $v \in V$ represent the stations which are served by a single railway line. An edge $(u, v) \in E$ represents a direct link served by the train along the railway line. Hence, the train serving this line starts its journey at station v_1, and traverses the graph through the directed edges. We assume that the train stops at every intermediate station $v_i, i \in \{2, \ldots, n-1\}$ and ends its journey at station v_n. At each stop, passengers may board or alight from the train. We refer to the passenger streams as passenger paths. Each passenger path boards the train at a station v_i, and alights at station $v_j, j > i$. We say the passenger paths enter at station v_i. These passenger paths represent passengers either starting their journey at station v_i, or having arrived at station v_i with another train, a feeder train, and wishing to connect to the train line. The passenger paths may continue their journey outside the train line, but this is not part of the considered problem. The passenger paths are known a priori, as well as the number of passengers travelling along each of them.

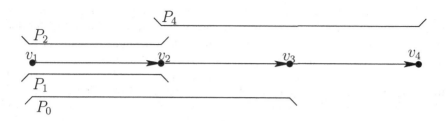

Fig. 1. An illustration of a railway line. The railway line has two intermediate stations (v_2 and v_3), and four paths: P_1, P_2, P_3 and P_4.

An illustration of a railway line and its paths is given in Figure 1. In the example, the passenger path P_1 connects to the train at station v_1 and alights from the train at station v_2. Similarly, path P_0 connects to the train at station v_1 and alights from the train at station v_3, where it continues its journey outside our track.

Some of the connecting passenger paths at the stations $v_i \in V$ might arrive there delayed. We say these paths enter the train line delayed, or that they have a source delay. This reflects the model of feeder trains arriving delayed at station v_i. For simplicity, we assume that all source delays equal δ time units. In order to allow such delayed paths to connect to the train, the train must wait for the delayed paths. Once delayed, the train is not able to catch up on any of this delay, and will hence arrive at its target station v_n with a delay of δ time units. Furthermore, the next train departs T time units later, and will run on time. Hence, should a passenger path miss its connection, it will catch the next train along the line and arrive at its destination with a total delay of T time units. We refer to passenger paths which have missed a connection as dropped paths, and to those facing an arrival delay as delayed paths.

It is not a priori known which passenger paths entering at station $v_i \in V$ start delayed. The delays occur in an online fashion. The online setting, which can be thought of as an adversary, notifies the online algorithm which of the entering paths at station $v_i \in V$ are delayed only when the train arrives at station v_i. Hence, only at that point in time can an online algorithm figure out how many passengers reach the station delayed, and how many are still on time.

When the train reaches station $v_i \in V$, the traffic control must decide whether the train departs on time, or whether it waits for the delayed passenger paths boarding the train in v_i. If the train departs on time, the delayed entering passengers miss their connection and will have to board the next train along this route. If the train waits for the delayed passengers, it will be delayed by δ time units. As the delays of the passenger paths are of the same size, a decision to wait will immediately guarantee the connections for all future delayed entering passenger paths. Hence, the decision to be taken is from which station on the train will wait, from then on allowing transfers of all future delayed entering passenger paths. Naturally, non-delayed passengers paths which are influenced by this decision thereby face a delay of δ time units.

Our objective is to minimize the total passenger delay occurring on the train line, i.e. the sum of the delays over all passengers which travel with the line. Notice that, in general, the objective includes the unavoidable delay of δ time units per passenger of the passenger paths that enter the line delayed.

As mentioned, we analyze the online setting of this problem. To solve the offline problem we merely have to decide where to start waiting. Therefore, we can efficiently enumerate the n different waiting policies. The adversary can decide which of the passenger paths to delay. When the train arrives at station v_i, the adversary must notify which passenger paths entering at that station are delayed. This implies that the online algorithm, when arriving at station v_i, knows exactly which passenger paths were delayed at stations $v_j, j \in \{1, \ldots, i\text{-}1\}$, and which passenger paths are delayed at station v_i. As giving the least information to the online algorithm is advantageous for the adversary, we assume the adversary does not reveal the delayed passenger paths connecting at stations $v_k, k > j$ until we reach these stations.

3 Competitive Online Algorithms

This section presents a family of 2-competitive online algorithms for the described delay management problem. First, we introduce some notation and some inequalities needed to prove the competitiveness of the online algorithms. Next, we point out the similarity between the presented online delay management problem and the Ski-Rental problem. Finally, we move the focus to the online algorithms and their analysis.

3.1 Notation and A-Priori Knowledge

As mentioned earlier, we assume the passengers paths on the train line to be known. Table 1 introduces a set of variables representing the number of passengers having a specified status at their connecting station (delayed, on time). These variables reflect the a-posteriori knowledge of the delays, i.e. the number of passenger in a specific state when the delay configuration is entirely known. If the train starts to wait for delayed entering passenger paths at station j, the value of our objective $\Delta(j)$ is:

$$\Delta(j) = (T - \delta) \sum_{i<j} D^i + \delta \sum_{i=1}^{n} D^i + \delta O^{\geq j}$$

We wish to find j such, that $\Delta(j)$ is minimal. Notice that we include the delay $\delta \sum_{i=1}^{n} D^i$, which we cannot optimize, in the objective. In Section 4 we show that if we do not include this delay, no online algorithm can be more competitive than applying a trivial strategy having unbounded competitiveness.

We define the variables of Table 1 also for the point of view of an online algorithm, see Table 2. The variables are identified by the same letters used for the offline algorithm, but by using lowercases. The following relations hold:

Table 1. The passenger variables for the offline setting

$O^{\geq i}$ The number of passengers which stay on the train at station i plus on time connecting passengers at and after station i, i.e. passengers newly subject to a delay if the train waits at station i.

D^i The number of connecting passengers which arrive delayed at station i, i.e. passengers subject to be dropped at station i if the train departs on time from station i.

Table 2. The passenger variables for the online setting

$o^{\geq i}$ The sum of the number of passengers which are on the train at station i, the number of on time passengers connecting at i, and the number of connecting passengers at future stations. Note that the online algorithm believes that the latter passengers will be on time. These are the passengers subject to be newly delayed if the train waits at station i.

d^i The number of connecting passengers which arrive delayed at station i. These will be dropped if the train departs on time from station i.

$$O^{\geq i+1} \leq O^{\geq i} \tag{1}$$

$$o^{\geq i+1} \leq o^{\geq i} \tag{2}$$

$$O^{\geq i} \leq o^{\geq i} \tag{3}$$

$$o^{\geq i} = O^{\geq i} + \sum_{j>i} D^j \leq O^{\geq i} + \sum_j D^j \tag{4}$$

Inequalities (1) and (2) hold, because at each station passengers may alight from the train. Hence, the number of passengers influenced by a delay monotonically decreases if the train starts to wait at a later station. Inequality (3) holds, as the online algorithm does not know which passengers will arrive delayed after station i. Hence, $o^{\geq i}$ is an upper bound on the number of passengers that will be delayed by a waiting decision. Finally, this overestimate equals the actual number of passengers influenced by this decision (i.e. with the a-posteriori knowledge of the delayed passenger paths), plus the number of passengers that will be delayed after station i. This number can naturally be bounded as shown in inequality (4).

When at a station i, the online algorithm knows the correct number of delayed passengers: $d^j = D^j, \forall j \leq i$.

3.2 Relation to the Ski-rental Problem

In the Ski-Rental problem, a skier wishes to go skiing. He does not known how many times he will actually go skiing, because he will only ski if the weather is nice and there is enough snow. This reflects the online situation. Initially, the skier does not own a pair of skis. Each time he goes skiing, he can either rent the skis at a fixed price or buy the skis. Obviously, buying is more expensive than renting. With the a-posteriori knowledge on the number of times he went skiing, it is clear that it only makes sense to buy the skis if the overall renting costs exceed the price of buying the skis. Furthermore, in that case the skier should buy the skis the first time he goes skiing. Indeed, the only decision to be taken is whether to buy the skis or to rent them. A well known online strategy is to buy the skis as soon as the overall renting costs would exceed the costs of buying the skis. This online strategy is 2-competitive, as analyzed in [6].

We introduce a slight variant of the above Ski-Rental problem. As the skier rents the skis always at the same shop, the shop owner gives him a discount on buying the skis. The discount is proportional to the amount of money the skier has already paid for renting skis, and the proportionality factor is α. Next, we allow the renting price to be variable, but known each time the skier wishes to rent the skis. We call this problem the Discounted Ski-Rental problem. The usual strategy for the Ski-Rental problem can also be applied to the Discounted Ski-Rental problem. The skier buys the skis as soon as the overall renting costs would exceed the actual costs of buying the skis, i.e. the price of the skis with the discount.

In the following, we present a one to one correspondence between a restricted version of the online delay management problem and the Discounted Ski-Rental problem. We restrict the online delay management problem as follows: all passenger paths have as destination v_n. The objective is still to minimize the overall passenger delay, including the delay δ of passenger paths entering delayed at the stations along the train line. Note that the latter part of the objective cannot be optimized. In this setting, the contribution to the objective of $\delta \sum_i D^i + \delta O^{\geq j}$, i.e. the delay of the paths arriving with a delay δ at their destination, plus the δ delay of all dropped paths, is independent of the waiting decision.

We map this constant sum of delays to the original, undiscounted, price of buying the skis in the Discounted Ski-Rental problem. Moreover, the sum of the costs of the dropped paths (i.e. T time units per passenger) at station v_j corresponds to the renting price of the skis on that day. Therefore, there is a bijection between waiting at station v_j and buying the skis on day j. Hence, the cost of buying the skis on day j corresponds to the delay caused by waiting at station v_j, whereas the cost of renting the skis on day j corresponds to the cost of dropping the delayed paths at station v_j. By setting the proportional discount factor to $\alpha = \frac{\delta}{T}$ we complete the mapping. Indeed, the cost of waiting at station v_j is the same as buying the skis on day j:

$$\text{ski-cost}(j) = T \sum_{i<j} D^i + \delta \sum_i D^i + \delta O^{\geq j} - \frac{\delta}{T} T \sum_{i<j} D^i = \Delta(j)$$

Our analysis of the online delay management problem hence also provides a 2-competitive algorithm for the above Discounted Ski-Rental problem. Further, the mapping also provides some intuition for the online algorithm in Section 3.3. In the 2-competitive algorithm for the Discounted Ski-Rental problem, the skier buys the skis as soon as the overall renting costs exceed the costs of buying the skis. In the next section we show that a similar strategy is 2-competitive for the online delay management problem.

We point out that the general setting where passengers can alight from the train at any station along the path, is still related to a Ski-Rental problem. However, the mapping is much less intuitive, and therefore we omit it. Finally, differently from the Ski-Rental problem, the maximum number of times the decision is to be taken is known a priori in the online delay management problem, as the number of stations on the train line is known in each instance.

3.3   A Family of 2-Competitive Online Algorithms

In this section, we describe a family of 2-competitive online policies for the delay management problem described in Section 2. Recall that paths may end before the last station of the line. Hence, the problem is structurally different from both the Discounted Ski-Rental problem and the general Ski-Rental problem. In fact, in most Ski-Rental problems the key decision for an optimal adversary is whether or not to buy skis. Hence, the decision is boolean. On the contrary, for an optimal adversary of the online delay management problem, the decision is not only whether to wait or not, but additionally where the online algorithm should start to wait in order to achieve the optimal policy.

Nevertheless, the family of online algorithms resembles the classical online algorithm for the Ski-Rental problem. Loosely speaking, the train should wait at station v_j if the delay caused by dropping passengers up to and including station v_j exceeds the delay caused to on time passengers by waiting in v_j.

As we present a family of algorithms, we must be a little more precise. For $t \in [T - \delta, T]$, the online algorithm $\text{ALG}(t)$ of the family lets the train wait at station j if

$$t \sum_{i \leq j} d^i \geq \delta o^{\geq j}.$$

Note that the two extremal values of t lead to two extremal behaviors within the same policy. By setting $t = T - \delta$ we obtain the algorithm that starts to wait as late as possible. By setting $t = T$, we obtain the algorithm of the family that starts to wait as early as possible. Below, we show that both extremal algorithms are 2-competitive.

Intuitively, these algorithms achieve the competitive ratio of 2 by a similar argument as for the Ski-Rental problem: the algorithms drop all source delayed passenger paths, until the accumulated delay balances the delay which would occur if the train started to wait. At this point, an adversary could leave all other remaining passenger paths to be on time, thus causing again the same amount of delay to the online algorithm, whereas it would have been optimal not to wait at all. On the other hand, should the online algorithm have waited earlier, the analysis shows that the adversary must also have had a delay equal to the one of the online algorithm.

Theorem 1. *The family of online algorithms* $\text{ALG}(t)$ *which start to wait at station* j *if* $t \sum_{i \leq j} d^i \geq \delta o^{\geq j}, t \in [T - \delta, T]$, *is 2-competitive on the single train line with fixed passenger paths.*

Proof. Let j^* be the station where the optimal offline algorithm waits and j be the station where the online algorithm $\text{ALG}(t)$ waits. The analysis is subdivided into two cases. In fact, if not optimal, the online algorithm either started waiting too early ($j < j^*$) or too late ($j^* < j$). In a worst-case scenario, for analyzing the first case, we take the algorithm of the family which waits the earliest, i.e. $\text{ALG}(T)$. Similarly, for analyzing the case where the algorithm waits too late, we take the algorithm of the family which waits the latest, i.e. $\text{ALG}(t - \delta)$.

Case $j^* < j$: we compare the objective value of the two solutions:

$$\Delta(j) = (T - \delta) \sum_{i<j} D^i + \delta \sum_i D^i + \delta O^{\geq j}$$

$$= \underbrace{(T - \delta) \sum_{i<j^*} D^i + \delta \sum_i D^i + \delta O^{\geq j^*}}_{\Delta(j^*)} - \delta O^{\geq j^*} + \delta O^{\geq j} + (T - \delta) \sum_{i=j^*}^{j-1} D^i$$

First, we note that as $j^* < j$, inequality (1) implies the following: $-\delta O^{\geq j^*} + \delta O^{\geq j} \leq 0$.

Second,

$$\sum_{i=j^*}^{j-1} D^i \leq \sum_{i\leq j-1} D^i.$$

Since the train did not wait at station $j - 1$, and we use $t = T - \delta$, the following inequalities hold:

$$(T - \delta) \sum_{i\leq j-1} D^i \leq \delta O^{\geq j-1} \overset{(4)}{\leq} \delta \left(O^{\geq j-1} + \sum_i D^i \right)$$

$$\leq \delta \left(O^{\geq j^*} + \sum_i D^i \right) \leq \Delta(j^*).$$

Concluding,

$$\Delta(j) \leq \Delta(j^*) - \delta O^{\geq j^*} + \delta O^{\geq j} + \delta \left(O^{\geq j^*} + \sum_i D^i \right) \leq 2\Delta(j^*).$$

Case $j^* > j$: similarly to the previous case, we compare the values of the two solutions:

$$\Delta(j) = \Delta(j^*) - \delta O^{\geq j^*} + \delta O^{\geq j} - (T - \delta) \sum_{i=j}^{j^*-1} D^i.$$

Inequality (3) and waiting at j with $t = T$ imply

$$\delta O^{\geq j} \leq \delta o^{\geq j} \leq T \sum_{i\leq j} D^i \leq T \sum_{i\leq j^*} D^i.$$

Since

$$T \sum_{i \leq j^*} D_i \leq (T - \delta) \sum_{i \leq j^*} + \delta \sum_i D_i \leq \Delta(j^*),$$

we finally have

$$\Delta(j) \leq \Delta(j^*) - \delta o^{\geq j^*} + \Delta(j^*) - (T - \delta) \sum_{i=j}^{j^*-1} D^i \leq 2\Delta(j^*). \qquad \Box$$

We have shown that the family of online algorithms $\text{ALG}(t)$, $t \in [T - \delta, T]$ is 2-competitive.

Corollary 1. *The analysis of the family of online algorithms* $\text{ALG}(t)$, *with* $t \in [T - \delta, T]$, *is tight.*

Proof. We show that the analysis is tight, both when the online algorithm starts waiting earlier or later than optimum. For each case, we analyze the worst-case scenario for the family $\text{ALG}(t)$: for starting to wait too late, we analyze the criterion which will wait the latest, i.e. $(T - \delta) \sum_{i \leq j} d_i \geq \delta o^{\geq j}$; for starting to wait too early, we analyze the criterion of the family which will wait the earliest, i.e. $T \sum_{i \leq j} d_i \geq \delta o^{\geq j}$.

We start by analyzing the case where the online algorithm starts waiting after the optimal solution. Consider the simple train line built by three stations, $V = \{v_1, v_2, v_3\}$, $E = \{(v_1, v_2), (v_2, v_3)\}$. We introduce two passenger paths $P_0 = \{v_1, v_2\}$ and $P_1 = \{v_2, v_3\}$, both connecting from other feeder trains. Potentially, these passenger paths could be delayed. Let p_0 be the number of passengers following path P_0, $p_1 = \frac{(T-\delta)}{\delta} p_0 + \epsilon$ the number of passengers following path P_1. At station v_1, the adversary declares P_0 to be delayed and P_1 to be on time. As $(T - \delta)p_0 < \delta p_1 = (T - \delta)p_0 + \delta\epsilon$, the online algorithm will not wait. Upon the arrival of the train at v_2, the adversary also delays P_1. Then, the online algorithm will certainly wait in v_2, as it will not delay any other passengers. The optimal offline solution would already have waited at v_1, hence online algorithm started to wait after the optimum. The ratio between the two solutions is:

$$\frac{\text{Online}}{\text{Opt}} = \frac{Tp_0 + \delta p_1}{\delta(p_0 + p_1)} = 2\frac{Tp_0 + \delta\epsilon}{Tp_0 + \delta\epsilon} - \frac{\delta p_0 + \delta\epsilon}{Tp_0 + \delta\epsilon} = 2 - \frac{\delta p_0 + \delta\epsilon}{Tp_0 + \delta\epsilon} \qquad (5)$$

For $\epsilon \to 0$ the ratio converges to $r_1 = 2 - \frac{\delta}{T}$. For $\frac{\delta}{T} = \epsilon'$, and by letting $\epsilon' \to 0$, we get arbitrarily close to 2. For analyzing the case where the online algorithm starts waiting too early, we consider a train line similar to the above, but with different passenger paths. Let $P_0 = \{v_1, v_2, v_3\}$, $P_1 = \{v_1, v_2\}$, $P_2 = \{v_2, v_3\}$, carrying p_0, p_1 and $p_2 = \frac{T}{\delta} p_0 - \epsilon$ passengers, respectively. Initially, the adversary shows P_0 to be delayed. The online algorithm does not wait at station v_0, as the other two paths are assumed to be on time: indeed, $\delta(p_1 + p_2) > Tp_0$, for $\delta p_1 > \delta\epsilon$. When the train arrives at station v_1, the adversary leaves P_2 on time.

But then, $Tp_0 > \delta p_2 = Tp_0 - \delta\epsilon$, hence the online algorithm starts to wait at v_2, interestingly enough, for nobody. The optimal algorithm would not have waited anywhere, and would only have dropped the path P_0. This gives the competitive ratio of

$$\frac{\text{Online}}{\text{Opt}} = \frac{Tp_0 + \delta p_2}{Tp_0} = \frac{2Tp_0 - \delta\epsilon}{Tp_0} \stackrel{\epsilon \to 0}{=} 2. \qquad (6)$$

This case is thus independent from the ratio $\frac{\delta}{T}$, and the analysis directly shows its tightness. $\qquad\square$

4 Competitiveness of Online Algorithms

In the following, we show two bounds on the competitive ratio for all online algorithms on the train line. First, we discuss that, if the objective accounts only for the delay which can be optimized, we cannot be better than $\frac{T}{\delta}$-competitive. This actually implies that we cannot do better than applying the trivial strategy of waiting as soon as there is a delayed entering passenger path, and to stay on time otherwise. We then analyze the objective discussed in Section 3.1 and show that no online algorithm can be better than Φ-competitive, where $\Phi = \frac{\sqrt{5}+1}{2}$ is the Golden Ratio.

4.1 Competitiveness with Additional Delay Objective

In this section, we consider the so-called additional delay objective function, which accounts only for the delay which can be optimized on the network, i.e. without the unavoidable delay $\delta \sum_i D^i$. With the previously introduced notation, the objective value occurring if the train waits at station j is defined as

$$\Delta_{\text{ADD}}(j) = (T - \delta) \sum_{i<j} D^i + \delta O^{\geq j}$$

Theorem 2. *No online algorithm on a single train line with fixed passenger paths can be better than $\frac{T}{\delta}$-competitive when minimizing only the additional delay.*

Proof. We analyze the network shown in Figure 2. The line we wish to optimize travels between stations A and C, and has an intermediate stop in B. We introduce two passenger paths, P_1 connecting to the train line in A and carrying $p_1 = 1$ passengers, P_2 connecting to the train line in B and carrying $p_2 = \frac{T(T-\delta)}{\delta^2}$ passengers. When in A, the adversary announces that the passenger path P_1 is delayed by δ time units. He can still choose if or not he will delay the passenger path P_2.

If the online algorithm decides to wait in A, the adversary leaves P_2 on time. Thus, the optimal offline policy is to stay on time to the end of the trip. The delay accumulated by the online algorithm is $\frac{T(T-\delta)}{\delta^2}\delta = \frac{T(T-\delta)}{\delta}$, the optimal strategy accumulates only $(T - \delta)$ delay. Hence, the online algorithm is $\frac{T}{\delta}$-competitive.

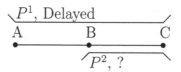

Fig. 2. The train network used for showing the non-competitiveness of the additional delay-objective

If the online algorithm decides to leave A on time, the adversary will delay P_2 as well. The optimal offline policy in this case is to wait in A, which produces a zero valued objective. The online algorithm produces an additional delay of at least T, hence the competitive ratio in this case is infinite.   □

The setting for proving the $\frac{T}{\delta}$-competitiveness might seem peculiar, as in one case the train travels empty between stations A and B. This can be resolved by introducing an on time passenger path between A and C carrying just one passenger. The same setting can then be used to prove a lower bound on the competitiveness of $\frac{T}{\delta} - 1$. This is asymptotically the same as above, and in practice it does not change significantly if $\frac{T}{\delta}$ is large.

4.2 Φ-Competitiveness for Total Delay Objective

In the following section, we prove that no online algorithm can be better than Φ-competitive if we use the objective function accounting for all delays introduced in Section 3.1.

Theorem 3. *No online algorithm on a train line with fixed passenger paths can be better than Φ-competitive when minimizing the total delay, where $\Phi = \frac{\sqrt{5}+1}{2}$ is the Golden Ratio.*

Proof. We introduce a network similar to the one of the previous proof (see Figure 3). This time, we introduce three passenger paths: P_1, carrying p_1 passengers and connecting to the train line at station A; P_2, carrying p_2 passengers, starting their journey in A; P_3, carrying p_3 passengers, connecting to the train line in B. At the beginning, the adversary declares P_1 to be delayed and P_2 to be on time.

Now, an online algorithm must decide whether to wait in A or not. The situation the adversary wants to enforce is the following: whatever decision the online algorithm takes, it is c-competitive. Then, he chooses the parameters such that c is maximal. The following mathematical program describes the adversary's parameter choices:

$$\max c$$
$$\delta(p_1 + p_2 + p_3) \geq cTp_1 \tag{7}$$
$$Tp_1 + \delta(p_2 + p_3) \geq c\delta(p_1 + p_2 + p_3) \tag{8}$$
$$T(p_1 + p_3) \geq c\delta(p_1 + p_2 + p_3) \tag{9}$$

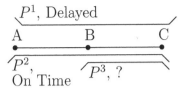

P^1, Delayed

A   B   C

P^2,
On Time P^3, ?

Fig. 3. The simple network used in the proof for not competitiveness of the all-delay objective

Inequality (7) reflects the situation when the online algorithm waits in A, thus delaying all three paths, but it would have been better not to wait, as P_3 was on time. The left hand side (LHS) reflects the costs of the online algorithm, the right hand side (RHS) the costs of the optimal offline solution, weighted with the competitive ratio c. Inequality (8) reflects the situation where the online algorithm does not wait in A but waits in B, as the adversary then delays P_3, and it would have been better to wait in A. Again, the LHS describes the costs of the online algorithm, the RHS the costs of the optimal delay policy weighted with the competitive ratio. Finally, inequality (9) describes the situation where the online algorithm decides not to wait at all even if P_3 delays, and it would have been better to wait in A. The LHS and RHS of the inequality describe the costs as before. For this last online policy we do not consider the case where the optimal offline strategy waits in B. Were this strategy better than waiting in A, we would only make the bound on the competitive ratio bigger than what we show here.

For simplicity, we normalize all passenger numbers with respect to p_1, and the delays with respect to δ. Hence, P_1 carries 1 passenger, and $\delta = 1$. Due to this normalization, in the following we should formally refer to the drop delay as T' and to the passenger numbers as p'_2 and p'_3. To improve readability, we omit the primes. The mathematical program becomes:

$$\max c$$

$$1 + p_2 + p_3 \geq cT \tag{10}$$

$$T + p_2 + p_3 \geq c(1 + p_2 + p_3) \tag{11}$$

$$T(1 + p_3) \geq c(1 + p_2 + p_3) \tag{12}$$

We restrict our attention to the case where $p_2 \leq (T-1)p_3$. In this case, (11) is tighter than (12), so we can omit the latter equation. As we are constructing a specific solution to the mathematical program, we let inequality (10) be tight. Note that by choosing (11) to be tight, we can construct an example with the same competitive ratio as shown below. Now we can set $p_3 = cT - 1 - p_2$. Thus, substituting into (11):

$$T + p_2 + cT - 1 - p_2 \geq c + cp_2 + c^2T - c - cp_2$$

$$T(1 + c) - 1 \geq c^2T$$

Let $c = \Phi - \epsilon$, and recall that $\Phi + 1 = \Phi^2$:

$$T(1 + \Phi) - T\epsilon - 1 \geq \Phi^2 T - 2\Phi\epsilon T + \epsilon^2 T$$
$$(2\Phi - 1 - \epsilon)\epsilon T \geq 1 \tag{13}$$

As long as $(2\Phi - 1 - \epsilon)\epsilon \geq 0$, we can choose T such that (13) is satisfied. The condition is satisfied for $0 < \epsilon \leq 2\Phi - 1$, and we can set $T = \frac{1}{(2\Phi - 1 - \epsilon)\epsilon}$. By letting $\epsilon \to 0$, we can get arbitrarily close to Φ. In all, this shows that the competitive ratio of any online algorithm cannot be better than Φ.

Notice that a closer inspection of the constructed instance shows that within the setting of this example, we cannot prove the competitive ratio to be greater than Φ, as choosing a negative ϵ leads to a contradiction.   □

5 Conclusion

We considered the online delay management problem for a single train line with passengers transferring from other feeder trains. Since such a feeder train may arrive at a transfer station with an arrival delay, the connecting passengers may be delayed as well. In this online setting, the train line only knows the delays of the entering passengers at its current station, as well as at the previous stations on the line.

As such, we provided a natural next step to the research in [1], who considered the online situation of delays at a single station. We proposed a family of Ski Rental-like 2-competitive online algorithms, and presented lower bounds on the competitive ratio that hold for any online algorithm for the single train line. As we do not know of any other theoretical work on online delay management problems, our results provide a first step in the direction of online delay management for more general networks and with more realistic assumptions.

Indeed, the extension of our results to two crossing train lines, or to a railway network with a tree topology are interesting topics for further research. Other directions for future research include different arrival delays for the connecting passengers, and the inclusion of timetable buffer times.

References

1. Anderegg, L., Penna, P., Widmayer, P.: Online train disposition: to wait or not to wait? In: Wagner, D. (ed.) Electronic Notes in Theoretical Computer Science, vol. 66, Elsevier, Amsterdam (2002)
2. Biederbick, C., Suhl, L.: Improving the quality of railway dispatching tasks via agent-based simulation. In: Allan, J., Brebbia, C.A., Hill, R.J., Sciutto, G., Sone, S. (eds.) Computers in Railways IX, Proceedings of the 9th International Conference on Computer Aided Design, Manufacture and Operation in the Railway and Other Mass Transit Systems (COMPRAIL), Dresden, Germany, pp. 785–795. WIT Press, Southampton, Boston (2004)

3. Gatto, M., Glaus, B., Jacob, R., Peeters, L., Widmayer, P.: Railway delay management: Exploring its algorithmic complexity. In: Hagerup, T., Katajainen, J. (eds.) SWAT 2004. LNCS, vol. 3111, pp. 199–211. Springer, Heidelberg (2004)
4. Gatto, M., Jacob, R., Peeters, L., Schöbel, A.: The computational complexity of delay management. In: Kratsch, D. (ed.) WG 2005. LNCS, vol. 3787, pp. 227–238. Springer, Heidelberg (2005)
5. Heimburger, D., Herzenberg, A., Wilson, N.: Using simple simulation models in the operational analysis of rail transit lines: A case study of the MBTA's red line. Transportation Research Record 1677, 21–30 (1999)
6. Karlin, A., Manasse, M., Rudolph, L., Sleator, D.: Competitive snoopy caching. Algorithmica 3(1), 79–119 (1988)
7. O'Dell, S., Wilson, N.: Optimal real-time control strategies for rail transit operations during disruptions. In: Computer-Aided Transit Scheduling. Lecture Notes in Economics and Math. Sys. pp. 299–323. Springer, Heidelberg (1999)
8. Schöbel, A.: A model for the delay management problem based on mixed-integer-programming. In: Zaroliagis, C. (ed.) Electronic Notes in Theoretical Computer Science, vol. 50, Elsevier, Amsterdam (2001)
9. Schöbel, A.: Optimization in Public Transportation: Stop Location, Delay Management and Tariff Zone Design in a Public Transportation Network. In: Optimization and Its Applications, vol. 3, Springer, Heidelberg (2007)

Author Index

Lecture Notes in Computer Science

Sublibrary 1: Theoretical Computer Science and General Issues

For information about Vols. 1– 4494
please contact your bookseller or Springer

Vol. 4665: J. Hromkovič, R. Královič, M. Nunkesser, P. Widmayer (Eds.), Stochastic Algorithms: Foundations and Applications. X, 167 pages. 2007.

Vol. 4664: J. Durand-Lose, M. Margenstern (Eds.), Machines, Computations, and Universality. X, 325 pages. 2007.

Vol. 4661: U. Montanari, D. Sannella, R. Bruni (Eds.), Trustworthy Global Computing. X, 339 pages. 2007.

Vol. 4649: V. Diekert, M.V. Volkov, A. Voronkov (Eds.), Computer Science – Theory and Applications. XIII, 420 pages. 2007.

Vol. 4647: R. Martin, M.A. Sabin, J.R. Winkler (Eds.), Mathematics of Surfaces XII. IX, 509 pages. 2007.

Vol. 4646: J. Duparc, T.A. Henzinger (Eds.), Computer Science Logic. XIV, 600 pages. 2007.

Vol. 4644: N. Azémard, L. Svensson (Eds.), Integrated Circuit and System Design. XIV, 583 pages. 2007.

Vol. 4641: A.-M. Kermarrec, L. Bougé, T. Priol (Eds.), Euro-Par 2007 Parallel Processing. XXVII, 974 pages. 2007.

Vol. 4639: E. Csuhaj-Varjú, Z. Ésik (Eds.), Fundamentals of Computation Theory. XIV, 508 pages. 2007.

Vol. 4638: T. Stützle, M. Birattari, H. H. Hoos (Eds.), Engineering Stochastic Local Search Algorithms. X, 223 pages. 2007.

Vol. 4630: H.J. van den Herik, P. Ciancarini, H.H.L.M.(J.) Donkers (Eds.), Computers and Games. XII, 283 pages. 2007.

Vol. 4628: L.N. de Castro, F.J. Von Zuben, H. Knidel (Eds.), Artificial Immune Systems. XII, 438 pages. 2007.

Vol. 4627: M. Charikar, K. Jansen, O. Reingold, J.D.P. Rolim (Eds.), Approximation, Randomization, and Combinatorial Optimization. XII, 626 pages. 2007.

Vol. 4624: T. Mossakowski, U. Montanari, M. Haveraaen (Eds.), Algebra and Coalgebra in Computer Science. XI, 463 pages. 2007.

Vol. 4623: M. Collard (Ed.), Ontologies-Based Databases and Information Systems. X, 153 pages. 2007.

Vol. 4621: D. Wagner, R. Wattenhofer (Eds.), Algorithms for Sensor and Ad Hoc Networks. XIII, 415 pages. 2007.

Vol. 4619: F. Dehne, J.-R. Sack, N. Zeh (Eds.), Algorithms and Data Structures. XVI, 662 pages. 2007.

Vol. 4618: S.G. Akl, C.S. Calude, M.J. Dinneen, G. Rozenberg, H.T. Wareham (Eds.), Unconventional Computation. X, 243 pages. 2007.

Vol. 4616: A.W.M. Dress, Y. Xu, B. Zhu (Eds.), Combinatorial Optimization and Applications. XI, 390 pages. 2007.

Vol. 4614: B. Chen, M. Paterson, G. Zhang (Eds.), Combinatorics, Algorithms, Probabilistic and Experimental Methodologies. XII, 530 pages. 2007.

Vol. 4613: F.P. Preparata, Q. Fang (Eds.), Frontiers in Algorithmics. XI, 348 pages. 2007.

Vol. 4600: H. Comon-Lundh, C. Kirchner, H. Kirchner (Eds.), Rewriting, Computation and Proof. XVI, 273 pages. 2007.

Vol. 4599: S. Vassiliadis, M. Bereković, T.D. Hämäläinen (Eds.), Embedded Computer Systems: Architectures, Modeling, and Simulation. XVIII, 466 pages. 2007.

Vol. 4598: G. Lin (Ed.), Computing and Combinatorics. XII, 570 pages. 2007.

Vol. 4596: L. Arge, C. Cachin, T. Jurdziński, A. Tarlecki (Eds.), Automata, Languages and Programming. XVII, 953 pages. 2007.

Vol. 4595: D. Bošnački, S. Edelkamp (Eds.), Model Checking Software. X, 285 pages. 2007.

Vol. 4590: W. Damm, H. Hermanns (Eds.), Computer Aided Verification. XV, 562 pages. 2007.

Vol. 4588: T. Harju, J. Karhumäki, A. Lepistö (Eds.), Developments in Language Theory. XI, 423 pages. 2007.

Vol. 4583: S.R. Della Rocca (Ed.), Typed Lambda Calculi and Applications. X, 397 pages. 2007.

Vol. 4580: B. Ma, K. Zhang (Eds.), Combinatorial Pattern Matching. XII, 366 pages. 2007.

Vol. 4576: D. Leivant, R. de Queiroz (Eds.), Logic, Language, Information and Computation. X, 363 pages. 2007.

Vol. 4547: C. Carlet, B. Sunar (Eds.), Arithmetic of Finite Fields. XI, 355 pages. 2007.

Vol. 4546: J. Kleijn, A. Yakovlev (Eds.), Petri Nets and Other Models of Concurrency – ICATPN 2007. XI, 515 pages. 2007.

Vol. 4545: H. Anai, K. Horimoto, T. Kutsia (Eds.), Algebraic Biology. XIII, 379 pages. 2007.

Vol. 4533: F. Baader (Ed.), Term Rewriting and Applications. XII, 419 pages. 2007.

Vol. 4528: J. Mira, J.R. Álvarez (Eds.), Nature Inspired Problem-Solving Methods in Knowledge Engineering, Part II. XXII, 650 pages. 2007.

Vol. 4527: J. Mira, J.R. Álvarez (Eds.), Bio-inspired Modeling of Cognitive Tasks, Part I. XXII, 630 pages. 2007.

Vol. 4525: C. Demetrescu (Ed.), Experimental Algorithms. XIII, 448 pages. 2007.

Vol. 4514: S.N. Artemov, A. Nerode (Eds.), Logical Foundations of Computer Science. XI, 513 pages. 2007.

Vol. 4513: M. Fischetti, D.P. Williamson (Eds.), Integer Programming and Combinatorial Optimization. IX, 500 pages. 2007.

Vol. 4510: P. Van Hentenryck, L.A. Wolsey (Eds.), Integration of AI and OR Techniques in Constraint Programming for Combinatorial Optimization Problems. X, 391 pages. 2007.

Vol. 4507: F. Sandoval, A.G. Prieto, J. Cabestany, M. Graña (Eds.), Computational and Ambient Intelligence. XXVI, 1167 pages. 2007.

Vol. 4502: T. Altenkirch, C. McBride (Eds.), Types for Proofs and Programs. VIII, 269 pages. 2007.

Vol. 4501: J. Marques-Silva, K.A. Sakallah (Eds.), Theory and Applications of Satisfiability Testing – SAT 2007. XI, 384 pages. 2007.

Vol. 4497: S.B. Cooper, B. Löwe, A. Sorbi (Eds.), Computation and Logic in the Real World. XVIII, 826 pages. 2007.